RAMAN SPECTROSCOPY

Edited by **Gustavo Morari do Nascimento**

Raman Spectroscopy

http://dx.doi.org/10.5772/intechopen.68928
Edited by Gustavo Morari do Nascimento

Contributors

Marcin Pisarek, Jan Krajczewski, Marcin Holdynski, Tomasz Plocinski, Miroslaw Krawczyk, Andrzej Kudelski, Maria Janik-Czachor, Jiatao Zhang, Meng Xu, Prabhakar Misra, Tito Trindade, Sara Fateixa, Helena Nogueira, Alexandra Nunes, Miguel Ramirez-Elias, Francisco Javier González, Izabela Rzeznicka, Hideyuki Horino, Manohar Chirumamilla, Anisha Chirumamilla, Alexander S. Roberts, Andrea Cerea, Francesco De Angelis, Remo Proietti Zaccaria, Peter Kjær Kristensen, Roman Krahne, Sergey Bozhevolnyi, S. Sutherland S. Sutherland, Kjeld Pedersen, Andrea Toma, Esben Skovsen, Sabina Botti, Weiping Cai, Guangqiang Liu, Qian Zhao, Pietro Galinetto, Benedetta Albini, Marcella Bini, Maria Cristina Mozzati, Horia Iovu, Iuliana Biru, Kai Jiang, Jinzhong Zhang, Junhao Chu, Zhigao Hu, Liping Xu, Gustavo Morari Do Nascimento

Notice

First published in London, United Kingdom, 2018 by IntechOpen
IntechOpen is the global imprint of INTECHOPEN LIMITED, registered in England and Wales, registration number: 11086078, The Shard, 25th floor, 32 London Bridge Street
London, SE19SG – United Kingdom
Printed in Croatia

British Library Cataloguing-in-Publication Data
A catalogue record for this book is available from the British Library

Additional hard copies can be obtained from orders@intechopen.com

Raman Spectroscopy , Edited by Gustavo Morari do Nascimento
p. cm.
Print ISBN 978-1-78923-000-0
Online ISBN 978-1-78923-001-7

For EU product safety concerns:
IN TECH d.o.o., Prolaz Marije Krucifikse Kozulić 3, 51000 Rijeka, Croatia,
info@intechopen.com or visit our website at intechopen.com.

Meet the editor

Dr. Gustavo Morari do Nascimento holds a position of Professor at the Federal University of ABC. He has strong experience in many fields related to characterization of nanomaterials by using spectroscopic techniques associated to microscopic resolution. He obtained a doctoral degree from the University of São Paulo (USP) in 2004, writing a thesis about spectroscopic characterization of nanocomposites formed by conducting polymers and clays. Afterward, he received in 2007/2008 a postdoctoral degree at Massachusetts Institute of Technology (MIT) in the resonance Raman study of double-walled carbon nanotubes doped with halogens under the guidance of the legendary Mildred S. Dresselhaus. Back in Brazil, he has spent 3 years (2009–2011) at the Federal University of Minas Gerais (UFMG) working with synthesis of nanostructured carbon modified with molecular magnets. Nowadays, his research focus is on molecular characterization of modified carbon nanostructured materials and polymer nanocomposites by using different spectroscopic techniques. Resonance Raman and surface-enhanced Raman spectroscopy (SERS) coupled with microscopy techniques added to X-ray absorption techniques at National Synchrotron Light Laboratory have been the main techniques employed in the investigation.

Contents

Preface

This book brings some examples of the state-of-the-art applications of Raman spectroscopy in characterization of materials and biomaterials, mainly through intensification processes, such as resonance Raman (RR) and Surface-enhanced Raman spectroscopy (SERS). The main goal of this book is to open up to an extended audience the wide possibilities of applications of Raman spectroscopy for academic, industrial, biomedical, and environmental purposes. All authors and editors try to use fluent language in order to make the reading possible for a non-specialized public. In fact, this collective work will be beneficial to students, teachers, and researchers of many areas who are interested to expand their knowledge about Raman spectroscopy applied to nanotechnology, biotechnology, environmental science, inorganic chemistry, in situ and in vivo detection, and health sciences.

This book is organized starting from an introductory chapter that discusses basic aspects of conventional Raman Spectroscopy and also the special cases where the Raman scattering signal can be strongly amplified. Many examples are exploited. The book is organized in five main sections: (i) Introduction, (ii) Surface-Enhanced Raman Spectroscopy: Nanosubstrates and Applications, (iii) SERS and Raman Spectroscopy: Carbon Nanomaterials, (iv) Raman Spectroscopy of Ferrite Nanomaterials, and (v) Raman Spectroscopy Applied to Biomedical Sciences.

The second section (ii) has three chapters that focus on preparation of highly sensible nanomaterials to be used as efficient and reliable platforms for SERS measurements: (a) In the first chapter, the authors (M. Chirumamilla et al.) studied the synthesis and characterization of three-dimensional (3D) nanostructures with multiple branches (MB) as SERS substrates with breakthrough performances in hotspot-mediated ultra-sensitive detection; (b) in the following chapter (Marcin Pisarek et al.), the results of recent investigations into TiO_2 nanotubular oxide layers on Ti metal loaded with Ag nanoparticles are investigated. The efficiency of these materials is discussed as surface plasmon resonators for precise surface analytical investigations of numerous types of organic molecules at concentrations as low as, e.g., 10^{-9} M, and (c) finally, the synthesis of precisely controllable anisotropic noble metal nanoparticles (NPs) is reviewed (M. Xu and J. Zhang). This review has demonstrated the correlation of the key morphological parameters to achieve the strong E-field and ultra-sensitive SERS detection.

In fact, the SERS effect and also surface chemistry in general can be studied at molecular level when the conventional Raman system is joined to a scanning tunneling microscope (STM); in this case, the technique is appropriately named Tip-enhanced Raman spectroscopy (TERS-STM). The origin of the chemical enhancement has been the subject of much debate over the years. In this chapter (I. Rzeznicka and H. Horino), the effects of adsorption state of a molecule and its orientation over Raman signal are studied from the standpoint of surface chemistry at the nanoscale. In addition, two chapters deal with special applications of SERS technique: (a) the recent research on the development of SERS substrates based on polymer nanocomposites and their applications in different fields is reviewed (S. Fateixa et al.). In addition, the joint use of Raman imaging and SERS in nanocomposite development is discussed, and (b) the detection of organophosphorus molecules (W. Cai et al.) is the main focus of the last chapter; the authors showed a route that could be suitable for detection of some organophosphorus nerve agents and other molecules weakly interacted with the coin metal substrates by choosing the appropriate modifiers.

The third section (iii) brings chapters that used conventional Raman spectroscopy and SERS in the study of different carbon allotropes. The versatility of Raman spectroscopy is illustrated in the characterization of single (SWNT) and multi-walled (MWNT) carbon nanotubes, few layers of graphene, and its functionalized forms, with an emphasis on gas-sensing applications (P. Misra et al.). The characteristic features in Raman spectra of carbon allotropes are exposed, and the D and G band intensities are deeply investigated. In particular, one chapter is devoted to graphene and different types of graphene oxide and its nanocomposites (E.I. Bîru and H. Iovu). The last chapter (S. Botti and A. Rufoloni) in this section brings new results in the SERS studies of carbon nanotubes and graphene at pristine and also modified conditions. These materials can amplify the SERS signal mainly by chemical mechanism.

The fourth section (iv) brings two chapters that use Raman spectroscopy as a powerful tool for characterization of some important materials, such as ferrite nanoparticles and ferroelectric oxides. The first chapter gives a broad overview (P. Galinetto et al.) of the Raman spectroscopy results in the characterization of $ZnFe_2O_4$ nanoparticles. The sensitivity of the Raman signal to probe cation disorder favored the appearance of several detailed works on a rich variety of nanosized zinc ferrites. An overview on the experimental results is reported and discussed at variance with synthesis methods, grain dimensions, and dopants. The second chapter systematically studies the structure of ferroelectric oxides and rare-earth element-doped ferroelectric materials (Junhao Chu et al.). Structural transformations that alter the crystal symmetry often have a significant effect on the Raman signal. The Curie temperature (T_C), distortion degree, and phase structure of the ferroelectric materials have been monitored by temperature-dependent Raman spectroscopy.

The last section (v) is composed of two chapters dedicated to special applications of Raman spectroscopy in biomedical area. The use of Raman spectroscopy as a tool for biochemical investigation is the main focus of the first chapter in this section (A. Nunes and S. Magalhães). Raman spectroscopy, mainly SERS, was already applied to successfully diagnose several types of cancer and infections, and preliminary results are also promising in the context of Alzheimer's disease. The success of Raman spectroscopy in biomedical applications is based on the fact that the molecular composition of healthy tissue is different from diseased tissue; also several disease biomarkers can be identified in Raman spectra, which can be used to diagnose or monitor the progress of certain medical conditions (M. G. R-Elías and F. J. González). This chapter outlines an overview of the use of Raman spectroscopy for in vivo medical diagnostics and demonstrates the potential of this technique to address biomedical issues related to human health.Finally, I would like to give special thanks to all authors that contributed for this book (in alphabetical order): Alessandro Rufoloni, Alexander S. Roberts, Alexandra Nunes, Andrea Cerea, Andrea Toma, Andrzej Kudelski, Anisha Chirumamilla, Benedetta Albini, Daniel Casimir, Duncan S. Sutherland, Elena Iuliana Bîru, Esben Skovsen, Francesco De Angelis, Francisco Javier González, Guangqiang Liu, Hideyuki Horino, H.I.S.

Nogueira, Horia Iovu, Iman Ahmed, Izabela Rzeznicka, Jan Krajczewski, Jiatao Zhang, Jinzhong Zhang, Junhao Chu, Kai Jiang, Kjeld Pedersen, Liping Xu, Manohar Chirumamilla, Marcella Bini, Marcin Holdyński, Marcin Pisarek, Maria Cristina Mozzati, Maria Janik-Czachor, Meng Xu, Michael Stenbæk Schmidt, Miguel Ghebré Ramírez-Elías, Mirosław Krawczyk, Peter Kjær Kristensen, Pietro Galinetto, Prabhakar Misra, Qian Zhao, Raul Garcia-Sanchez, Remo Proietti Zaccaria, Roman Krahne, Sabina Botti, Sandra Magalhães, Sergey I. Bozhevolnyi, S. Fateixa, Tomas Rindzevicius, Tomasz Płociński, T. Trindade, Weiping Cai, and Zhigao Hu. Special thanks go to Ms. Marina Dusevic for supporting and assisting with the book edition.

Gustavo Morari do Nascimento
Federal University of ABC
Santo André, Brazil

Section 1

Introduction

Chapter 1

Introductory Chapter: The Multiple Applications of Raman Spectroscopy

Gustavo M. Do Nascimento

Additional information is available at the end of the chapter

http://dx.doi.org/10.5772/intechopen.75795

1. Introduction

The vibrational spectroscopies have been largely employed in the determination of the molecular structure and played a central role in many areas, from fundamental chemistry to materials engineering. Among these techniques, the Raman spectroscopy is essential today in the study of nanostructured materials and biological systems in *in situ* and *in vivo* conditions. In addition, by intensification process, very small quantities can be analyzed (for certain cases in the limit of single molecule regime). Detailed information about the molecular structure can be obtained, because each vibration of atoms shows a characteristic position and intensity. Both position and intensity are influenced by the chemical environment, or in other words by chemical bonds, inter- and intramolecular forces.

2. Raman scattering

The Raman spectra have a different origin compared to infrared absorption (IR); an inelastic scattering process is responsible for the appearance of the vibrational bands. In IR spectroscopy, the radiation is absorbed by available vibrational states (E_1, E_2, E_3,… E_N), it means that photon energy must be coincident with the energy difference between two available states. Thus, $E_N - E_1 = h\upsilon$, being h the Planck's constant. In Raman spectroscopy, a molecule can scatter a monochromatic (ν_o) radiation. Typically, lasers with photon energies ranging from UV to near-IR region are used in the Raman experiments. The elastic scattering is called Rayleigh scattering, and it is much more intense than the inelastic Raman scattering (approximately 10^{-8} lower than the intensity of the incident radiation). The Raman spectrum appears in a wavelength (λ_o) slightly higher or lower than the incident radiation [1–5].

Hence, this scattering process can be visualized as a perturbation of the molecule by incident photon, which *eigenstate* can be described as the linear combination of the *stationary-state* (*or time-independent*) wave functions of the molecule. The pertubated states of the molecule are called "*virtual*" because they are not states from the stationary condition (*eigenstates*). The photon released when the molecule returns to electronic ground state has, more often, the same incident photon energy, characterizing then the Rayleigh scattering. The Raman scattering, that contents the vibrational information, can be exemplified in the energy diagram represented in **Figure 1**. The photon energy $(\nu_o - \nu_m)$ appears in the region of the spectrum called Stokes region; in addition, the photons with energy $(\nu_o + \nu_m)$ appear in the anti-Stokes region. The bands observed in anti-Stokes region are less intense than those in the Stokes region due to the differences in population according to the *Maxwell-Boltzmann* distribution, making the bands from Stokes region more intense than the anti-Stokes Raman region [6–8].

A vibration (vibration coordinate ϱ) is Raman active when the polarizability (α) (or induced dipolar momentum) of the molecule is changed; being described as: $\left(\frac{\partial \alpha}{\partial \varrho}\right) \neq \varnothing$. Hence, as the nature of the Raman effect is physically different from the infrared, the selection rules are also different, which lead to different spectra. In fact, these two techniques are complementary, being used in many cases combined for structural elucidation.

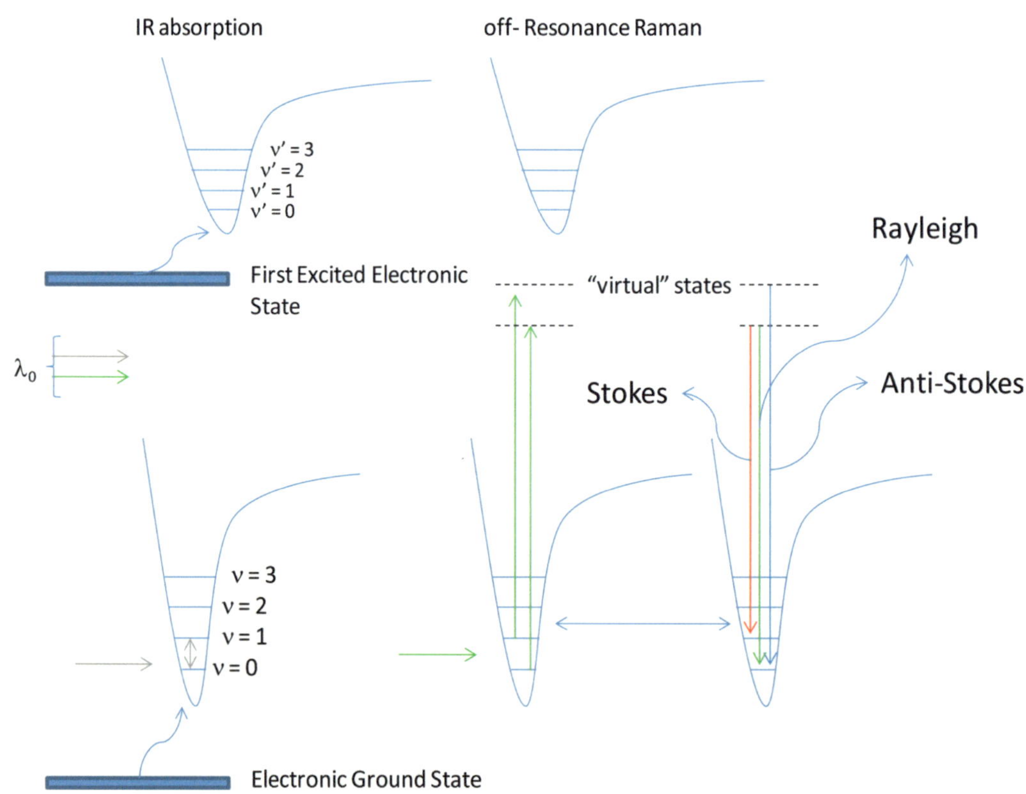

Figure 1. Schematic representation of the energy diagrams for a diatomic molecule. The figure represents the IR absorption and the Raman scattering in off-resonance or normal condition, by using a laser line with energy is far enough to excite the molecule to other electronic states.

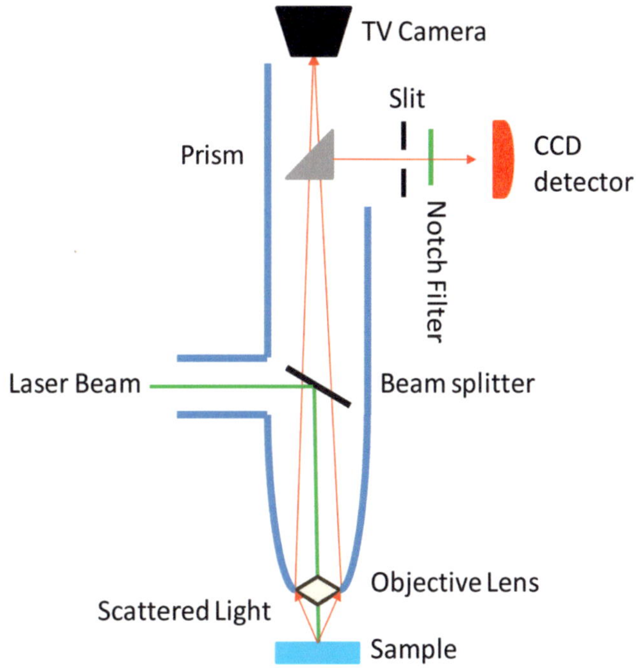

Figure 2. Conventional Raman microscope.

Nowadays, the Raman spectra can be collected in instruments having microscope (see **Figure 2**). The laser line is introduced to the sample on the microscope stage. In fact, more sophisticated microscopes can be attached in the Raman spectrograph, as STM or AFM microscopes and also electron microscopes [9, 10]. The scattered light is collected at 180°, and a high efficiency notch filter must be used to cutoff the incident radiation and also the elastic scattering because the Raman signal is very small (typically a cross section factor of 10^{-6} to 10^{-12} of the incident radiation). Raman microscopy is a nondestructive technique and usually has no requirement for sample preparation, and the measurements can be done in aqueous solution (not possible in conventional IR spectroscopy). The main advantage here is the ability to focus the laser on a very small part of the sample (1 μm approximately or smaller). The high lateral resolution and depth of field (the order of a few micrometers) are very useful, for instance, in the study of nanostructured materials and *in situ* measurements [11–15].

3. Resonance Raman

In the off-resonance or normal Raman spectroscopy, the intensities are linearly proportional to the intensity of the incident light (I_o, see **Figure 3**), proportional to the fourth power of the wavelength of the scattered light (λ_s^4 or ν_s in wavenumber units), and proportional to the square of the polarizability tensor ($[\alpha]^2$) [6–8, 16]. However, when the laser line has similar energy to a permitted electronic transition, the Raman signal is amplified for about 10^5 magnitude orders; this is what characterizes the resonance Raman effect. The theoretical

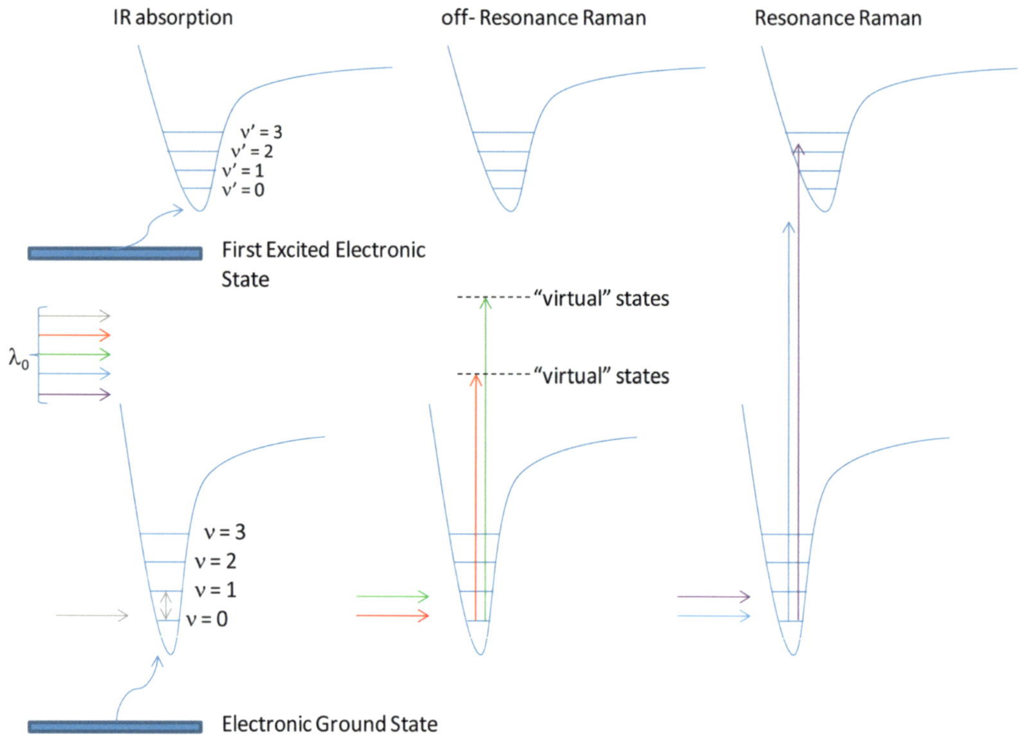

Figure 3. Schematic representation of two electronic states (ground and excited) and their respective vibrational levels. The arrows indicated the possible transitions. In figure v_o and v_s (the scattered frequency is composed of: $v_{ev,gm}$ and $v_{ev,gn}$, the stokes and anti-stokes components, respectively) are the laser line and the scattered frequencies. The equation that describes the resonance Raman scattering is formed in the numerator part by transition dipole moment integrals between the electronic ground state (g, for the vibrational m or n states) and an excited electronic state (e, for any vibrational v states). The sum is done over all possible (e,v) states. The denominator part is the difference or sum of the scattered and incident light, added by the dumping factor ($i\Gamma_{ev}$) that contains information about the lifetime of the transition states.

formalism developed by Albrecht et al. is commonly employed in the interpretation of the resonance Raman [6–8, 16].

Generally, the tensor of polarizability is described as shown in **Figure 3**. The equation is formed in the numerator part by transition dipole moment integrals between the electronic ground state (g, for the vibrational m or n states) and an excited electronic state (e, for any vibrational v states). The sum is done over all possible (e,v) states. In the denominator part is the difference or sum of the scattered and incident light, added by the dumping factor ($i\,\Gamma_{ev}$) that contents information about the lifetime of the transition states. This enormous intensification makes, in principle, the Raman spectrum easy to be acquired. But, in a state of resonance, a lot of radiation is absorbed, leading to a local heating and frequently can be observed a decomposition of the sample. Despite this problem, the RR spectroscopy has been largely used in the study of the different chromophore units present in many compounds varying from conducting polymers, nanocomposites, carbon allotropes to DNA [17–21].

4. Surface-enhanced Raman spectroscopy

The Raman signal can also be amplified by the adsorption of molecules in certain metallic surfaces. This behavior was discovered in 1974 by Fleishmann, Hendra and McQuillan [22]. Initially, the phenomena was associated to the increase of metallic surface area during oxidation–reduction cycles, but soon, it was evident that the strong increase of the Raman signal cannot be explained only considering the changes in metallic surface [23–25]. Finally, this new intensification process is known as surface enhanced Raman spectroscopy (SERS), evidently this process can be combined to RR effect and the acronym for this behavior is SERRS.

There are two main mechanisms to explain the SERS effect: (1) the electromagnetic model and (2) the charge transfer or chemical model. In the electromagnetic model, the interactions among the incident electromagnetic radiations (laser lines) with the surfaces are the main concerns. In this model, the form, size, alignment and type of metallic substrate are the main characteristics to shape the absorption of the surface plasmon resonance. The enhancement of the local electromagnetic field promoted by surface plasmon resonances is more efficient in metallic regions named hot-spots (nanoscale gaps or nanoholes), and as a consequence, the SERS signal from molecules adsorbed in hot-spots dominates the overall detected intensity [26, 27].

In the charge-transfer or chemical model, the interactions between the adsorbed molecules (mainly molecular orientation) with the metallic atoms involved in adsorption (atoms organization) are considered. The adsorption increases the polarizability of the molecule, which can be further increased by charge transfer between the adsorbed molecule and the metallic surface induced by the incident radiation [28, 29]. The applied electrical potential can also be considered in cases where the molecules are adsorbed in an electrode. The SERS technique permits the study of surface interactions, adsorption process, electrode reactions, biological systems, single molecule detection, and so on [30–33].

5. Outlook

Hence, this introduction tries to summarize the principles of main mechanisms of intensification of the Raman signal. These two effects (Resonance Raman and SERS) can be used separately or combined in the structural studies. In addition, the attachment of powerful microscopes with atomic/molecular resolution can also amplify the Raman signal, and this is the case of Tip-enhanced Raman spectroscopy (TERS). As a consequence, this combination opens the opportunity to study the Raman signal at unparalleled spatial resolution. In the present book, some examples of the state-of-the-art applications of Raman spectroscopy in characterization of materials and biomaterials, mainly through resonance Raman (RR) and surface-enhanced Raman spectroscopy (SERS) are deeply discussed.

Author details

Gustavo M. Do Nascimento

Address all correspondence to: morari@yahoo.com; gustavo.morari@ufabc.edu.br

Federal University of ABC-CCNH, Brazil

References

[1] Nakamoto N, Ferraro JR. Introduction to Raman Spectroscopy. London: Academic Press, Inc.; 1994

[2] Rostron P, Gerber D. Raman Spectroscopy, a review. International Journal of Engineering and Technical Research. 2016;**6**:50-64

[3] Ospitalli F, Sabetta T, Tullini F, Nannetti MC, Di Lornardo G. The role of Raman microspectroscopy in the study of black gloss coatings on roman pottery. Journal of Raman Specroscopy. 2005;**36**:18

[4] Smith BC. Fundamentals of Fourier Transform Infrared Spectroscopy. USA: CRC Press, Inc.; 1996

[5] Milosevic M. Applied Spectroscopy Reviews. 2004;**39**:365

[6] Batchelder DN. In: Brässler H, editor. Optical Techniques to Characterize Polymer Systems. Amsterdam: Elsevier; 1987

[7] Batchelder DN, Bloor D. Advances in Infrared and Raman Spectroscopy. London: Wiley-Heyden; 1984

[8] Clark JH, Dines TJ. Resonance Raman spectroscopy, and Its application to inorganic chemistry. New analytical methods (27). Angewandte Chemie (International Ed. in English). 1986;**25**:131

[9] Kumar N, Su W, Vesely M, Weckhuysen BM, Pollard AJ, Wain AJ. Nanoscale chemical imaging of solid–liquid interfaces using tip-enhanced Raman spectroscopy. Nanoscale. 2018;**10**:1815

[10] Black L, Brooker A. SEM–SCA: Combined SEM – Raman spectrometer for analysis of OPC clinker. Advances in Applied Ceramics. 2007;**106**:327

[11] Do Nascimento GM, Constantino VRL, Temperini MLA. Spectroscopic characterization of doped poly(benzidine) and its nanocomposite with cationic clay. The Journal of Physical Chemistry. B. 2004;**108**:5564

[12] Do Nascimento GM, Constantino VRL, Landers R, Temperini MLA. Aniline polymerization into montmorillonite clay: A spectroscopic investigation of the intercalated conducting polymer. Macromolecules. 2004;**25**:9373

[13] Do Nascimento GM, Temperini MLA. Structure of polyaniline formed in different inorganic porous materials: A spectroscopic study. European Polymer Journal. 2008;**44**:3501

[14] Sestrem RH, Ferreira DC, Landers R, Temperini MLA, Do Nascimento GM. Synthesis and spectroscopic characterization of polymer and oligomers of ortho-phenylenediamine. European Polymer Journal. 2010;**46**:484

[15] Do Nascimento GM, De Oliveira RC, Pradie NA, Gessolo Lins PR, Worfel PR, Martinez GR, Di Mascio P, Dresselhaus MS, Corio PJP, Photobio A. Single-wall carbon nanotubes modified with organic dyes: Synthesis, characterization and potential cytotoxic effects. Chemistry. 2010;**211**:99

[16] McHale JL. Molecular Spectroscopy. USA: Prentice-Hall; 1999

[17] Furukawa Y, Ueda F, Hydo Y, Harada I, Nakajima T, Kawagoe T. Vibrational spectra and structure of polyaniline. Macromolecules. 1988;**21**:1297

[18] Quillard S, Louarn G, Lefrant S, MacDiarmid AG. Vibrational analysis of polyaniline: A comparative study of leucoemeraldine, emeraldine, and pernigraniline bases. Physical Review B. 1994;**50**:12496

[19] Do Nascimento GM, Hou T, Kim YA, Muramatsu H, Hayashi T, Endo M, Akuzawa N, Dresselhaus MS. Double-wall carbon nanotubes doped with different Br2 doping levels: A resonance Raman study. Nano Letters. 2008;**8**:4168

[20] Do Nascimento GM, Silva TB, Corio P, Dresselhaus MS. Charge-transfer behavior of polyaniline single wall carbon nanotubes nanocomposites monitored by resonance Raman spectroscopy. Journal of Raman Specroscopy. 2010;**41**:1587

[21] Benevides JM, Overman SA, Thomas GJ Jr. Raman, polarized Raman and ultraviolet resonance Raman spectroscopy of nucleic acids and their complexes. Journal of Raman Specroscopy. 2005;**36**:279

[22] Fleischmann M, Hendra PJ, McQuillan AJ. Raman spectra of pyridine adsorbed at a silver electrode. Chemical Physics Letters. 1974;**26**:163-166

[23] Jeanmaire DL, Van Duyne RP. Surface Raman spectroelectrochemistry: Part I. Heterocyclic, aromatic, and aliphatic amines adsorbed on the anodized silver electrode. Journal of Electroanalytical Chemistry and Interfacial Electrochemistry. 1977;**84**:1-20

[24] Albrecht MG, Creighton JAJ. Anomalously intense Raman spectra of pyridine at a silver electrode. American Chemical Society. 1977;**99**:5215-5217

[25] Kneipp K, Wang Y, Kneipp H, Perelman LT, Itzkan I, Dasari RR, Feld MS. Single molecule detection using surface-enhanced Raman scattering (SERS). Physical Review Letters. 1997;**78**:1667-1670

[26] Corni S, Tomasi J. Surface enhanced Raman scattering from a single molecule adsorbed on a metal particle aggregate: A Theoretical Study. The Journal of Chemical Physics. 2002;**116**:1156

[27] Schatz GC, Young MA, Van Duyne RP. Electromagnetic mechanism of SERS. Topics in Applied Physics. 2006;**103**:19-46

[28] Huang Y-F, Wu D-Y, Zhu H-P, Zhao L-B, Liu G-K, Ren B, Tian Z-Q. Surface-enhanced Raman spectroscopic study of p-aminothiophenol. Physical Chemistry Chemical Physics. 2012;**14**:8485-8497

[29] Lombardi JR, Birke RLAJ. A unified approach to surface-enhanced Raman spectroscopy. Physical Chemistry C. 2008;**112**:5605-5617

[30] Cotton TM, Kim JH, Chumanov GD. Application of surface-enhanced Raman spectroscopy to biological systems. Journal of Raman Specroscopy. 1991;**22**:729

[31] Pettinger B, Wetzel H. Surface-enhanced Raman scattering from crystal violet adsorbed on a silver electrode. Chemical Physics Letters. 1981;**78**:398

[32] Campion A, Kambhampati P. Surface-enhanced Raman scattering. Chemical Society Reviews. 1998;**27**:241

[33] Nie S, Emory SR. Probing single molecules and single nanoparticles by surface-enhanced Raman scattering. Science. 1997;**275**:1102

Section 2

Surface Enhanced Raman Spectroscopy: Nanosubstrates and Applications

Chapter 2

Engineering 3D Multi-Branched Nanostructures for Ultra-Sensing Applications

Anisha Chirumamilla, Manohar Chirumamilla,
Alexander S. Roberts, Andrea Cerea, Esben Skovsen,
Francesco De Angelis, Remo Proietti Zaccaria,
Peter Kjær Kristensen, Roman Krahne,
Duncan S. Sutherland, Sergey I. Bozhevolnyi,
Kjeld Pedersen and Andrea Toma

Additional information is available at the end of the chapter

http://dx.doi.org/10.5772/intechopen.74066

Abstract

The fabrication of plasmonic nanostructures with sub-10 nm gaps supporting extremely large electric field enhancement (hot-spot) has attained great interest over the past years, especially in ultra-sensing applications. The "hot-spot" concept has been successfully implemented in surface-enhanced Raman spectroscopy (SERS) through the extensive exploitation of localized surface plasmon resonances. However, the detection of analyte molecules at ultra-low concentrations, i.e., down to the single/few molecule level, still remains an open challenge due to the poor localization of analyte molecules onto the hot-spot region. On the other hand, three-dimensional nanostructures with multiple branches have been recently introduced, demonstrating breakthrough performances in hot-spot-mediated ultra-sensitive detection. Multi-branched nanostructures support high hot-spot densities with large electromagnetic (EM) fields at the interparticle separations and sharp edges, and exhibit excellent uniformity and morphological homogeneity, thus allowing for unprecedented reproducibility in the SERS signals. 3D multi-branched nanostructures with various configurations are engineered for high hot-spot density SERS substrates, showing an enhancement factor of 10^{11} with a low detection limit of 1 fM. In this view, multi-branched nanostructures assume enormous importance in analyte detection at ultra-low concentrations, where the superior hot-spot density can promote the identification of probe molecules with increased contrast and spatial resolution.

Keywords: hot-spot engineering, 3D multi-branched nanostructures, nanostars, top-down fabrication methods, ultra-sensitive SERS substrates

1. Introduction

Nanoplasmonics is an emerging area of scientific research with a variety of applications in spectroscopy, metamaterials engineering, biosensing, lasing, photocatalysis, nonlinear and quantum optics [1–8]. This discipline deals with the study of collective oscillations of conduction electrons at metal-dielectric interfaces, which can be resonantly excited upon external irradiation. Coupling of electromagnetic (EM) fields to the free-electron motion leads to enhanced optical near fields, confined in subwavelength regions and localized in close proximity of metallic nanoparticles (localized surface plasmon resonances—LSPRs), which are very sensitive to the local dielectric environment [9, 10]. Within this context, nanostructures endowed with sharp tips/edges and sub-10 nm interparticle separation (IPS) are ideal candidates for nanoscale manipulation of optical energy, promoting nanofocusing of EM radiation into hot-spots [11–15]. In recent years, a great variety of nanostructure geometries, for instance, nanospheres, nanocubes, nanocones, nanoantennas, nanoaggregates and nanostars, have been fabricated by bottom-up and top-down approaches [16–24], in order to engineer their plasmonic resonances and increase the field enhancement and hot-spot densities. Control over nanostructure morphology shows promising applications in bio/chemical sensing using the synergistic combination of LSPR and surface-enhanced Raman spectroscopy (SERS) as detection paradigm [9, 25–26].

As shown in **Figure 1**, the interaction of the incident light with a plasmonic nanostructure can promote resonant oscillation of the free electron cloud, which, for particles smaller than the exciting wavelength, can give rise to standing waves, i.e., to LSPRs. Noble metals, such as gold, silver and copper, are the best candidates for supporting plasmon activity, due to their low electron losses, high carrier densities and high field amplitudes on the particle surface. The nanostructure morphology (size, shape and arrangement) together with the surrounding dielectric environment plays a key role in the excitation of plasmonic resonances [10]. In LSPR-based sensing devices, the analytes adsorbed on the nanostructure

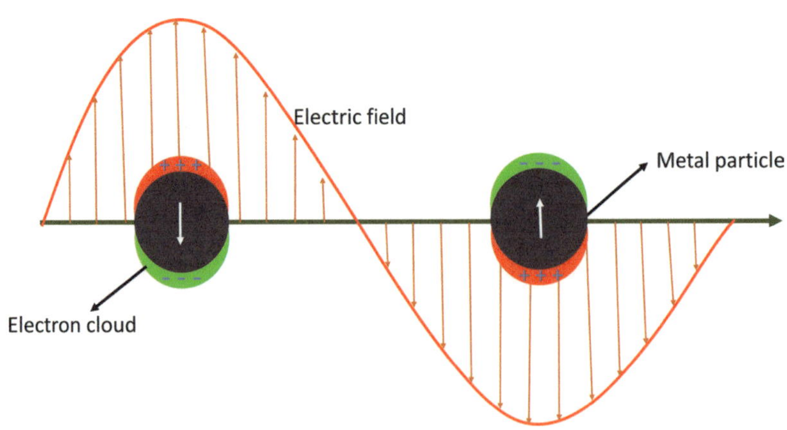

Figure 1. Schematic representation of the localized surface plasmon resonance on a metal particle.

surface influence the surrounding dielectric medium, thus resulting in an effective refractive index change and subsequent resonance shift. Although this technique is very powerful for molecule detection, it lacks analyte specificity, and it suffers from reduced efficiency at very small concentrations.

On the contrary, SERS enables the detection of biological and chemical analytes with high specificity and sensitivity even at ultra-low concentrations, exploiting the electromagnetic enhancement offered by plasmonic nanostructures [3, 27–32]. The huge enhancement typical of SERS substrates is mainly based on chemical and electromagnetic phenomena. In chemical enhancement [33], the charge transfer between electronic energy levels of the metal and the adsorbed molecules increases the Raman scattering cross-section up to a factor of 10^2. The electromagnetic contribution, instead, is crucially depending on the near-field intensity associated with the nanostructure plasmonic activity. In SERS detection, the intrinsically weak Raman signals can be enhanced by many orders of magnitude (a factor of 10^8 has been observed in [34, 35]), due to the interaction between the adsorbed molecules and the electromagnetic near-fields. In this respect, nanostructures featuring closely spaced gaps and/or sharp protrusions are of great interest in detecting analytes at ultra-low concentration [8, 36, 37]. However, the tiny availability of molecules at the single/few entities limit can dramatically reduce the effectiveness of a plasmon-based sensitive device. It is therefore imperative for overcoming this practical limitation to design and engineer nanostructure architectures with sufficiently high hot-spot densities. **Figure 2** shows the local EM field mapping on a multi-branched (MB) nanoparticle using electron energy loss spectroscopy: strong near-field intensities are clearly promoted and well-confined by the sharp apexes. In view of that, nanostructures with different layouts, multiple branches and single-digit IPS will be studied in the present work.

Various nanofabrication methods, applying both bottom-up and top-down approaches, have been used to fabricate SERS-based platforms for sensing applications [38–46]. Among them, colloidal techniques provide a wide variety of nanoparticles with sharp protrusions, although they are still suffering for poor control over uniformity and arrangement [35]. The positioning of the analyte molecule in the vicinity of the hot-spot is critical for improving the SERS enhancement factor and hence the detection limit. Due to the lack of reproducible

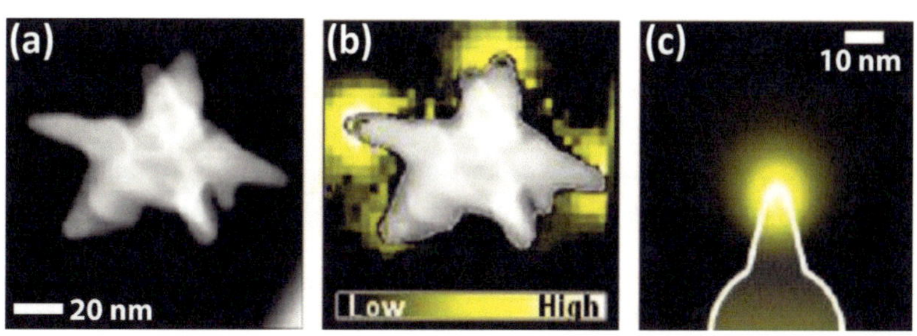

Figure 2. (a) Scanning transmission electron microscopy dark-field image of an individual gold nanostar. (b and c) Electron energy loss spectroscopy intensity mapping and the calculated intensity map of the plasmon resonances around a nanostar apex, respectively. Reprinted with permission from [35]. Copyright (2009) American Chemical Society.

SERS signals and homogeneity of the structures, alternative fabrication methods are needed. In this context, lithographic techniques can overcome these limitations, thus providing a feasible strategy towards the realization of uniformly patterned nanostructures over large areas. In the present chapter, fabrication and characterization of three-dimensional (3D) multi-branched plasmonic architectures realized by means of electron-beam lithography (EBL) and reactive ion etching (RIE) techniques will be investigated. Numerical calculations, Raman and optical characterization will be used for demonstrating outstanding performances in analyte detection at ultra-low concentrations.

2. Experimental techniques

2.1. Fabrication of multi-branched nanostructures

Figure 3a–c shows the fabrication protocol of typical six-branched nanostructures with 2D and 3D layout (without and with perforated metal (PM) topologies). A combination of electron beam lithography (EBL, Raith 150-Two), and reactive ion etching (RIE, SENTECH) has been successfully employed for the uniform production of 3D and 3D PM nano-architectures. As highlighted in **Figure 3a** planar and 3D nanostructures share a common EBL step. A 250 nm polymethyl-methacrylate (PMMA, 950 kDa) layer was spin coated at 3000 rpm for 60 s onto a p-type c-Si (100) wafer. The substrate has been heated at 180 °C for 9 min to get a homogenous PMMA film. After e-beam exposure (electron energy 30 keV and beam current 130 pA), the substrate has been developed in a 3:1 mixture of isopropanol and methyl isobutyl ketone at 4 °C for 3 min. For 2D structures, 3 nm Ti and 18 nm Au were deposited, and the unexposed PMMA removed by ultrasonically assisted lift-off in acetone. In the case of 3D multi-branched nanostar (MBNS) structures (**Figure 3b**), a 20 nm chromium layer was deposited on top of 3 nm Ti and 18 nm Au to act as an etch mask. The excess metal was removed using an ultrasonically assisted lift-off process. Thereafter, substrates were reactive-ion etched (with an etch rate of ≈100 nm min^{-1}) in an atmosphere of SF_6 (30 Standard Cubic Centimeters per Minute—SCCM) and C_4F_8 (32 SCCM) at 1 mTorr, where temperature, power and etching time were held at 4°C, 18 W and 25 s, respectively. After the RIE procedure, the chromium layer was removed by wet-etching in a ceric ammonium nitrate-based mixture (Sigma-Aldrich). 3D PM structures (**Figure 3c**) were obtained by depositing 20 nm of chromium after developing the PMMA. Excess chromium has been removed in acetone lift-off, leaving behind multi-branched shape chromium patterns on silicon, serving as etching masks. RIE was employed to produce the underlying pedestals. Subsequently, the chromium mask was removed leaving behind silicon stars on silicon posts. 3 nm Ti and 18 nm Au were evaporated at a deposition rate of 0.3 Å/s in order to form the Au MBNS as well as the perforated film on the underlying substrate.

2.2. Numerical calculations

Finite integration technique FIT (computer simulation technology-microwave, CST-MW) was used to calculate the near-field properties of the MBNS structures while rigorous coupled

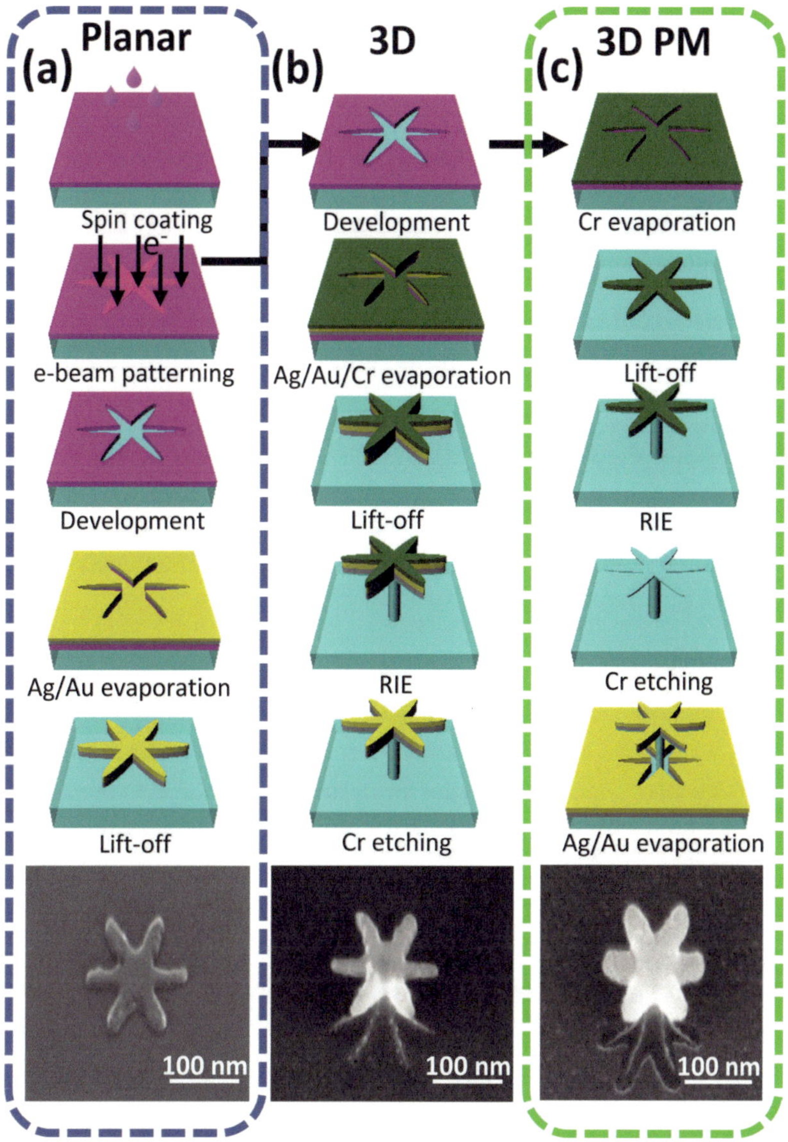

Figure 3. (a–c) Schematic illustration of the protocol, for the fabrication procedure of six-branched nanostructure with different topologies. The corresponding SEM images for six-branched nanostructures are shown below. Reprinted with permission from reference [6]. Copyright 2017 John Wiley and Sons.

wave analysis (Synopsys' Optical Solutions, RSoft) was employed to calculate the far-field response of the nanostructures.

2.3. Optical characterization

The spectral response of the MB nanostructures under normal incidence irradiation was measured using an Olympus IX-73 research microscope. A 100 W halogen lamp (Olympus) with a broadband illumination source in the visible and near-infrared spectral range has been used. The linearly polarized light was obtained by a Glan-Taylor polarizer. Reflection spectra were measured with a 50× objective of numerical aperture 0.5. The collected light has been acquired through a spectrometer with a Peltier-cooled charge-coupled-device from Ocean Optics (QE65000 and NIRquest512 for visible and near-infrared measurements, respectively).

2.4. Surface-enhanced Raman spectroscopy

SERS spectra were recorded with a Renishaw inVia micro-Raman spectrometer equipped with laser excitations at 830, 785, 633 and 532 nm, and a thermo-electrically cooled charge-coupled device (CCD) as detector. A 150× LEICA HCX PL APO objective (numerical aperture 0.95) was used. The diameter of the laser spot was around 680 nm, 800 nm, 1 μm and 1.07 μm for excitation wavelengths at 532, 633, 785 and 830 nm, respectively. The first order silicon peak at 520 cm^{-1} was used to calibrate the instrument; all the spectra were recorded at room temperature in the backscattering geometry. Wire 3.0 software was used to correct the baseline with a third-order polynomial fit. The probe molecules, p-aminothiophenol (p-MA) used in this study, were purchased from Sigma-Aldrich.

2.5. Analyte preparation

A stock solution of p-MA at 1×10^{-3} M concentration was prepared by dissolving the solid analyte into ethanol. Afterward, 1×10^{-6} M to 1×10^{-15} M solutions were prepared by further dilution. Molecules were deposited onto the substrate by chemisorption process. The samples were dipped for 20 min and then rinsed in ethanol in order to remove the excess molecules that were not covalently bounded to the metallic surface. Finally, the substrates were dried with nitrogen gas.

3. Results and discussion: engineering 3D MBNS structures

Raman efficiency and SERS signal intensity have been systematically investigated in different plasmonic platforms, elucidating the effect of substrate, IPS, polarization, metal composition, number of branches and geometrical arrangement, on EM near-field localization and enhancement.

3.1. Effect of geometry, IPS and polarization on SERS enhancement

The effect of 2D and 3D geometry on electric field and SERS enhancements was initially studied using five-branched nanostructure dimers with sub-10 nm IPS as a test-bench. **Figure 4a** shows the schematic presentation of the five-branched 3D PM dimer structure, where L = 150 nm, h = 150 nm, B_w = 50 nm and P_D = 40 nm denote star size, height, branch width and Si pillar diameter, respectively. The effect of nanostructure height on E-field enhancement is shown in **Figure 4b–d**, where the structure size and IPS were kept constant

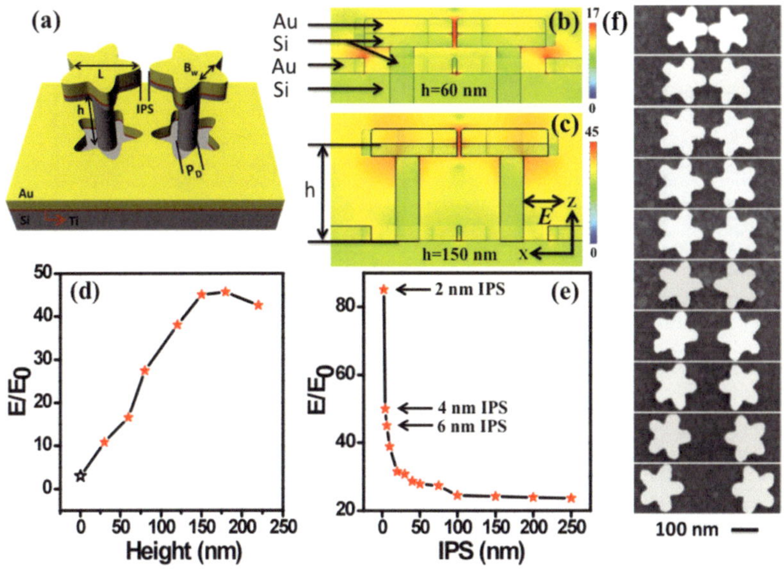

Figure 4. (a) Schematic representation of five-branched 3D PM nanostructure dimer with 150 nm structure size and 6 nm IPS. (b and c) E-field distribution of the nanostructures at h = 60 and 150 nm, respectively. The excitation source is set to 830 nm. Calculated E-field enhancement with respect to nanostructure height (d), and as a function of IPS (e). (f) Normal-incidence SEM images of five-branched 3D PM nanostructure dimer with IPS varying from 6 to 200 nm (top to down, respectively). Reprinted with permission from reference [36]. Copyright 2014 John Wiley and Sons.

for all the samples. A low E-field enhancement (E/E_0, where E and E_0 are the local and the incident electric fields) is observed for the MB nanostructures directly laid on the bulk Si substrate (for h = 0 nm), owing to the strong overlapping of the local fields within the high refractive index Si material. For h = 60 nm, a 6× improvement of local E-field enhancement compared to the 2D structure is observed due to the reduction of the overlap between local E-fields and Si substrate (**Figure 4b**). At h = 150 nm (**Figure 4c**), the local E-field enhancement is 15× that of the corresponding nanostructure with "planar" geometry. Thereafter, a slight reduction is observed with a further increase of h. The role of IPS on E-field enhancement is investigated (**Figure 4e**) with nanostructures of fixed L, h and different IPS ranging from 2 to 250 nm. For IPS of around 2 nm, an E-field enhancement of 85 is observed, and it decreases exponentially with increasing interparticle distances. The large E-fields at low IPS are due to the strong interaction of the LSPRs supported by the nanostructures, thus resulting in the strong localization of intense E-fields (hot-spots). **Figure 4f** shows the SEM images of the nanostructures with 6–200 nm IPS, top-down, respectively.

In order to evaluate the effect of height and IPS experimentally, SERS measurements were performed with p-MA molecules chemisorbed from a solution at 10 μM concentration, **Figure 5**. The incident laser wavelength, power and acquisition time were set to 830 nm, 1.4 mW and 10 s, respectively. The incident light polarization was kept parallel to the IPS axis. **Figure 5a** shows the SERS spectrum of p-MA molecules on five-branched nanostructure dimers with 150 nm height and 6 nm IPS. Prominent modes of p-MA were clearly visible: strong bands

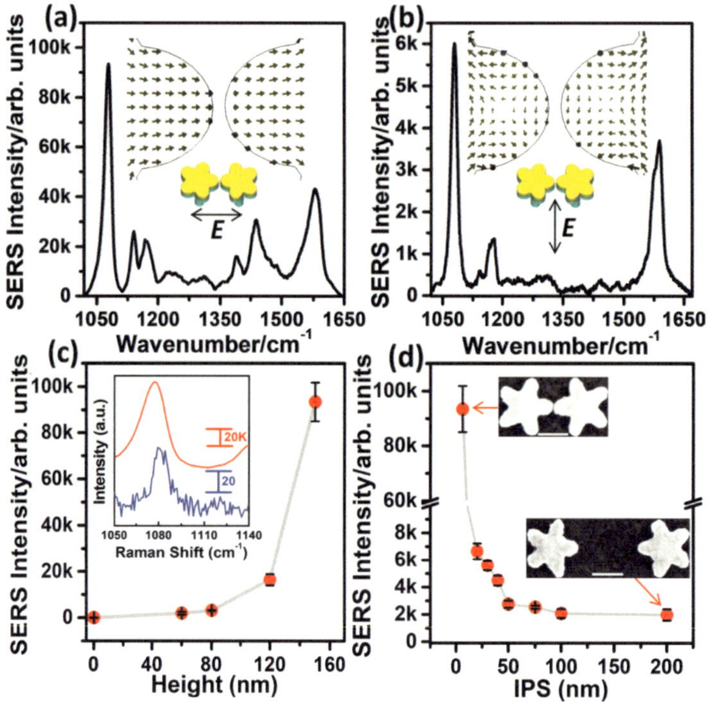

Figure 5. (a and b) Typical SERS spectra of p-MA molecules (chemisorbed at 10 μM concentration) on five-branched 3D PM nanostructure dimer with 6 nm IPS. The excited polarization is set parallel (a) and perpendicular (b) to IPS axis. The corresponding calculated surface-charge distribution is represented by direction arrows, shown in the inset. (c) Experimental SERS signal intensity at 1077 cm^{-1} as a function of nanostructure height, where the IPS is fixed at 6 nm. The inset shows the SERS spectra of five-branched nanostructures at h = 0 (blue spectrum) and 150 nm (red spectrum). (d) Variation of the SERS signal intensity with respect to IPS with constant h at 150 nm. Reprinted with permission from Ref. [36]. Copyright 2014 John Wiley and Sons.

are centered at 1077 and 1590 cm^{-1} while low-intense bands correspond to 1140, 1179, 1390 and 1438 cm^{-1} [6, 8]. Polarization-dependent SERS signal enhancement of MB nanostructures is shown in **Figure 5a** and **b**, where the polarization was set parallel and perpendicular to IPS axis, respectively. Insets represent the calculated surface-charge distributions in correspondence of the IPS region. A dipolar-like distribution of surface charges was observed when the incident light is parallel to the IPS axis (inset of **Figure 5a**). The in-phase dipole moments generate an intense E-field (hot-spot) in the IPS region due to the strong coupling of the sub-10 nm gapped nanostructures. In the case of perpendicular polarization (inset of **Figure 5b**), the effective dipoles are aligned across the interparticle nanocavity leaving a low-intensity local E-field in the IPS region. The corresponding SERS spectra (**Figure 5a** and **b**) discriminate the polarization-induced SERS signal intensities with a factor of 15. The C-S stretching mode located at 1077 cm^{-1} was used to calculate the SERS enhancement factor. **Figure 5c** displays the SERS intensity dependence on nanostructure height of the specific strong band located at 1077 cm^{-1}, showing an exponential-like growth. The inset confirms the high signal-to-noise (SNR) ratio of the 3D PM structures compared to the planar case, for h = 0 (blue spectrum) and h = 150 nm (red spectrum), respectively. The impact of IPS on SERS signal enhancement is shown in **Figure 5d**. An exponential increment in the SERS intensity (around 50×) has been observed upon reduction of the IPS from 200 nm to 6 nm. The experimental findings, i.e., SERS signal dependence on nanostructure height and interparticle separation, are in good agreement with the numerical calculations reported in **Figure 4d** and **e**.

3.2. Effect of bimetal layer, and recycling of SERS substrates

To evaluate the impact of the metal layer composition on SERS signal enhancement, five-branched nanostructure dimers in a ring structure with Au, Ag and AgAu metal layers were investigated. The schematic representation of the AgAu architecture is presented in **Figure 6a**, while SEM images of the corresponding structure are reported in **Figure 6b–d**. The near-field distribution of Au and AgAu nanostructures has been summarized in **Figure 7a–d**. Due to

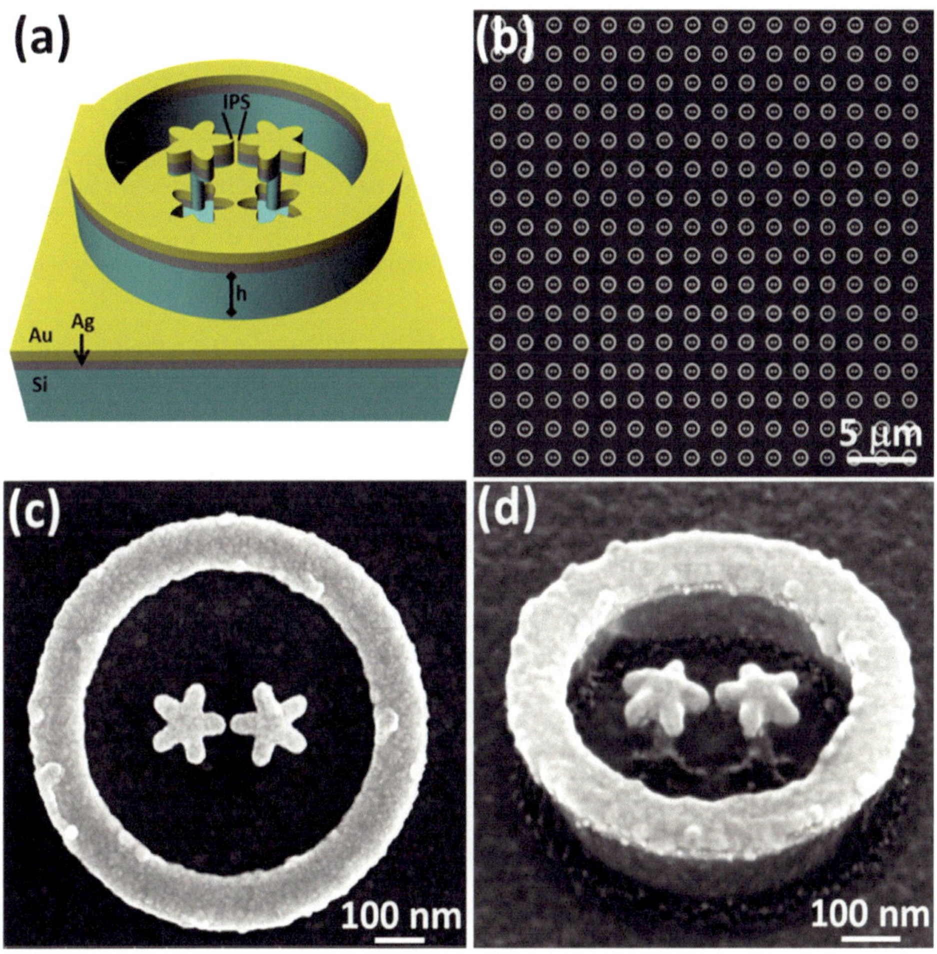

Figure 6. (a) Schematic representation of five-branched 3D PM nanostructure dimer. (b) Normal-incidence SEM image of the nanostructures over a large area. (c and d) Top and tilted view images of individual nanostructures. Reprinted with permission from Ref. [8]. Copyright 2014 American Chemical Society.

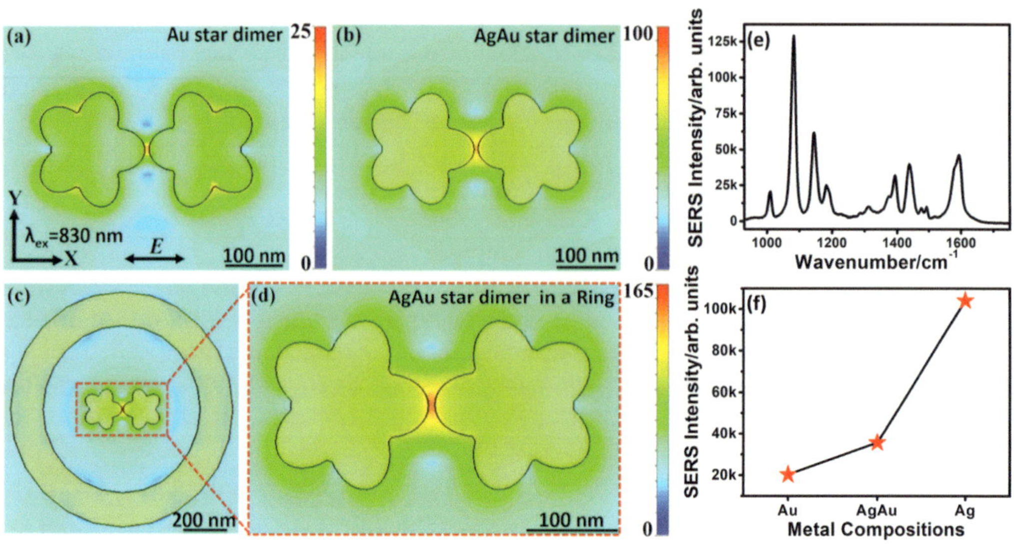

Figure 7. E-field distribution of five-branched 3D PM nanostructure dimer with Au (a) and AgAu (b) metal layers, in the x–y plane. (c and d) E-field distribution of five-branched 3D PM nanostructure dimer in a ring structure with AgAu metal layers. Reprinted with permission from reference [8]. Copyright 2014 American Chemical Society. (e) Typical SERS spectrum of p-MA molecule (at 10 μM concentration) on five-branched 3D PM nanostructure dimer in the ring structure. The exciting laser source, power and accumulation time were set to 830 nm, 1.4 mW and 15 s, respectively. The incident light polarization was set parallel to IPS axis. (f) SERS signal intensity at 1077 cm^{-1} as a function of metal layer compositions.

the strong plasmon resonance of Ag, bimetallic AgAu configurations are endowed with a 4 times higher EM field enhancement. The effect of the nanoring on the MB dimer is shown in **Figure 7c** and **d**. The scattered light is reflected back towards the ring centre where the nanostructures are placed, hence increasing the EM field enhancement in the IPS region by a factor 1.65. **Figure 7e** shows the typical Raman spectrum of p-MA (10 μM concentration) on AgAu five-branched dimers in the ring structure with an IPS of 5 nm. The polarization of the incident light was fixed parallel to the IPS axis. Characteristic Raman bands of p-MA can be clearly observed in the acquired spectrum. The impact of metal layer composition on SERS signal enhancement is plotted for the band located at 1077 cm^{-1} (**Figure 7f**), using 20 nm IPS MB dimer structures with 20 nm Au, 20/20 nm Ag/Au and 20 nm Ag metal layers.

The 3D nanostructure configuration presented so far allows recycling, long-term stability and reutilization of the SERS substrate, thus reducing fabrication costs. **Figure 8a** shows the regeneration protocol for the present architecture. However, as clearly highlighted by the cleaning steps, the process is fully compatible with any kind of 3D plasmonic configuration. In 3D geometry, the nanostructure morphology is conserved by the underlying Si template, while the plasmon active layer is simply recovered by a mask-less wet etching process, followed by metal redeposition. The chemisorbed analyte molecules used for SERS measurements will be completely removed together with the pre-existent metal layer, thus allowing employment of different molecular species after recycling. In order to test the effectiveness of the regeneration process, we have recycled the AgAu nanostructures up to five times (**Figure 8b**) and investigated their SERS response. The corresponding SEM images are shown in the top panel.

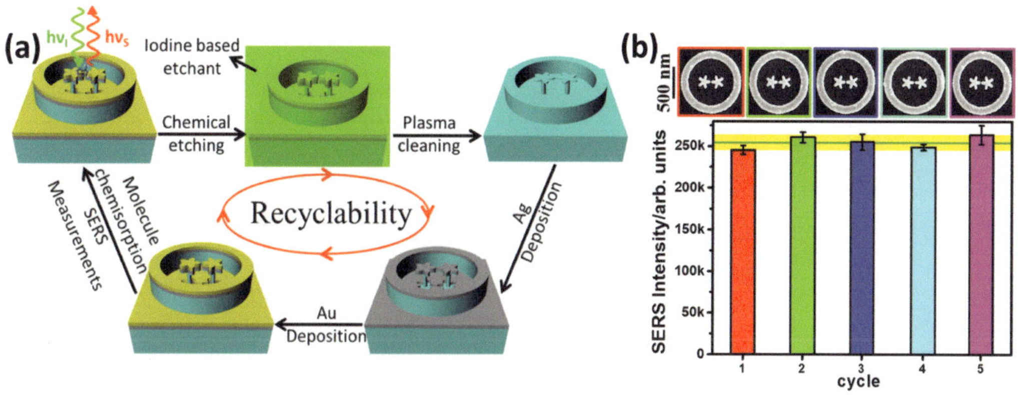

Figure 8. (a) Schematic representation of the maskless recycling process. (b) SERS signal intensity of p-Ma at 1077 cm^{-1} vs. number of regeneration cycles and corresponding SEM images at each recycling step in the top panel. Reprinted with permission from Ref. [8]. Copyright 2014 American Chemical Society.

The SERS signal intensity of the C—S stretching band of p-MA measured after each regeneration cycle is reported in **Figure 8b**. A good correspondence between different recycling steps is clearly observed, showing an average SERS signal deviation below 10%.

3.3. Effect of branch number, perforated metal layer and overall nanostructure arrangement

So far, we have engineered the five-branched nanostructure dimers to improve the hot-spot intensity as a function of nanostructure height, IPS, incoming light polarization and metal composition. Nevertheless, at very low concentrations (down to the femtomolar scale), the number of molecules available per hot-spot is decreased, and thus it is necessary to endow the nanostructures with high hot-spot densities in order to improve their detection limit. In view of that, single plasmonic nanostructures with multiple branches (4–10) and sharp protrusions (tip radius < 10 nm) were investigated.

The schematic representation of a typical eight-branched bimetallic AgAu nanostructure with 3D PM geometry and 200 nm IPS is shown in **Figure 9a**. Nanostructures with 4–10 branches, 20 nm Ag and 20 nm Au metal layers, 140 nm width, 150 nm high and tip radius of 10 nm were used to investigate the effect of hot-spot density on SERS signal enhancement. Normal-incidence SEM image of eight-branched MB nanostructure is shown in **Figure 9b**, where the inset depicts a 54° tilted view. Besides, normal-incidence SEM images of individual MB structures with 4–10 branches are reported in the bottom panel of **Figure 9b**. A typical reflection spectrum of eight-branched 3D PM nanostructures is shown in **Figure 9c**, where multiple plasmonic resonances are clearly observable. To evaluate the effect of branch number on the LSPRs and hot-spot generation, far-field optical spectra and near-field distribution were calculated using RSoft and CST-MW numerical approaches. Distribution of E-field enhancement (E/E_0) for 4–10 branched 2D nanostructures at their respective LSPRs is shown in **Figure 10a–g**. The corresponding data points are shown in **Figure 10h**, along with their experimental counterparts. The incident polarization is set parallel to the x-axis. In the case of 4 branched nanostructures, two

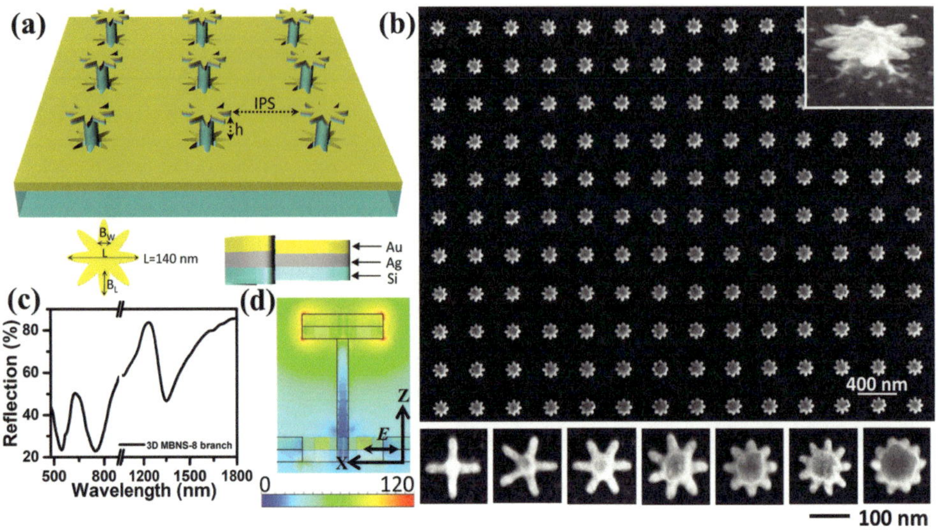

Figure 9. (a) Schematic of bimetallic 3D PM nanostructure for eight-branched nanostructure geometry. (b) Normal incidence SEM image of 3D MB nanostructures with eight-branches arranged in a quadratic array of 200 nm IPS. The inset represents the 54° tilted view of the MBNS with 150 nm silicon pillar height. The bottom panel shows SEM images of MBNS with 4–10 branches, where each micrograph shares the same scale bar. (c) A typical reflection spectrum of eight-branched 3D PM nanostructure. (d) Electric field distribution of eight-branched 3D PM nanostructure at its characteristic LSPR position of 685 nm.

hot-spots are observed parallel to the incident light polarization direction. The hot-spots density increases according to the number of branches, inducing a clear blue shift in LSPR spectral position (**Figure 10h**). This behaviour can be explained using a simple dipole theory. An increasing hot-spot density, that goes along with the number of branches, results in a higher total restoring force which blue shifts the nanostructure LSPR. Tailoring of the branch morphology enables precise control over the generation and spatial distribution of the hot-spots on the single nanostructure, thus opening new perspectives in reproducible SERS signal detection from large biomolecules, where the analyte size is many times larger than the individual hot-spot volume.

The influence of 2D, 3D and 3D PM geometries on the optical response and SERS enhancement was experimentally validated, as reported in **Figures 11** and **12**, respectively. Numerically calculated reflectance spectra of the eight-branched nanostructures with 2D, 3D and 3D PM topologies (see schematics in **Figure 11b–d**) are shown in **Figure 11a**. The corresponding experimental spectra are summarized in **Figure 11e**. In both cases, the optical spectra are normalized with respect to the flat/unpatterned area of the same sample. The experimental spectra show good correlation with their numerical counterparts. The E-field distribution of the nanostructures at their LSPR position is shown in **Figure 11f–h**. In the 2D case, near-field distribution maps (**Figure 11f**) clearly show the E-field confinement at the metal and bulk silicon interface. Due to the overlap of the E-field profiles into the bulk silicon a low-intense E-field enhancement (around 20) is observed. In general, surface plasmon resonances are tightly confined in high refractive index materials (e.g. silicon), which results in low E-field enhancement, low extinction-cross section, large propagation losses, broadening and red shifting of resonances compared to low-index materials [47]. As highlighted in the previous sections, 3D nanostructures decouple the hot-spot confinement from the substrate and enhance its strength by reducing the effective refractive

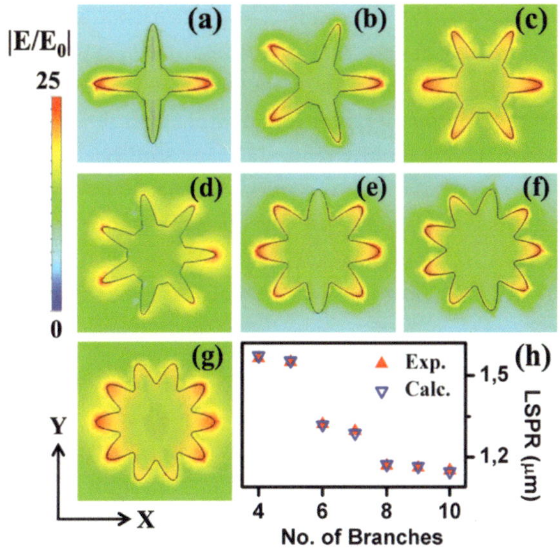

Figure 10. (a–g) E-field distribution of 2D nanostructures with varying branch number (4–10). (h) Corresponding LSPR positions for experimental and calculated data points. Reprinted with permission from Ref. [6]. Copyright 2017 John Wiley & Sons.

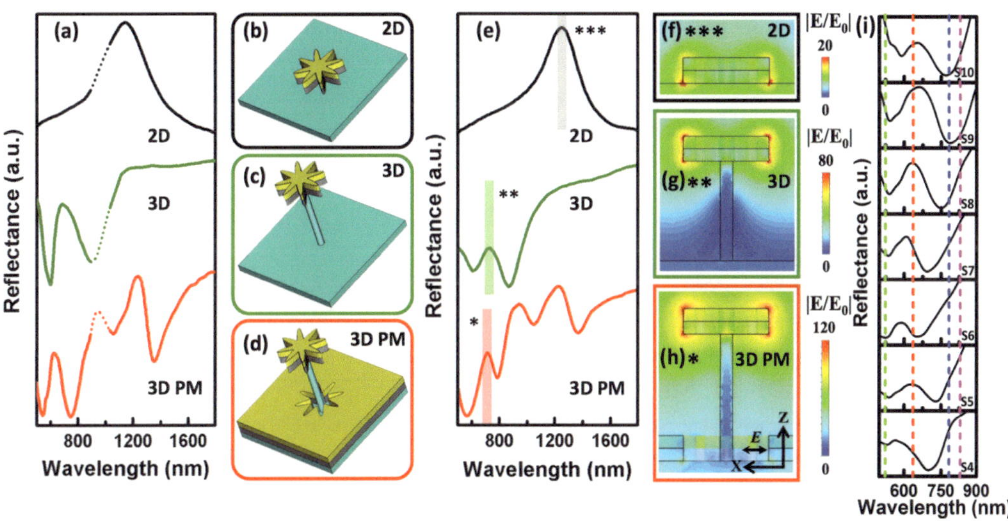

Figure 11. (a) Experimental reflectance spectra of eight-branched nanostructures with 2D, 3D and 3D PM geometries. (b–d) Schematic representations of the structure geometries simulated in (a). (e) Experimentally measured reflection spectra of the nanostructures shown in (a). (f–h). Near-field profiles of E-field distribution in x–z plane for the LSPRs of the nanostructures with 2D, 3D and 3D PM topologies. Reprinted with permission from Ref. [6]. Copyright 2017 John Wiley & Sons.

index of the embedding medium. The reflectance spectrum of the 3D nanostructure (**Figure 11a**) shows a significant blue shift of LSPR (around 485 nm), with a resonance maximum at around 680 nm. **Figure 11g** clearly shows that the generated E-fields (with an enhancement of around 80) are decoupled from the substrate by means of a dielectric nanopedestal. These strong E-field regions can be easily accessible to the analytes in Raman measurements if compared to the planar configuration. In the case of 3D PM geometry, the numerical spectrum (**Figure 11a**) shows various modes corresponding to the LSPR of the nanostructure and the underlying nanohole, and surface plasmon polaritons (SPP) at metal/air and metal/substrate interfaces. The LSPR position of the 3D PM structure (650 nm) is slightly shifted (a blue shift of 30 nm) from the 3D geometry, and a high E-field enhancement of 120 is observed. A deeper analysis of the perforated metal contribution has been discussed elsewhere [6]. The spectral features corresponding to the LSPRs of the 3D PM nanostructure and star-shaped hole cavity are located at about 620 and 1350 nm, while the 543 and 1050 nm resonances can be associated to the SPPs of the PM layer.

In order to assess the impact of hot-spot density on SERS enhancement, Raman measurements were performed on 4–10 branched nanostructures with p-MA molecules chemisorbed at 1 μM concentration. Typical SERS spectra of 8 branched nanostructures with 2D, 3D and 3D PM geometries are shown above. The incident laser wavelength, acquisition time and power were set to 785 nm, 30 s and 1 mW, respectively, and the incident light polarization was fixed along the x-axis. Characteristic Raman modes of p-MA are clearly visible in the spectra acquired on 3D PM MB nanostructures. In the planar case, the peaks centered around 1077 and 1590 cm^{-1} are experimentally observable, but the other bands are buried into the background noise. When the nanostructures are decoupled from the substrate via a dielectric nanopedestal (3D case), a significant rise in SERS signal intensity is observed along with the presence of all characteristic p-MA Raman modes. Furthermore, an additional enhancement in SERS signal intensity is observed for the 3D PM nanostructure geometry due to the coupling of the MBNS with the reflected light coming from the perforated metal layer underneath. The 3D PM MB nanostructure shows an absolute SERS enhancement in the order of 10^{11} obtained from the evaluation of the peak intensity at 1077 cm^{-1}, with reference to Raman spectra of p-MA molecules on a planar SERS active gold film. Details on the corresponding calculations can be found in [6].

In order to understand the wavelength-dependent hot-spot generation and SERS enhancement, 4–10 branched 3D PM nanostructures were excited with four different laser sources (**Figure 12b**). The SERS signal intensity of the 1077 cm^{-1} band is plotted with respect to the number of branches and excitation lasers (532, 633, 785 and 830 nm). Due to different spectral power densities of the excited lasers, the SERS intensities are normalized independently with respect to the highest peak intensity obtained in the series. A monotonic increment of the SERS signal is observed by raising the number of branches for both 785 and 830 nm laser excitations, which are off-resonance with respect to the nanostructure LSPRs (see, **Figure 11i**). In this scenario, the SERS signal intensity can be associated to the increment of the number of hot-spots with multiple branches. For 633 nm laser excitation, the eight-branched nanostructure shows highest SERS signal intensity owing to the overlap between the LSPR and the laser source. A similar trend has been observed for 532 nm excitation source.

Ultra-sensitive detection of analyte molecules was probed on eight-branched 3D PM nanostructures with p-MA molecules at concentrations ranging from 1 μM to 1 fM (**Figure 12c**). The incident wavelength, power and acquisition time were set to 785 nm, 1 mW and 3 s, respectively. At 1 μM concentration, the SERS spectra show the characteristic Raman bands of p-MA with good SNR. A decrease in SERS signal intensity is observed upon reduction of the molecular concentration from 1 μM to 1 fM, due to a lower number of adsorbed molecules on the nanostructure

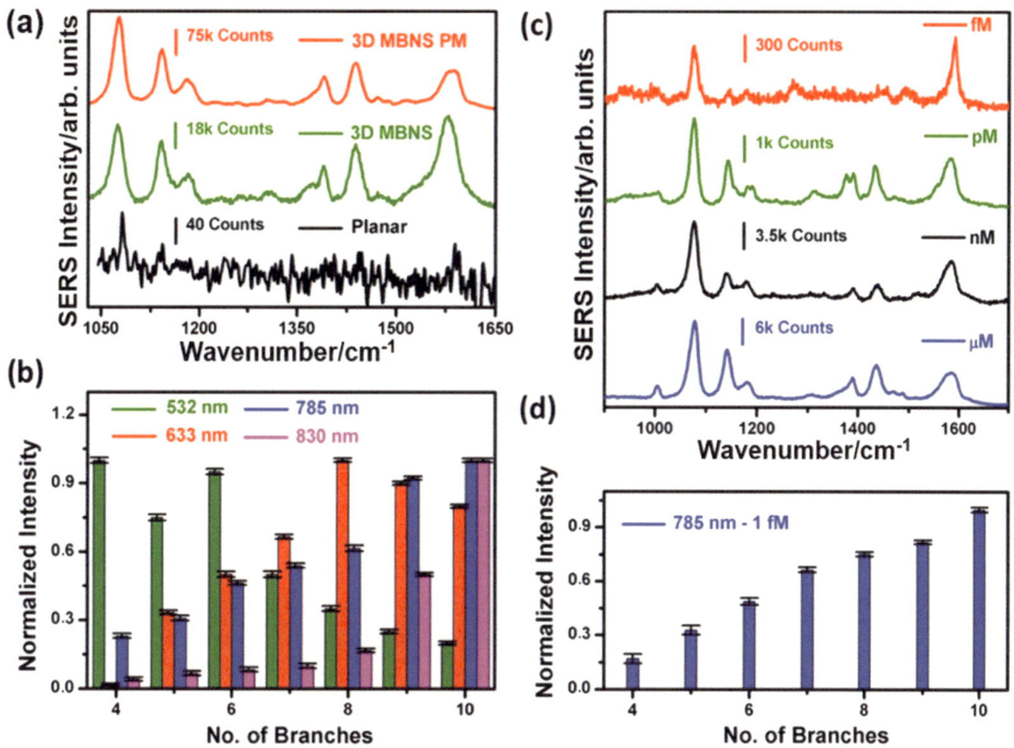

Figure 12. (a) SERS spectra of p-MA molecules chemisorbed at 1 µM concentration on eight-branched nanostructures with 2D, 3D and 3D PM geometries. (b) SERS signal intensity at 1077 cm^{-1} as a function of a number of branches and different exciting wavelengths (532, 633, 785 and 830 nm). (c) SERS spectra of p-MA molecules at 1 µM, 1 nM, 1 pM and 1 fM concentrations taken on eight-branched 3D PM nanostructures for 785 nm laser excitation. (d) SERS signal intensity at 1077 cm^{-1} versus number of branches at ultra-low concentration, 1 fM. Reprinted with permission from Ref. [6]. Copyright 2017 John Wiley & Sons.

surface. For 1 fM concentration, only the Raman bands at 1077 and 1590 cm^{-1} are clearly visible, while the other features lie in the background noise. Despite of this, analyte detection at 1 fM showed the ultra-sensing capability of the substrate towards single/few molecules detection. In order to assess the effect of branch number on the detection limit, SERS measurements were performed on 4–10 branched 3D PM nanostructures with p-MA chemisorbed from a 1 fM concentrated solution (**Figure 12d**). An increment in SERS signal intensity at 1077 cm^{-1} is observed by raising the number of branches, owing to the larger number of generated hot-spots.

Reproducible identification and detection of biological samples/chemicals at ultra-low concentrations remains a huge challenge due to lack of high hot-spot density substrates. At ultra-low concentrations, single/few molecules adsorbed in the vicinity of the hot-spot sites provide the majority of the SERS signal intensity. In this situation, if the molecules are not absorbed in the proximity of the hot-spot, the molecular fingerprint of the analyte cannot be identified. Thus, development of plasmonic nanostructures with a high hot-spot density that enables reproducible detection at ultra-low concentrations is of paramount importance in the field of molecular sensing. Multi-branched nanostructure designs hold concrete promises in this direction. In view of that, eight-branched 3D PM nanostructures arranged in the form of single, dimer, 3 × 3 array of clusters and chain of nanostructures, as shown in **Figure 13**, were investigated by keeping the interparticle distance fixed at 200 nm.

Near-field distributions of 2D single, dimer, 3 × 3 periodic array and chain of eight-branched nanostructures at their characteristic LSPRs, are shown in the **Figure 14a–d**, respectively. In order to reduce the computational time, 2D structures were used to compare the hot-spot density and E-field enhancement. The corresponding LSPR positions (experimental and calculated) are shown in **Figure 15**. A red-shift in the plasmon resonance has been observed with different layouts, as a consequence of the interaction between adjacent nanostructures. It is clearly visible that the hot-spot density is increased with respect to the arrangement schemes and, the highest E-field enhancement is observed for nanostructures arranged in the form of a chain.

SERS spectra were acquired in order to address the effect of different geometrical configurations on the measured signal. **Figure 16a** shows a typical SERS spectrum of p-MA (at 1 μM concentration) on eight-branched 3D PM structures arranged in the form of a chain. The exciting laser, acquisition time and power were set to 785 nm, 10 s and 1 mW, respectively, while the impinging light polarization was fixed along the x-axis. **Figure 16b** shows the SERS signal intensity for the band at 1077 cm^{-1} with respect to the nanostructure arrangement. Dimer configuration presents higher field enhancement in comparison to the isolated geometry, as

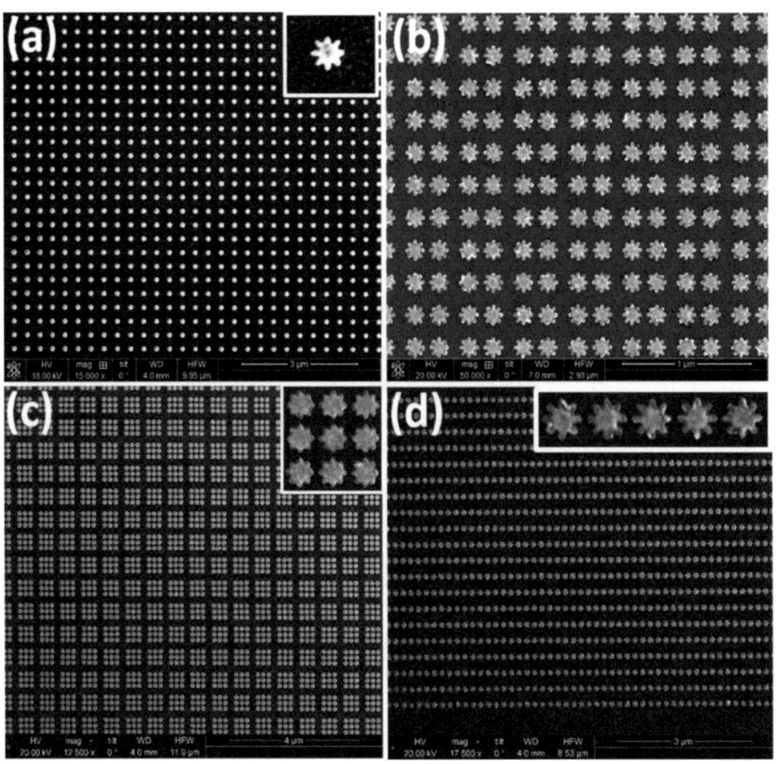

Figure 13. (a–d). Normal incidence SEM images of the eight-branched 3D PM MB nanostructures in the form of single, dimer, 3 × 3 periodic array of clusters and chain of nanostructures, respectively. The inset shows the magnified view of the nanostructures.

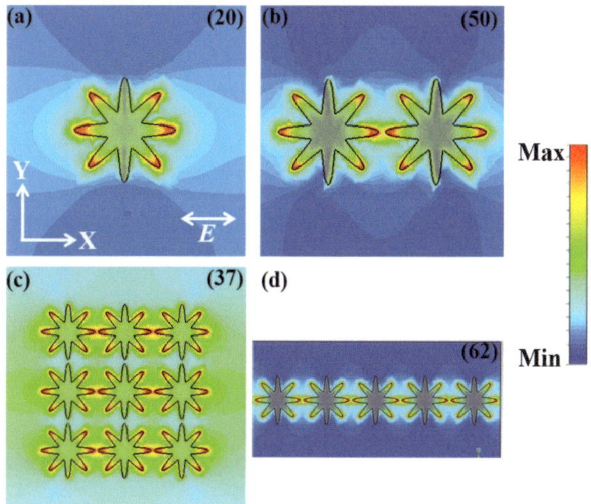

Figure 14. E-field distribution of single, dimer, 3 × 3 periodic array and chain of MBNS (a–d) at their characteristic LSPRs. The incident light is polarized along the x-axis.

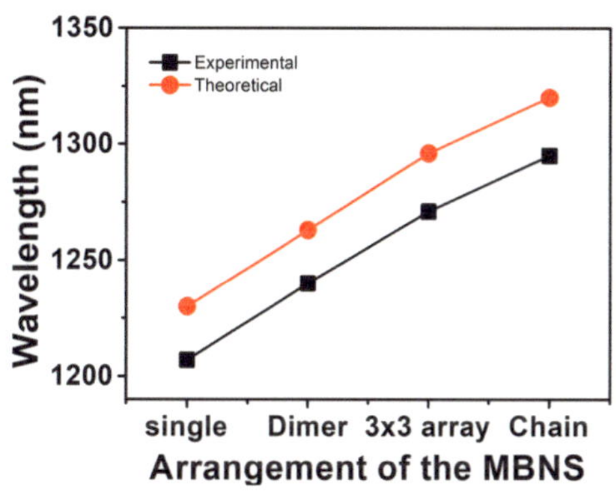

Figure 15. LSPR positions (theoretical and experimental) of 2D eight-branched nanostructures arranged in the form of single, dimer, 3 × 3 periodic array of clusters and chain of nanostructures.

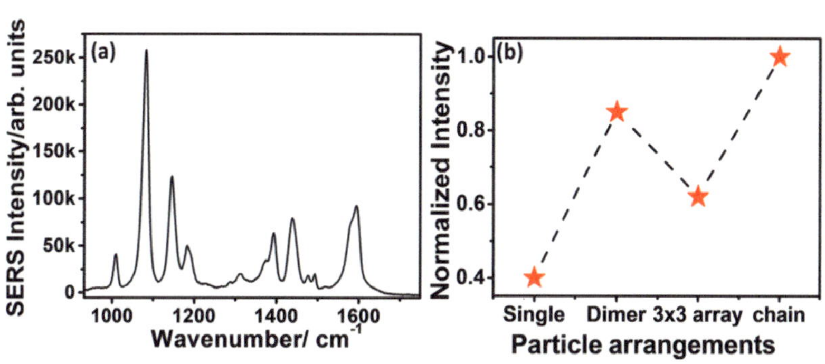

Figure 16. (a) The SERS spectrum of p-MA at 1 μM concentration taken on eight-branched 3D PM nanostructures arranged in the form of a chain. (b) Normalized SERS signal intensity variation at 1077 cm^{-1} with respect to different arrangements of the nanostructures at 1 μM p-MA concentration.

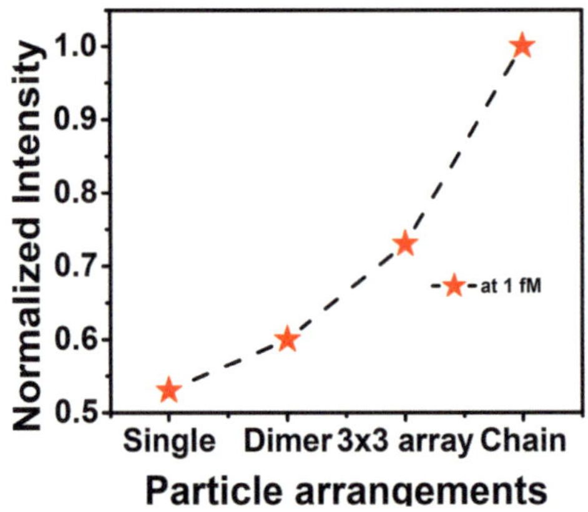

Figure 17. Normalized SERS signal intensity variation at 1077 cm^{-1} with respect to different arrangements of the eight-branched 3D PM nanostructures at 1 fM p-MA concentration.

a consequence of the strong hot-spot confined in the IPS region. Periodic 3 × 3 arrays of nanostructures show higher E-field enhancement compared to single nanostructures, and a lower E-field enhancement with respect to the dimer layout, which is in good agreement with the E-field enhancements observed in **Figure 14**. Nanostructures arranged in the form of chain show the highest E-field enhancement (a factor of 62) in comparison to the other geometries. **Figure 17** shows the SERS signal intensity at 1077 cm^{-1} as a function of the arrangements of nanostructures at 1 fM p-MA concentration. As in the previous situation, the highest SERS intensities were observed for chains, as a result of the higher hot-spot density.

4. Conclusions and outlook

In summary, the engineering of 3D multi-branched (up to 10 branches) nanostructures for sensing of analyte molecules at ultra-low concentrations, down to 1fM, is demonstrated to be highly feasible. Numerical simulations were performed to understand the underlying physics of high electric field enhancement of the plasmonic nanostructures. The advancement of the 3D fabrication methods enables the realization of uniform, homogenous and reproducible SERS devices. Reflection and SERS measurements were carried out to evaluate the MB nanostructure performances. Within this context, we demonstrated the importance of the geometry, IPS and polarization on SERS signal enhancement using 3D five-branched nanostar dimers (with sub-10 nm IPS). The elevated 3D geometry shows the advantage of high E-field enhancement over 2D geometry due to decoupling from the underlying substrate of the strong optical-near fields localized at the metal/dielectric interface. In particular, the 3D geometry enables direct interaction of analytes with hot-spot spatial regions, which are severely affected by solid dielectric substrates in the 2D geometry case. This kind of SERS architectures is particularly important in miniaturized lab-on-chip Raman detection systems, thus allowing the exploitation of lower laser powers with no consequencie over the device sensitivity. Moreover, the low-cost recycling capability of the 3D geometry counterbalances the production cost and time defined by the lithographic process. The effect of metal layer composition on SERS signal enhancement of p-MA molecules, and recycling capabilities of

3D structures were investigated with five-branched nanostar dimers in the ring structures. The effect of the number of branches, varying from 4 to 10, on the hot-spot generation and SERS enhancement was evaluated by using individual nanostructures (separated by 200 nm IPS). Moreover, the arrangement of MB nanostructures in various configurations (single, dimer, 3 × 3 array of clusters and chain of nanostructures) was evaluated to improve the device detection limit towards the single-molecule regime. 3D multi-branched nanostructures exhibit enhancement factors in the order of 10^{11} with an extremely high sensing capability (down to 1fM concentration). In this view, engineering the aforementioned architectures for high hot-spot density paves the way towards commercial biosensing applications, which require single/few-molecule detection sensitivity, with scalable manufacturing methods and cost-effective approaches. The proposed devices do not require specific labeling of the investigated analytes and multiple testing can be evaluated on the same platform. A new class of biological experiments will be therefore feasible, including monitoring growth factors that are produced from cultured cells. Moreover, our plasmonic nanostructures can be employed for direct detection of proteins within biological samples and real-time monitoring of chemical reactions. When applied to biomedicine, the present results, combined with already available purification methods, suggest the possibility of improving the early detection of several diseases, including cancer, where the number of clinically significant molecules at the onset of the pathology is very small and often generated by a single cell.

Author details

Anisha Chirumamilla[1,2], Manohar Chirumamilla[2*], Alexander S. Roberts[3], Andrea Cerea[4], Esben Skovsen[2], Francesco De Angelis[4], Remo Proietti Zaccaria[4,5], Peter Kjær Kristensen[2], Roman Krahne[4], Duncan S. Sutherland[1], Sergey I. Bozhevolnyi[3], Kjeld Pedersen[2] and Andrea Toma[4*]

*Address all correspondence to: mch@nano.aau.dk and andrea.toma@iit.it

1 Interdisciplinary Nanoscience Center (iNANO), Århus University, Aarhus C, Denmark

2 Department of Physics and Nanotechnology, Aalborg University, Aalborg Øst, Denmark

3 Centre for Nano Optics, University of Southern Denmark, Odense M, Denmark

4 Istituto Italiano di Tecnologia, Genova, Italy

5 Cixi Institute of Biomedical Engineering, Ningbo Institute of Materials Technology and Engineering, Chinese Academy of Sciences, Ningbo, China

References

[1] Cao Y, Zhang J, Yang Y, Huang Z, Long NV, Fu C. Engineering of SERS substrates based on noble metal nanomaterials for chemical and biomedical applications. Applied Spectroscopy Reviews. 2015;**50**:499. DOI: 10.1080/05704928.2014.923901

[2] Xuming Z, Yu Lim C, Ru-Shi L, Din Ping T. Plasmonic photocatalysis. Reports on Progress in Physics. 2013;**76**:046401

[3] Lane LA, Qian X, Nie S. SERS nanoparticles in medicine: From label-free detection to spectroscopic tagging. Chemical Reviews. 2015;**115**:10489. DOI: 10.1021/acs.chemrev.5b00265

[4] Fateixa S, Nogueira HIS, Trindade T. Hybrid nanostructures for SERS: Materials development and chemical detection. Physical Chemistry Chemical Physics. 2015;**17**:21046. DOI: 10.1039/C5CP01032B

[5] Huang J-A, Zhang Y-L, Ding H, Sun H-B. SERS-enabled lab-on-a-chip systems. Advanced Optical Materials. 2015;**3**:618. DOI: 10.1002/adom.201400534

[6] Chirumamilla M, Chirumamilla A, Roberts AS, Proietti Zaccaria R, De Angelis F, Kjær Kristensen P, Krahne R, Bozhevolnyi SI, Pedersen K, Toma A. Hot-spot engineering in 3D multi-branched nanostructures: Ultrasensitive substrates for surface-enhanced Raman spectroscopy. Advanced Optical Materials. 2017;**5**:1600836. DOI: 10.1002/adom.201600836

[7] Chirumamilla M, Chirumamilla A, Yang Y, Roberts AS, Kristensen PK, Chaudhuri K, Boltasseva A, Sutherland DS, Bozhevolnyi SI, Pedersen K. Large-area Ultrabroadband absorber for solar thermophotovoltaics based on 3D titanium nitride nanopillars. Advanced Optical Materials. 2017;**5**:1700552. DOI: 10.1002/adom.201700552

[8] Gopalakrishnan A, Chirumamilla M, De Angelis F, Toma A, Proietti Zaccaria R, Krahne R. Bimetallic 3D nanostar dimers in ring cavities: Recyclable and robust surface-enhanced Raman scattering substrates for signal detection from few molecules. ACS Nano. 2014;**8**:7986. DOI: 10.1021/nn5020038

[9] Willets KA, Duyne RPV. Localized surface Plasmon resonance spectroscopy and sensing. Annual Review of Physical Chemistry. 2007;**58**:267. DOI: 10.1146/annurev.physchem.58.032806.104607

[10] Kelly KL, Coronado E, Zhao LL, Schatz GC. The optical properties of metal nanoparticles: The influence of size, shape, and dielectric environment. The Journal of Physical Chemistry B. 2002;**107**:668. DOI: 10.1021/jp026731y

[11] Moskovits M. Imaging: Spot the hotspot. Nature. 2011;**469**:307

[12] Zhang R, Zhang Y, Dong ZC, Jiang S, Zhang C, Chen LG, Zhang L, Liao Y, Aizpurua J, Luo Y, Yang JL, Hou JG. Chemical mapping of a single molecule by plasmon-enhanced Raman scattering. Nature. 2013;**498**:82. DOI: 10.1038/nature12151

[13] Ahmed A, Gordon R. Single molecule directivity enhanced Raman scattering using nanoantennas. Nano Letters. 2012;**12**:2625. DOI: 10.1021/nl301029e

[14] Radziuk D, Moehwald H. Prospects for plasmonic hot spots in single molecule SERS towards the chemical imaging of live cells. Physical Chemistry Chemical Physics. 2015;**17**:21072. DOI: 10.1039/C4CP04946B

[15] Das G, Chirumamilla M, Toma A, Gopalakrishnan A, Proietti Zaccaria R, Alabastri A, Leoncini M, Di Fabrizio E. Plasmon based biosensor for distinguishing different peptides mutation states. Scientific Reports. 2013;**3**:1-6. DOI: 10.1038/srep01792

[16] Toma A, Das G, Chirumamilla M, Saeed A, Proietti Zaccaria R, Razzari L, Leoncini M, Liberale C, De Angelis F, Di Fabrizio E. Fabrication and characterization of a nanoantenna-based Raman device for ultrasensitive spectroscopic applications. Microelectronic Engineering. 2012;**98**:424. DOI: 10.1016/j.mee.2012.07.066

[17] Das G, Chirumamilla M, Gopalakrishnan A, Toma A, Panaro S, Proietti Zaccaria R, De Angelis F, Di Fabrizio E. Plasmonic nanostars for SERS application. Microelectronic Engineering. 2013;**111**:247. DOI: 10.1016/j.mee.2013.04.028

[18] Chirumamilla M, Das G, Toma A, Gopalakrishnan A, Proietti Zaccaria R, Liberale C, De Angelis F, Di Fabrizio E. Optimization and characterization of au cuboid nanostructures as a SERS device for sensing applications. Microelectronic Engineering. 2012;**97**:189. DOI: 10.1016/j.mee.2012.05.004

[19] Hanif M, Juluri RR, Chirumamilla M, Popok VN. Poly(methyl methacrylate) composites with size-selected silver nanoparticles fabricated using cluster beam technique. Journal of Polymer Science Part B: Polymer Physics. 2016;**54**:1152. DOI: 10.1002/polb.24021

[20] Accardo A, Proietti Zaccaria R, Candeloro P, Gentile F, Coluccio ML, Das G, Krahne R, Liberale C, Toma A, Panaro S, Miele E, Chirumamilla M, Rajamanickam V, Di Fabrizio E. In: Bhushan B, Luo D, Schricker SR, Sigmund W, Zauscher S, editors. Handbook of Nanomaterials Properties. Heidelberg: Springer; 2014. p. 615. DOI: 10.1007/978-3-642-31107-9_42

[21] Toma A, Das G, Proietti Zaccaria R, Chirumamilla M, Gentile F, Mecarini F, Coluccio ML, Leoncini M, Liberale C, Francardi M, De Angelis F, Di Fabrizio E. In: Lamy de la Chapelle M, Pucci A, editors. Nanoantenna: Plasmon-Enhanced Spectroscopies for Biotechnological Applications. USA: Pan Stanford; 2013. p. 267. DOI: 10.1201/b14594-9

[22] Wang ZY, Zhang RJ, Wang SY, Lu M, Chen X, Zheng YX, Chen LY, Ye Z, Wang CZ, Ho KM. Broadband optical absorption by tunable Mie resonances in silicon nanocone arrays. Scientific Reports. 2015;**5**:7810. DOI: 10.1038/srep07810

[23] Wei H, Hossein Abtahi SM, Vikesland PJ. Plasmonic colorimetric and SERS sensors for environmental analysis. Environmental Science: Nano. 2015;**2**:120. DOI: 10.1039/C4EN00211C

[24] Roberts AS, Søndergaard T, Chirumamilla M, Pors A, Beermann J, Pedersen K, Bozhevolnyi SI. Light extinction and scattering from individual and arrayed high-aspect-ratio trenches in metals. Physical Review B. 2016;**93**:075413. DOI: 10.1103/PhysRevB.93.075413

[25] Nie S, Emory SR. Probing single molecules and single nanoparticles by surface-enhanced Raman scattering. Science. 1997;**275**:1102. DOI: 10.1126/science.275.5303.1102

[26] Proietti Zaccaria RSP, Toma A, Chirumamilla M, Giugni A, Das G, Krahne R, Di Fabrizio E. In: Lamy de la Chapelle M, Giuseppe Gucciardi P, Lidgi-Guigui N, editors. Handbook of Enhanced Spectroscopy. USA: Pan Stanford; 2015. p. 141. DOI: 10.4032/9789814613330

[27] Etchegoin PG, Le Ru EC. A perspective on single molecule SERS: Current status and future challenges. Physical Chemistry Chemical Physics. 2008;**10**:6079. DOI: 10.1039/b809196j

[28] Harmsen S, Huang R, Wall MA, Karabeber H, Samii JM, Spaliviero M, White JR, Monette S, O'Connor R, Pitter KL, Sastra SA, Saborowski M, Holland EC, Singer S, Olive KP, Lowe SW, Blasberg RG, Kircher MF. Surface-enhanced resonance Raman scattering nanostars for high-precision cancer imaging. Science Translational Medicine. 2015;**7**:271ra7. DOI: 10.1126/scitranslmed.3010633

[29] Alvarez-Puebla RA, Liz-Marzán LM. SERS-based diagnosis and biodetection. Small. 2010;**6**:604. DOI: 10.1002/smll.200901820

[30] Lin C-C, Yang Y-M, Chen Y-F, Yang T-S, Chang H-C. A new protein A assay based on Raman reporter labeled immunogold nanoparticles. Biosensors and Bioelectronics. 2008; **24**:178. DOI: 10.1016/j.bios.2008.03.035

[31] Lee S, Kim S, Choo J, Shin SY, Lee YH, Choi HY, Ha S, Kang K, Oh CH. Biological imaging of HEK293 cells expressing PLCγ1 using surface-enhanced Raman microscopy. Analytical Chemistry. 2007;**79**:916. DOI: 10.1021/ac061246a

[32] Kneipp K, Wang Y, Kneipp H, Perelman LT, Itzkan I, Dasari RR, Feld MS. Single molecule detection using surface-enhanced Raman scattering (SERS). Physical Review Letters. 1997;**78**:1667. DOI: 10.1103/PhysRevLett.78.1667

[33] Reilly TH, Chang S-H, Corbman JD, Schatz GC, Rowlen KL. Quantitative evaluation of plasmon enhanced Raman scattering from nanoaperture arrays. The Journal of Physical Chemistry C. 2007;**111**:1689. DOI: 10.1021/jp066802j

[34] Xu H, Bjerneld EJ, Käll M, Börjesson L. Spectroscopy of single Hemoglobin molecules by surface enhanced Raman scattering. Physical Review Letters. 1999;**83**:4357. DOI: 10.1103/PhysRevLett.83.4357

[35] Rodríguez-Lorenzo L, Álvarez-Puebla RA, Pastoriza-Santos I, Mazzucco S, Stéphan O, Kociak M, Liz-Marzán LM, García de Abajo FJ. Zeptomol detection through controlled ultrasensitive surface-enhanced Raman scattering. Journal of the American Chemical Society. 2009;**131**:4616. DOI: 10.1021/ja809418t

[36] Chirumamilla M, Toma A, Gopalakrishnan A, Das G, Proietti Zaccaria R, Krahne R, Rondanina E, Leoncini M, Liberale C, De Angelis F, Di Fabrizio E. 3D nanostar dimers with a sub-10-nm gap for single-/few-molecule surface-enhanced Raman scattering. Advanced Materials. 2014;**26**:2353. DOI: 10.1002/adma.201304553

[37] Di Fabrizio E, Gentile F, Donnorso MP, Chirumamilla M, Miele E, Coluccio ML, La Rocca R, Brescia R, Krahne R, Das G, De Angelis F, Liberale C, Toma A, Razzari L, Manna L, Proietti Zaccaria R. In: Hashim AA, editor. The Delivery of Nanoparticles. Rijeka: Intech; 2012. p. 293. DOI: 10.5772/2647

[38] Ou FS, Hu M, Naumov I, Kim A, Wu W, Bratkovsky AM, Li X, Williams RS, Li Z. Hot-spot engineering in polygonal nanofinger assemblies for surface enhanced Raman spectroscopy. Nano Letters. 2011;**11**:2538. DOI: 10.1021/nl201212n

[39] Jung K, Hahn J, In S, Bae Y, Lee H, Pikhitsa PV, Ahn K, Ha K, Lee J-K, Park N, Choi M. Hotspot-engineered 3D multipetal flower assemblies for surface-enhanced Raman spectroscopy. Advanced Materials. 2014;**26**:5924. DOI: 10.1002/adma.201401004

[40] Forestiere C, Pasquale AJ, Capretti A, Miano G, Tamburrino A, Lee SY, Reinhard BM, Dal Negro L. Genetically engineered Plasmonic Nanoarrays. Nano Letters. 2012;**12**:2037. DOI: 10.1021/nl300140g

[41] Kleinman SL, Frontiera RR, Henry A-I, Dieringer JA, Van Duyne RP. Creating, characterizing, and controlling chemistry with SERS hot spots. Physical Chemistry Chemical Physics. 2013;**15**:21. DOI: 10.1039/C2CP42598J

[42] Fabris L. Bottom-up optimization of SERS hot-spots. Chemical Communications. 2012; **48**:9346. DOI: 10.1039/C2CC34068B

[43] Chirumamilla M, Gopalakrishnan A, Toma A, Proietti Zaccaria R, Krahne R. Plasmon resonance tuning in metal nanostars for surface enhanced Raman scattering. Nanotechnology. 2014;**25**:235303. DOI: 10.1088/0957-4484/25/23/235303

[44] Yuan H, Wilson CM, Xia J, Doyle SL, Li S, Fales AM, Liu Y, Ozaki E, Mulfaul K, Hanna G, Palmer GM, Wang LV, Grant GA, Vo-Dinh T. Plasmonics-enhanced and optically modulated delivery of gold nanostars into brain tumor. Nanoscale. 2014;**6**:4078. DOI: 10.1039/C3NR06770J

[45] Fredriksson H, Alaverdyan Y, Dmitriev A, Langhammer C, Sutherland DS, Zäch M, Kasemo B. Hole–mask colloidal lithography. Advanced Materials. 2007;**19**:4297. DOI: 10.1002/adma.200700680

[46] Proietti Zaccaria R, Toma A, Das G, Giugni A, Tuccio S, Panaro S, Chirumamilla M, Gopalakrishnan A, Saeed A, Li H, Krahne R, Di Fabrizio E. In: Boriskina SV, Zheludev NI, editors. Singular and Chiral Nanoplasmonics. USA: Pan Stanford; 2014. p. 451. DOI: 10.4032/9789814613187

[47] Huck C, Toma A, Neubrech F, Chirumamilla M, Vogt J, De Angelis F, Pucci A. Gold nanoantennas on a pedestal for plasmonic enhancement in the infrared. ACS Photonics. 2015;**2**:497. DOI: 10.1021/ph500374r

Chapter 3

Titanium (IV) Oxide Nanotubes in Design of Active SERS Substrates for High Sensitivity Analytical Applications: Effect of Geometrical Factors in Nanotubes and in Ag-n Deposits

Marcin Pisarek, Jan Krajczewski, Marcin Hołdyński,
Tomasz Płociński, Mirosław Krawczyk,
Andrzej Kudelski and Maria Janik-Czachor

Additional information is available at the end of the chapter

http://dx.doi.org/10.5772/intechopen.72739

Abstract

In this chapter, we summarize the results of recent investigations into TiO_2 nanotubular oxide layers on Ti metal loaded with Ag nanoparticles, which act as efficient surface plasmon resonators. These Ag-n/TiO_2 NT/Ti composite layers appear to be useful as platforms for precise surface analytical investigations of minute amounts of numerous types of organic molecules: pyridine (Py), mercaptobenzoic acid (MBA), 5-(4-dimethylaminobenzylidene) rhodamine (DBRh) and rhodamine (R6G); such investigations are known as surface enhanced Raman Spectroscopy (SERS). Geometrical factors related to the nanotubes and the silver deposit affect the SERS activity of the resulting composite layers. The results presented here show that, for a carefully controlled amount of Ag-n deposit located mainly on the tops of titania nanotubes, it is possible to obtain high-quality, reproducible SERS spectra for probe molecules at an enhancement factor of 10^5–10^6. This achievement makes it possible to detect organic molecules at concentrations as low as, e.g., 10^{-9} M for R6G molecules. SEM investigations suggest that the size of the nanotubes, and both the lateral and perpendicular distribution of Ag-n (on the tube tops and walls), are responsible for the SERS activity. These features of the Ag-n/TiO_2 NT/Ti composite layer provide a variety of cavities and slits which function as suitable resonators for the adsorbed molecules.

Keywords: TiO_2 nanotubes (TiO_2 NT), Ag nanoparticles (Ag-n), PVD methods of Ag-n deposition, SERS measurements, enhancement factor (E_F)

1. Introduction

Searching for new and effective substrates for surface enhanced Raman Spectroscopy (SERS) applications is a subject of research at many scientific centers. This is because SERS spectroscopy is used to quickly characterize many functional materials important in advanced technologies, analytical chemistry and biology [1]. An important practical problem when carrying out analytical SERS measurements is to obtain suitable substrates containing a sufficient number of electromagnetic nanoresonators. The SERS effect is strongly dependent on an enhancement of the Raman scattering intensity by molecules adsorbed on a nanostructured metallic surface. The SERS enhancement factor is related to the size and shape of the nanostructures, which causes electromagnetic enhancement [2, 3]. Usually, the average value of SERS enhancement is around 10^5–10^6, but localized enhancement may reach values of 10^{10} at certain, highly efficient sub-wavelength regions of the surface [4]. It is therefore desirable for the resulting substrates to render a very large SERS spectra enhancement factor, where the enhancement factor thus obtained does not depend on where the measurement is made (the same enhancement factor should be present at different locations on the sample surface [5]).

A promising substrate for SERS measurements that meets these two most important practical requirements is regular TiO_2 nanotubes loaded with metal nanoparticles, which support surface plasmon resonance. One way of modifying the nanotubular layer and attaining high SERS activity is to enhance the surface with a small quantity of Ag or other plasmonic metal (e.g. Au or Cu) nanoparticles (plasmonic nanoparticles) [1, 6]. Such plasmonic metals themselves are known for their high SERS activity, which results only after a proper roughening of their surfaces [7]. The dielectric constant for these metals consists of a negative real part and a small positive imaginary part. When a nanoparticle made of these metals interacts with electromagnetic radiation, collective oscillations of the surface plasmons are induced [8]. These oscillations result in an enhanced electromagnetic field in close proximity to the surface of the nanoparticles. The surface of these metals, when properly roughened, provides suitable slits and cavities which serve as surface plasmon resonators that strongly enhance the intensity of the electromagnetic field. For Raman bands with a small Raman shift, the increase in the efficiency of Raman scattering is roughly proportional to the fourth power of the field enhancement [2, 9]. A detailed description of the SERS effect can be found in literature [2, 9–12].

It is well known that the enhancement factor (E_F) of SERS spectra depends on the metal morphology/topography. Only a suitable surface roughness of the SERS substrates can produce a stronger Raman signal. The optimum value of the size and shape of the noble metal particles may lead to a maximum E_F (which also depends on the nature of the metal, the excitation laser wavelength, and special experimental conditions) [13]. A highly active substrate provides superior conditions for measuring the SERS spectra of an adsorbate. Therefore, to gain some insight into the substrate–adsorbate interactions at the molecular level, and to detect different kinds of organic adsorbates, detailed knowledge of how to fabricate a highly sensitive and reproducible SERS substrate is of considerable importance [1, 5].

The high regularity of the TiO_2 nanotubular structures obtained ensures that the enhancement factors are high reproducible [14–16]. The specific morphology of the resulting structures (the large side wall surfaces of the nanotubes grown perpendicular to the substrate) makes it possible to prepare metallic nanograin systems with a large number of gaps between the

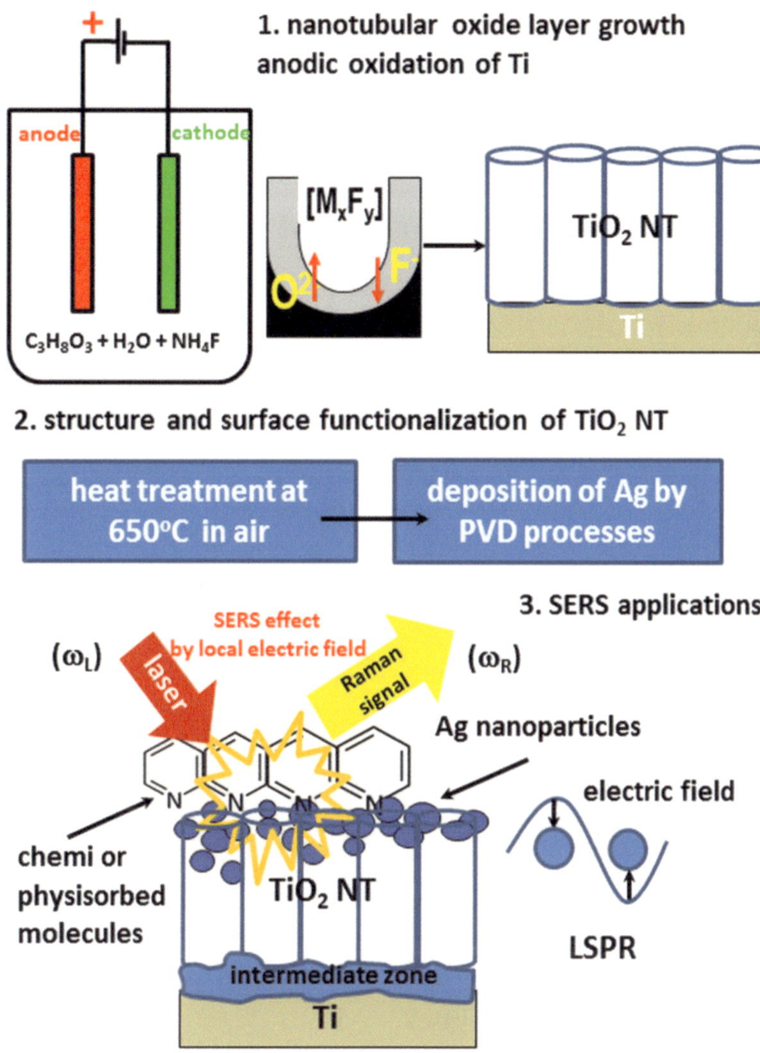

Figure 1. Schematic presentation of fabrication of TiO_2 nanotube layer and functionalization process for SERS application.

nanotubes (i.e. the metal nanoparticle systems are close to each other on the side walls of TiO_2 nanotubes that are perpendicular to the macroscopic Ti surface). As shown experimentally, the narrow gaps between the metal nanoparticles supporting surface plasmon resonance produce the largest SERS signal enhancement factors [17–19] (see **Figure 1**).

In this chapter we discuss in more detail the effect of a TiO_2 nanoporous structure with Ag nanoparticles deposited by different PVD methods (which provide precise control over the amount of silver sputtered onto the nanotubular substrate) on the SERS activity of the substrates prepared. We focus on the geometrical effects on SERS activity: nanotube diameter, size and distribution of Ag-n, and morphology of the Ag-n agglomerates. TiO_2 nanotubes are particularly suitable for such investigations because the strictly controlled electrochemical procedures make it possible to produce well-formed, nanotubular substrates that are "homogeneous," statistically having the same nanotube diameter and nanotube wall thickness (see **Figure 2**).

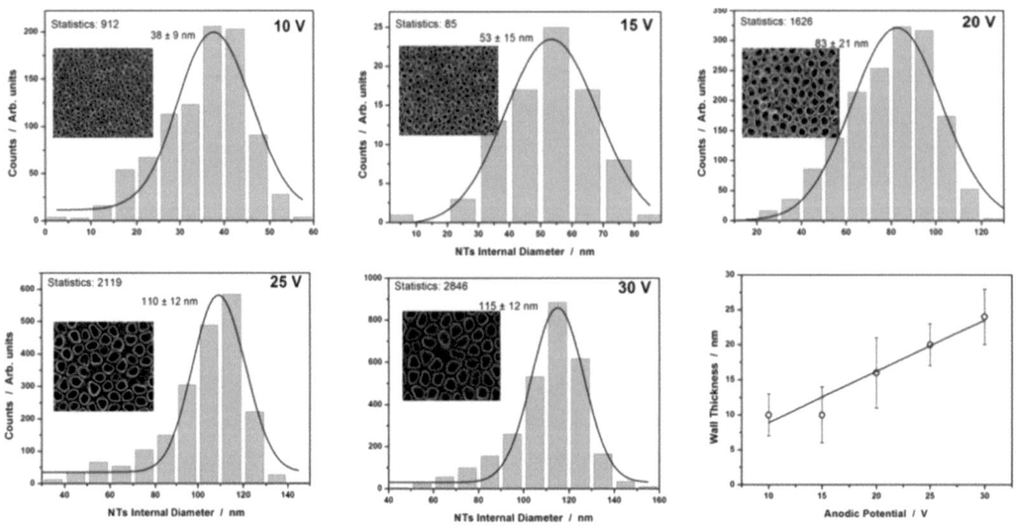

Figure 2. Effect of anodic voltage V_{max} on average diameter and wall thickness of nanotubes. The inserts show top views of TiO_2 nanotubes fabricated at different voltages [21].

As shown in **Figure 2**, the specific morphology of TiO_2 nanotubes can be relatively easily modified by controlling the conditions of anodic polarization (type of electrolyte, voltage and anodization time), because there is a direct linear relationship between anodization voltage and the average diameter of the nanotubes formed. In general, the diameter and wall thickness of the nanotubes increase with anodic voltage [20, 21]. The possibility of preparing nanotubes of different size, shape and wall thickness provides control over the geometrical surface area and specific surface area, which are important parameters when developing new substrates for SERS applications.

2. Fabrication, surface and structure characterization of SERS substrates based on TiO_2 nanotubes

Innovative SERS active platforms based on TiO_2 NT with noble metal deposits (Ag) were used for investigating various organic probe molecules such as pyridine (Py), mercaptobenzoic acid (MBA), organic dye 5-(4-dimethylaminobenzylidene) rhodamine (DBRh) and rhodamine 6G (R6G). For this purpose, simple electrochemical methods were applied: anodic oxidation of Ti foil (0.25 mm-thick, 99.5% purity, Alfa Aesar) in an optimized electrolyte: a glycerol/water mixture (volume ratio 50:50) with 0.27 M NH_4F under different constant voltages from 10 up to 30 V. They led to the formation of nanoporous titanium oxide structures, and subsequently, to the preparation of specific metal nanostructures on surfaces thereof during PVD processes (magnetron sputtering, evaporation at high and low vacuum). Before the Ag deposition, all of the samples were annealed in air at 650°C for 3 h in order to transform the structure of the TiO_2 NT from amorphous to crystalline [19, 22].

Ag nanoparticles (from 0.01 up to 0.03 mg/cm^2) were deposited using the sputter deposition technique: the evaporation method in a low vacuum ($p = 3 \times 10^{-3}$ Pa) with a JEE-4X JEOL device, and the DC magnetron sputtering technique using a Leica EM MED020 apparatus in a configuration perpendicular to the surface of the samples. More details are given elsewhere [18, 23]. We strictly controlled in situ the average amount of metal deposited per cm^2 using a quartz microbalance. One has to consider, however, that the true local amount of the metal deposits may vary substantially from site to site. The highly developed specific surface area of the nanotube arrays and its brush-like morphology may strongly affect Ag local distribution, resulting in considerable non-uniformity. For this type of process, silver targets of 99.9% purity (Kurt J. Lesker Company) were used. To better control the silver sputtering deposition process on the surface of the nanotubes, we applied the thermal evaporation method (0.01 and 0.05 mg/cm^2) using an EF 40C1 effusion cell inside an UHV preparation chamber (PREVAC, Poland). The cell was maintained at a temperature of 900°C during the process of resistive evaporation. Silver (2 mm-diameter wire, 99.999%, Alfa Aesar) was evaporated onto the room temperature-surface of the TiO_2 NT at a pressure of $1–2 \cdot 10^{-6}$ Pa at a constant evaporation rate of 0.13 nm/min. The evaporation rate from the silver effusion cell was calibrated and monitored using a TM-400 quartz crystal thickness monitor (Maxtek Inc.). The process of fabricating the SERS substrates is presented schematically in **Figure 1**.

The SERS platforms thus fabricated were characterized using various analytical methods. Electron Microscopy was applied for the morphological and structural characterization of the nanotubes after each preparation stage. The SEM observations (FEI NovaNanoSEM 450, Hitachi S70, Hitachi S-5500) were carried out at an accelerating voltage of 5 or 10 kV and an SE detector to reveal topographic contrast at the tops of the tubes. The thin sample for STEM observations was prepared by using the lift-out technique with a Hitachi NB-5000Focused Ion Beam system. The internal structure of the nanotubes was examined with a dedicated STEM Hitachi HD2700 in BF and atomic mass contrast. Observations at 200 kV revealed a columnar structure of the nanotubes and the transition zone. The surface chemical composition and the chemical state of the surface species (Ti, O, Ag) of the fabricated SERS platforms were examined using XPS spectroscopy. The XPS spectra were measured with a Microlab 350 Thermo Electron spectrometer at 300 W non-monochromatic Al K_α radiation and 1486.6 eV energy. AES spectroscopy was applied to determine the local chemical composition of the SERS substrates obtained. All of the spectra were recorded at an energy of 10 kV. The appropriate standards for AES and XPS reference spectra were also used. Finally, the Raman spectra of 2-mercaptoethanesulfonate (5×10^{-3} M), pyridine (5×10^{-2} M, in a mixture of pyridine and 0.1 M KCl), DBRh (10^{-4} M, in a mixture of water and ethanol (1:2)), and rhodamine from a 10^{-7} and 10^{-9} M aqueous solution were collected with a Horiba Jobin–Yvon Labram HR800 using a He-Ne laser (632.8 nm) as the excitation source. For each sample, 20 measurements were performed locally at various points chosen randomly.

3. Results and discussion

For our studies, we used an optimized electrolyte based on a mixture of glycerol and water (volume ratio 50:50) with 0.27 M NH_4F. The growth of nanotubes under a constant voltage is perpendicular to the metal substrate, as shown in **Figure 3**. STEM images reveal nanotubes that are hollow in shape, separated from each other, and that feature a characteristic

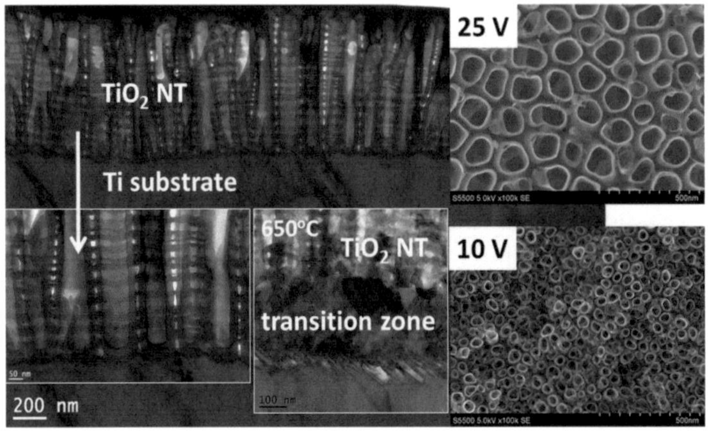

Figure 3. STEM images of a cross-section of a TiO_2 nanotubular layer before and after heat treatment at 650°C (20 V); insert shows a top view of the nanotubes obtained at 25 and 10 V.

"columnar structure." A typical surface morphology of the nanotubes is shown in the inset to **Figure 3** (10 and 25 V). Usually, the as-grown porous anodic layers exhibit poor adhesion to the Ti substrate, and so, to improve their adhesion and mechanical stability, the samples were annealed in air at 650°C for 3 h. The heat treatment in this temperature range does not cause any visible changes in the diameter or shape of the TiO_2 nanotubes, but in the annealing process an interfacial region is formed between the Ti substrate and the TiO_2 nanotubes that stabilizes the entire TiO_2 NT/Ti substrate system. Moreover, the heat treatment results in a change in the nanotube structure, from amorphous (directly after anodization) to crystalline: anatase [19, 21, 24]. All these factors are crucial for the suitability of the nanotubes as substrates when preparing SERS-active adsorbates.

After annealing, the above structures proved to be important for designing active substrates for SERS spectroscopy, where a large surface area and a stabilized structure are required [19]. The free-standing nanotubes adorned with Ag nanoparticles formed a layer of natural nanoresonators (antennas), which repeatedly enhanced Raman scattering [14, 18]. However, the geometrical factors of the Ag-n deposit and their relation to the geometry of the nanotubes are not yet well understood. Therefore, the details of the SERS mechanism should be carefully considered.

In general, there are two important mechanisms underlying SERS. The first, and the dominant, mechanism toward large SERS enhancement factors (E_F) is that of electromagnetic field enhancement, where localized surface plasmons (LSPs) in the metallic nanostructure increase the Raman signal intensity. The other contribution to SERS E_F is the chemical enhancement mechanism, where the charge transfer between the adsorbed molecule and the metal plays a critical role in enhancing and modifying the modes of molecular vibration [25]. Both enhancement mechanisms can operate simultaneously when using TiO_2 nanoporous structures in the form of freestanding nanotubes with a suitable, carefully prepared deposit of silver nanoparticles.

Figure 4 shows SEM images of TiO_2 nanotube layers (top-views) formed at 25 V and loaded with 0.01 mg/cm^2 of Ag. A careful inspection of **Figure 4a** reveals that the magnetron-deposited Ag-n is located on the tops and side walls of the nanotubes, while the silver particles become

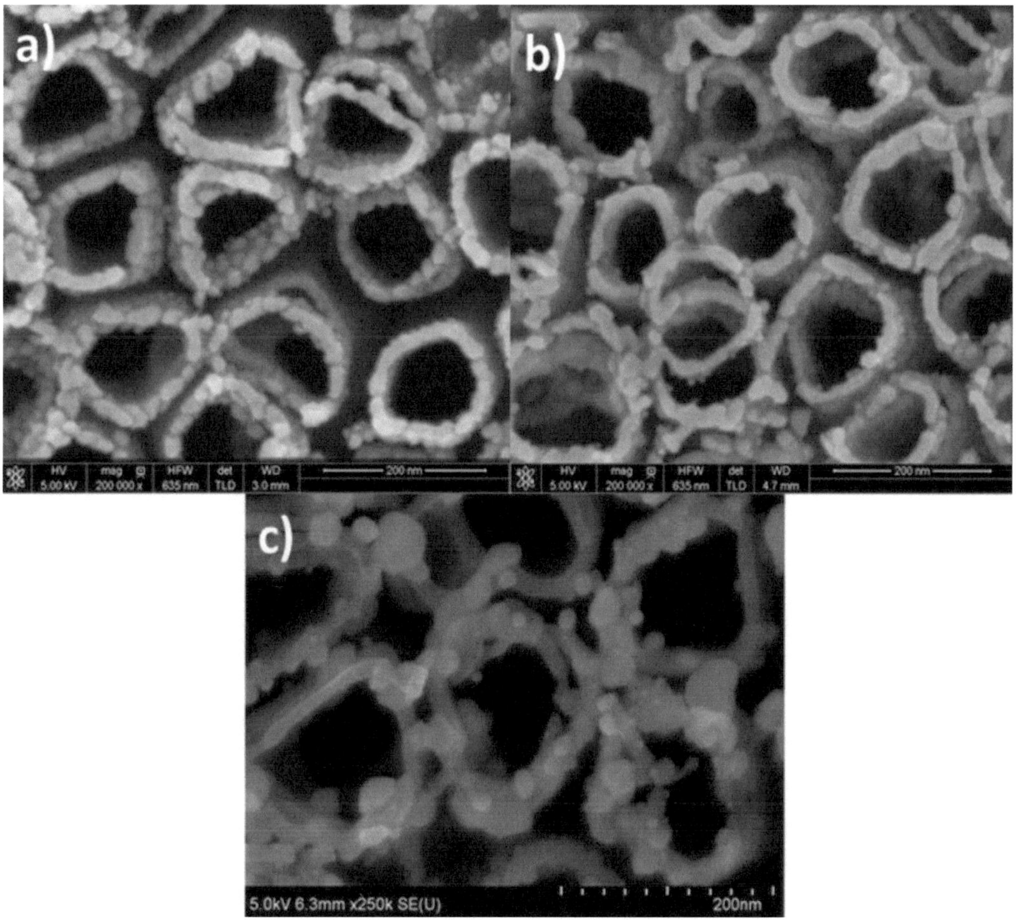

Figure 4. SEM images of TiO_2 nanotube layers (top-views) formed at 25 V and loaded with 0.01 mg Ag/cm^2 after different PVD processes: (a) magnetron sputtering, (b) evaporation at a low vacuum, and (c) evaporation at a high vacuum.

agglomerated and form rings. As a result of silver evaporation at a low vacuum, the plasmonic nanoparticles are distributed homogenously in the TiO_2 nanotube layer, tightly covering the walls of the nanotubes and forming a thin, solid coating around the tubes (**Figure 4b**). After evaporation at a high vacuum, where the process of silver deposition is very slow (0.13 nm/min), spherical Ag nanoparticles are formed (see **Figure 4c**). The diameter of these particles is below 50 nm. Densely-packed TiO_2 nanotubes coated with a plasmonic metal (Ag) could act as antenna-nanoresonators having a strictly defined geometry and surface development. This would ensure the formation of highly active places—"hot spots," i.e., gaps and cavities serving as surface plasmon resonators that significantly increase the intensity of the electromagnetic field.

Figure 5 shows typical AES spectra for samples with an Ag deposit (0.01 mg/cm^2) after vacuum evaporation at low and high vapor pressures. Signals from the Ag MNN, Ti LMM and O KLL Auger transitions are clearly visible. These results suggest that O is bound to Ti, and there are areas where the oxidized Ti substrate is not fully covered by the Ag deposit, which is consistent with the SEM microscopic observations.

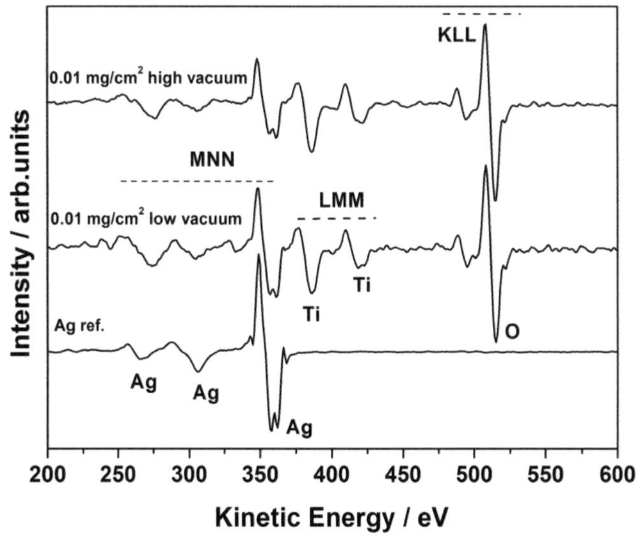

Figure 5. Typical Auger survey spectra taken at the surface of Ag/TiO_2 NT layers obtained at V_{max} = 25 V and covered with the same amount of Ag deposit (0.01 mg/cm^2). The Ag MNN signal for a pure Ag reference sample is also shown.

In order to gain further insight into the chemical state of the samples before and after the silver deposition processes, XPS measurements were performed for those samples having a small amount of Ag (0.01 mg/cm^2) (see **Table 1**). The deconvolution of the main XPS signals for $Ti2p_{3/2}$ and O1s suggests that titanium is bound to oxygen and forms titanium oxide IV. High-resolution XPS spectra confined to the Ag range gave the binding energies of $Ag3d_{5/2}$ peaks located at 367.9 (high vacuum), 368.3 (magnetron sputtering) and 368.4 eV (low vacuum), respectively, which is consistent with the literature [26–30] and also with our reference data. This implies that the Ag agglomerates and single nanoparticles located on the tops and in the deeper parts of the TiO_2 nanotubes are metallic silver. A small shift in the $Ag3d_{5/2}$ and $Ti2p_{3/2}$ peaks (see **Table 1**) for the samples after functionalization by PVD methods (±0.1–0.4 eV) may suggest that the XPS signals of Ag are modified by an interaction with the TiO_2 nanoporous substrate in relation to the position of the Ag standard peak. This interaction may induce a shift in the Fermi level in the deposited silver, in particular, if single nanoparticles are supported on the oxide carrier – the SMSI effect. Such effects have been reported by Goodman et al. and Lopez et al. [31, 32] for gold nanoparticles on TiO_2 supports, and were also observed in our recent work, where a ZrO_2 nanoporous layer was covered with Ag nanoparticles [33].

Figure 6 shows the SERS spectra of pyridine (Py) adsorbed at TiO_2 NT/Ti platforms fabricated at 25 V and covered with the same average amount of Ag-n (0.01 mg/cm^2) by the three different procedures: magnetron sputtering, and vacuum evaporation at low and high vapor pressures. The spectra presented were averaged from 20 measurements on each sample tested. The SERS spectra of pyridine are dominated by two bands: at ~1010 and at ~1034 cm^{-1} originating from the aromatic ring vibrations of this molecule. Moreover, on the spectra recorded, there are also others bands clearly visible at 1150 and 1220 cm^{-1}, which are also characteristic of Py adsorbed on a standard silver surface [34, 35]. The SERS measurements revealed that the distribution of Ag nanoparticles on the nanotubular substrate affects SERS intensity (compare

samples	Peak	BE / eV	Chemical Bond
25V TiO_2 NT / 650°C [in this work]	Ti2p$_{3/2}$	459.0	Ti-O (TiO_2)
	O1s	530.2	Ti-O (TiO_2)
25V TiO_2 NT / 650°C + thin Ag film (magnetron sputtering) [in this work]	Ti2p$_{3/2}$	459.3	Ti-O (TiO_2)
	O1s	530.6	Ti-O (TiO_2)
	Ag3d$_{5/2}$	368.3	Ag metal
25V TiO_2 NT / 650°C + thin Ag film (evaporation method at low vacuum) [in this work]	Ti2p$_{3/2}$	459.1 and 459.3	Ti-O (TiO_2)
	O1s	530.5 and 530.6	Ti-O (TiO_2)
	Ag3d$_{5/2}$	368.1 and 368.4	Ag metal
25V TiO_2 NT / 650°C + Ag single nanoparticles (evaporation method at high vacuum) [in this work]	Ti2p$_{3/2}$	459.1	Ti-O (TiO_2)
	O1s	530.4	Ti-O (TiO_2)
	Ag3d$_{5/2}$	367.9	Ag metal
Reference materials			
Titanium dioxide [20]	Ti2p$_{3/2}$	458.8	Ti-O (TiO_2)
Silver foil ref. material [in this work]	Ag3d$_{5/2}$	368.2	Ag metal
Silver metal [20]	Ag3d$_{5/2}$	368.3	Ag metal

Table 1. XPS results for titania nanotube layers before and after silver nanoparticles deposition by PVD methods at low and high vacuums.

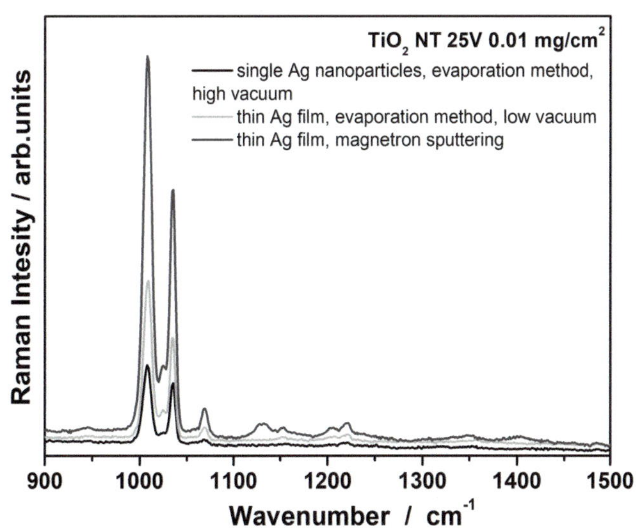

Figure 6. Average SERS spectra of Py adsorbed on Ag-n/TiO_2NT (25 V)/Ti substrates fabricated by: Ag magnetron sputtering, Ag evaporation method at a low vacuum, Ag thermal evaporation method at a high vacuum.

Figure 4). A roughly two-fold increase in SERS intensity occurs when the titania nanotubes are adorned with Ag-n after the magnetron sputtering process. Three kinds of Ag particles can be distinguished: those accumulated on the tops of the nanotubes, forming "rings," and single particles (see. **Figure 4**) separated from each other and/or those Ag particles produced in the high vacuum process, which are located on the tops ("mouth") of the nanotubes. The latter do not yield such a strong SERS effect. Apparently, the key factor in the SERS activity of Ag-n *is the size and mode of the specific surface area of the silver particles* formed during a specific vacuum process. While the SERS intensity reached was not very high, well-reproducible and good-quality SERS spectra were obtained.

The next step was to repeat the same measurements with a nanotubular substrate prepared at an anodization of 10 V to produce nanotubes having a smaller diameter, thereby increasing the specific surface area of the SERS-active silver. **Figure 7** shows typical SEM images of a platform for SERS measurements, based on TiO_2 nanotubes (10 V): a surface of nanotubes of titania on a Ti substrate after deposition of Ag nanoparticles by the evaporation method in a low vacuum of 0.01 (a), 0.02 (b), and 0.03 (d) mg Ag/cm^2. For the smallest amount of Ag (0.01 mg·cm^{-2}), the silver nanoparticles tend to gather on the tops of the nanotubes and on their side walls (c). Characteristic specific structures of Ag-n are formed around the nanotubes, consisting of agglomerates of Ag nanoparticles. Increasing the amount of silver (0.02 and 0.03 mg·cm^{-2}) leads further to a visible development of the Ag surface area, up to the formation of silver nanoparticle agglomerates having characteristic slits of from several to a few dozen nanometers.

Figure 8 shows the spectrum of pyridine adsorbed on Ag-n/TiO_2 NT (10 V) platforms from an aqueous solution of 0.05 M pyridine +0.1 M KCl. The spectra are dominated by two strong

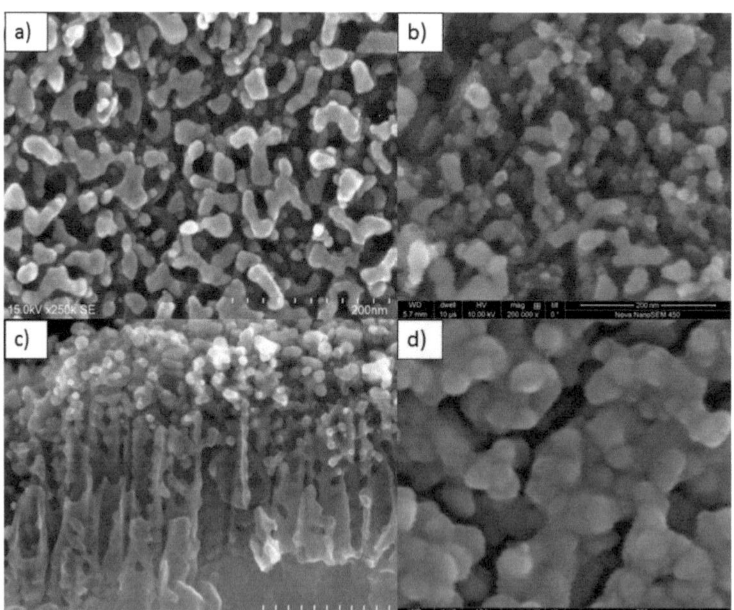

Figure 7. Top view of titania dioxide nanotubes (10 V) annealed for 2 h in air at a temperature of 650°C after silver deposition: 0.01 (a), 0.02 (b), 0.03 mg/cm^2 (d). A cross-sectional view of a nanoporous layer after deposition of Ag – 0.01 mg/cm^2 is also given, left side (c).

bands at 1004 and 1034 cm^{-1}. The band at 1004 cm^{-1} is due to the ring breathing mode (ν_1, Wilson notation), whereas that at 1034 cm^{-1} is due to symmetric triangular ring deformation (ν_{12}) [34, 35]. The SERS spectral intensity increases distinctly with the amount of Ag metal deposit, which correlates with a change in surface topography (compare **Figure 7**). This effect is most likely related to an increase in the number of narrow gaps between the silver particles themselves (locations particularly active in SERS spectroscopy) with an increasing amount of Ag deposit.

Our previous experiments show that the SERS enhancement factor for free-standing TiO_2 nanotubes adorned with silver nanoparticles is strictly related to the size of the nanotubes. The SERS enhancement factor (E_F) for the pyridine (Py) probe molecule was successfully estimated using the following formula (1):

$$E_F = I_{SERS}/I_{ref} \times hc_{ref}/N_{surf} \quad (1)$$

where I_{SERS} and I_{ref} are the Raman intensities obtained from the SERS and normal Raman (NR) investigations, respectively, c_{ref} stands for the concentration of pure Py in the NR measurements, and h is the depth-of-focus of the laser beam. The average number of adsorbed molecules of Py per geometrical surface area unit participating in the SERS measurements (N_{surf}) was calculated assuming that the adsorbed molecules are spheres closely packed on a plane to form a hexagonal lattice. AFM measurements were performed to determine the geometrical surface area, see **Figure 9a**. More details can be found in the publication [15].

For the same amount of Ag deposit (0.02 mg/cm^2, magnetron sputtering) on nanotubes fabricated at 10 V up to 25 V (see **Figure 2**), there was an increase in the SERS enhancement factor of from 10^5 to 10^6 (see **Figure 9b**).

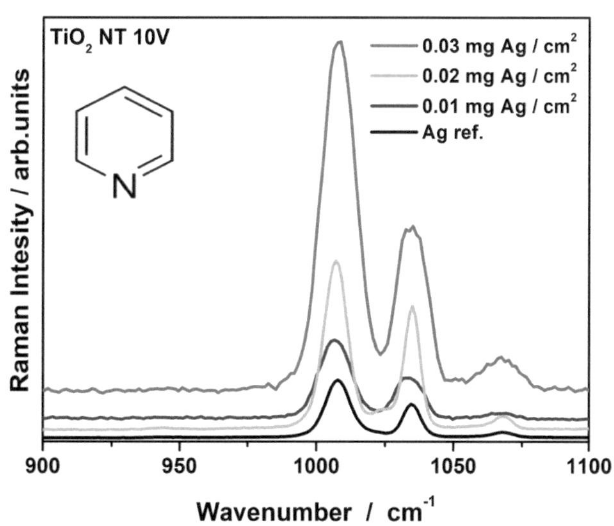

Figure 8. SERS spectra of pyridine adsorbed at the surface of nanotubes (10 V) annealed in air at 650°C for 2 h and coated with silver deposit: 0.01, 0.02, 0.03 mg/cm^2. A reference spectrum for pure Ag (electrochemically roughened silver surface) is also given.

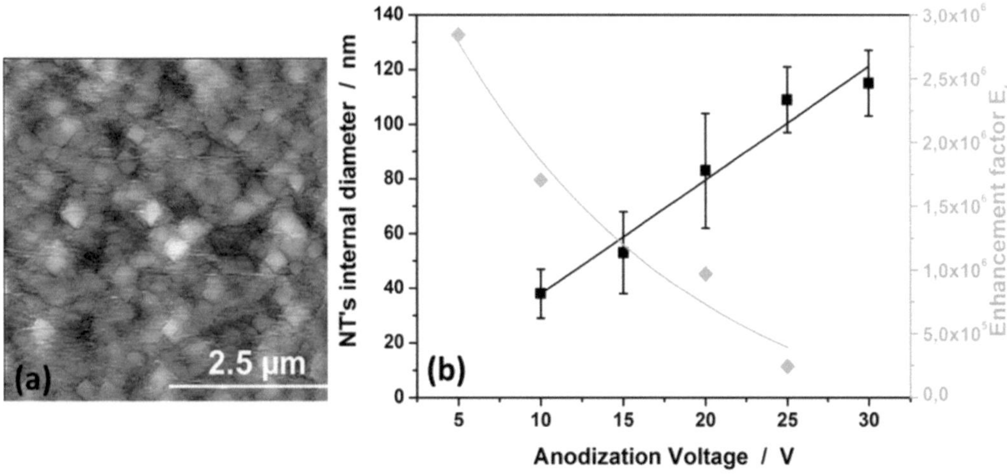

Figure 9. (a) Typical AFM image of silver nanoparticles (0.020 mg/cm²) deposited by the magnetron sputtering technique on TiO_2 oxide layers (20 V)—top view; (b) SERS enhancement factor E_F as a function of the final formation voltage V_{max} of the titania nanoporous layers. E_F estimated for Py band at ~1010 cm^{-1}, Ag deposited by magnetron sputtering technique—0.02 mg/cm².

The hundredfold increase in SERS enhancement shown in **Figure 9b** is apparently due to a combination of two factors: a change in the specific surface area of the nanotubes with formation voltage [15], and a change in the distribution and size of the Ag nanoparticles with nanotube size [23]. These geometrical factors affect the properties of the silver "nanoresonators" produced on the tops and side walls of the TiO_2 nanotubes. The present results confirm the importance of the size of the geometrical surface area of the TiO_2 nanotubes for the plasmonic properties of the Ag-n deposit itself and, consequently, for the properties of the Ag/TiO_2NT composite materials.

Figure 10 shows SERS spectra of two other probe molecules: p-mercaptobenzoic acid (a) and DBRh dye (b) recorded on the platforms, which were characterized by an enhancement factor larger than 10^6. In both cases, the spectra are of good quality and high intensity. They show that the background is not very high, and quite "flat" for both molecules. Our SERS experiments confirmed the good reproducibility of such substrates obtained by adorning TiO_2 nanotubes with Ag metal clusters and particles, which effectively support plasmon resonance [18].

Figures 11 and **12** show R6G spectra taken with our Ag-n/TiO_2 NT platform covered with a solution containing 10^{-7} or 10^{-9} mol/l R6G. A rhodamine molecule is often used when the relationship between the local electromagnetic field enhancement and a large SERS signal is explored, which makes it possible to measure Raman spectra from a single molecule located on an Ag particle or individual Ag nanoparticles [36–39]. The characteristic peaks at ~970, 1150, 1220, 1300, 1340, 1410, 1500 and 1600 cm^{-1} correspond to the Raman lines for R6G [3, 36]. In particular, the bands which are usually assigned to aromatic C-C stretching vibrations of R6G molecule are clearly visible. It can be seen that the SERS spectrum of R6G adsorbed on our Ag substrate exhibits enough intensity for this molecule to be detected even at a small concentration in an aqueous solution (below 10^{-7} mol/l). As a result of this reduced concentration of R6G, some bands are suppressed (compare **Figures 11** and **12**). This sensitivity of our

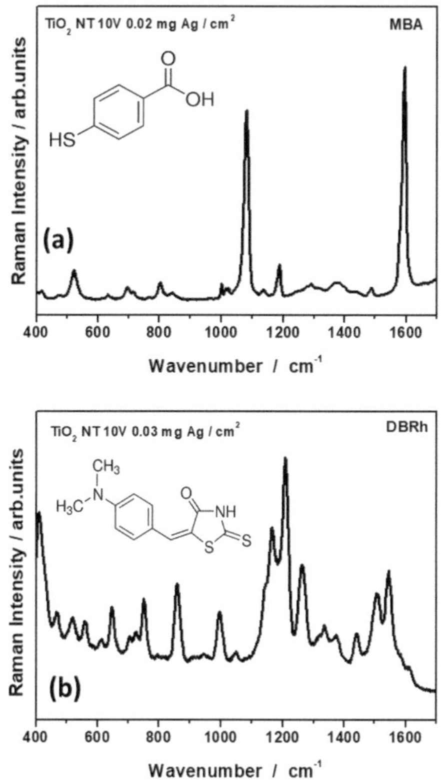

Figure 10. SERS spectra of p-mercaptobenzoic acid (a) and DBRh dye (b) recorded at the surface of 10 V nanotubes.

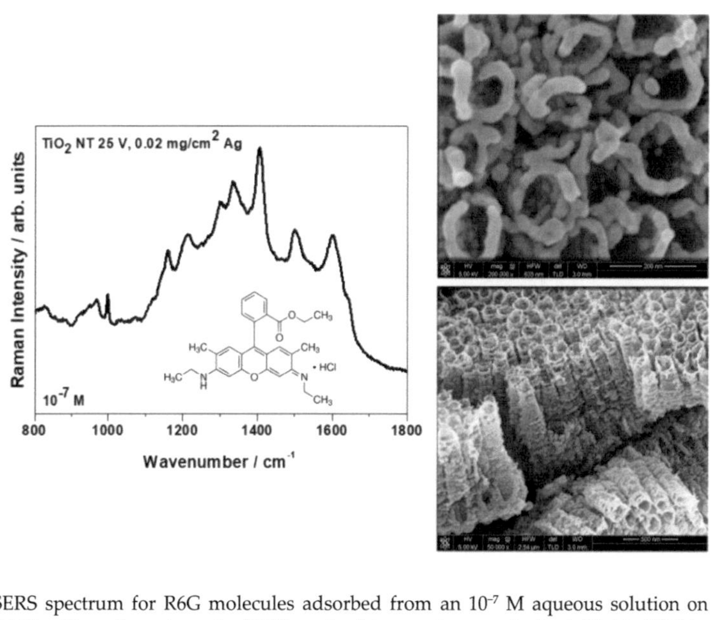

Figure 11. Left: SERS spectrum for R6G molecules adsorbed from an 10^{-7} M aqueous solution on a surface of TiO_2 NT nanotubes (25 V) with a silver deposit of 0.02 mg/cm^2 (magnetron sputtering). Right: SEM images showing the morphology and cross-sectional view of the SERS active platform.

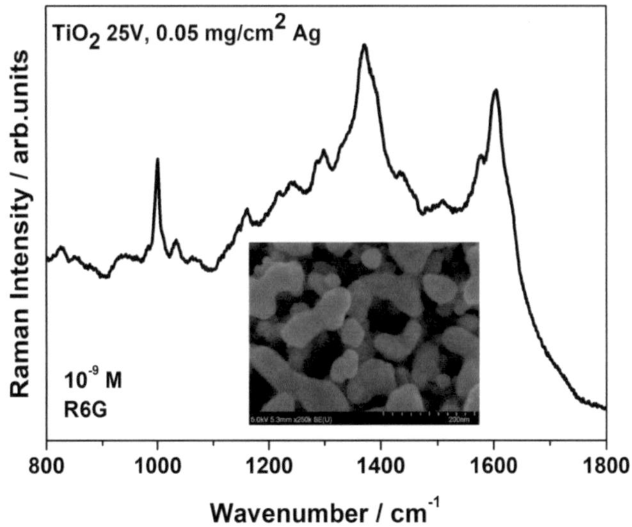

Figure 12. SERS spectrum of R6G molecules adsorbed from a 10^{-9} M aqueous solution on a surface of TiO_2 NT nanotubes (25 V) with a silver deposit of 0.05 mg/cm² (thermal evaporation method at a high vacuum).

fabricated SERS substrates is apparently related to the homogeneous Ag distribution on the surface of the regular nanoporous TiO_2 structures; this should make it possible to produce specific morphologies that are extremely useful in Raman investigations.

4. Conclusions and outlook

PVD methods (the sputtering technique) provide a strictly controlled, stable distribution of Ag-n on nanotubular titania, thus offering suitable platforms for SERS investigations and for measurements of high-quality SERS spectra. Such spectra were obtained for pyridine, p-mercaptobenzoic acid, rhodamine and rhodamine 6G dye adsorbed on Ag-n-functionalized nanoporous layers of titania. For the SERS measurements (with R6G), excellent spectra were obtained, even for a concentration of R6G in a solution as low as 10^{-9} mol/l. The E_F value was found to be the largest ($E_F = 2.8 \times 10^6$) for the smallest nanotube diameter ($d = 38 \pm 9$ nm). Stable, reproducible and highly active platforms for SERS investigations were obtained using relatively simple procedures and techniques.

The SERS enhancement induced by these new materials is associated primarily with two geometric factors: the specific surface area of the nanotubes, and an appropriate size and distribution of the Ag nanoparticles. These geometric factors (clearly visible using high-resolution microscopy) affect the properties of the metallic "nanoresonators" deposited on the TiO_2 nanotubes. Consequently, the probe molecules sense even slight changes on the surface, and these manifest themselves in the SERS intensity. However, the spectral differences observed result not only from the specific morphology, where the PVD techniques lead to the formation of cavities and slits that operate as adsorbed molecule resonators, but also from a specific interaction between the Ag nanoparticles and the TiO_2 nanotubes themselves. The further development of this type of SERS platform may involve using Ti of varying degrees of purity, and alloys of Ti, where the impurities and alloying elements will have an impact on the formation of nanoporous oxides layers by means of doping process (a change in the electron structure of the titanium oxide). As a consequence, this could lead to a significantly stronger enhancement of the intensity of the SERS spectra due to the occurrence of modified interactions between the doped titanium oxide and the metal nanoparticles, effectively supporting surface plasmons resonance (Ag, Au, Cu). All those factors should be taken into account when planning future analytical applications of new modified SERS substrates.

Acknowledgements

We are most grateful to Prof. Robert Nowakowski from the Institute of Physical Chemistry PAS (Poland) for his cooperation (AFM measurements) and many helpful discussions. This work was financially supported by the Institute of Physical Chemistry PAS.

Author details

Marcin Pisarek[1]*, Jan Krajczewski[2], Marcin Hołdyński[1], Tomasz Płociński[3], Mirosław Krawczyk[1], Andrzej Kudelski[2] and Maria Janik-Czachor[1]

*Address all correspondence to: mpisarek@ichf.edu.pl

1 Institute of Physical Chemistry, Polish Academy of Sciences, Warsaw, Poland

2 Faculty of Chemistry, University of Warsaw, Warsaw, Poland

3 Faculty of Material Science and Engineering, Warsaw University of Technology, Warsaw, Poland

References

[1] Sharma B, Frontiera RR, Henry AI, Ringe E, Van Duyne RP. SERS materials, applications, and the future. Materials Today. 2012;**15**:16-25. DOI: 10.1016/S1369-7021(12)70017-2

[2] Schlucker S. Surface-enhanced Raman spectroscopy: Concepts and chemical applications. Angewandte Chemie, International Edition. 2014;**53**:4756-4795. DOI: 10.1002/anie.201205748

[3] Tian F, Bonnier F, Casey A, Shanahan AE, Byrne HJ. Surface enhanced Raman scattering with gold nanoparticles: Effects of particle shape. Analytical Methods. 2014;**6**:9116-9123. DOI: 10.1039/c4ay02112f

[4] Fan M, Andrade GFS, Brolo AG. A review on the fabrication of substrates for surface enhanced Raman spectroscopy and their applications in analytical chemistry. Analytica Chimica Acta. 2011;**693**:7-25. DOI: 10.1016/j.aca.2011.03.002

[5] Cialla D, Marz A, Bohme R, Theil F, Weber K, Schmitt M, Popp J. Surface-enhanced Raman spectroscopy (SERS): Progress and trends. Analytical and Bioanalytical Chemistry. 2012;**403**:27-54. DOI: 10.1007/s00216-011-5631-x

[6] Krajczewski J, Kołątaj K, Kudelski A. Plasmonic nanoparticles in chemical analysis. RSC Advances. 2017;**7**:17559-17576. DOI: 10.1039/c7ra01034f

[7] Jeanmaire DL, Van Duyne RP. Surface raman spectroelectrochemistry: Part I. Heterocyclic, aromatic, and aliphatic amines adsorbed on the anodized silver electrode. Journal of Electroanalytical Chemistry. 1977;**84**:1-20. DOI: 10.1016/S0022-0728(77)80224-6

[8] Yamamoto YS, Ozaki Y, Itoh T. Recent progress and frontiers in the electromagnetic mechanism of surface-enhanced Raman scattering. Journal of Photochemistry & Photobiology, C: Photochemistry Reviews. 2014;**21**:81-104. DOI: 10.1016/j.jphotochemrev.2014.10.001

[9] Ding SY, You EM, Tian ZQ, Moskovits M. Electromagnetic theories of surface-enhanced Raman spectroscopy. Chemical Society Reviews. 2017;**46**:4042-4076. DOI: 10.1039/c7cs00238f

[10] Stiles PL, Dieringer JA, Shah NC, Van Duyne RP. Surface-enhanced Raman spectroscopy. Annual Review of Analytical Chemistry. 2008;**1**:601-626. DOI: 10.1146/annurev.anchem.1.031207.112814

[11] Chen X-J, Cabello G, Wu D-Y, Tian Z-Q. Surface-enhanced Raman spectroscopy toward application in plasmonic photocatalysis on metal nanostructures. Journal of Photochemistry and Photobiology C: Photochemistry Reviews. 2014;**21**:54-80. DOI: 10.1016/j.jphotochemrev.2014.10.003

[12] Kudelski A. Raman spectroscopy of surfaces. Surface Science. 2009;**603**:1328-1334. DOI: 10.1016/j.susc.2008.11.039

[13] Du Y, Shi L, He T, Sun X, Mo Y. SERS enhancement dependence on the diameter and aspect ratio of silver-nanowire array fabricated by anodic aluminium oxide template. Applied Surface Science. 2008;**255**:1901-1905. DOI: 10.1016/j.apsusc.2008.06.140

[14] Roguska A, Kudelski A, Pisarek M, Opara M, Janik-Czachor M. Surface-enhanced Raman scattering (SERS) activity of Ag, Au and Cu nanoclusters on TiO_2-nanotubes/Ti substrate. Applied Surface Science. 2011;**257**:8182-8189. DOI: 10.1016/j.apsusc.2010.12.048

[15] Pisarek M, Hołdyński M, Roguska A, Kudelski A, Janik-Czachor M. TiO_2 and Al_2O_3 nanoporous oxide layers decorated with silver nanoparticles – Active substrates for SERS measurements. Journal of Solid State Electrochemistry. 2014;**18**:3099-3109. DOI: 10.1007/s10008-013-2375-x

[16] Pisarek M, Roguska A, Kudelski A, Hołdyński M, Janik-Czachor M: Ch.17: Self-organized TiO_2, Al_2O_3 and ZrO_2 nanotubular layers: Properties and applications. In: Aliofkhazraei M editor. Comprehensive Guide for Nanocoatings Technology, Vol.3. New York Nova Science Publishers, Inc. 2015. p. 435-462, ISBN:978-1-63482-647-1

[17] Lamberti A, Virga A, Chiado A, Chiodoni A, Bejtka K, Rivolo P, Giorgis F. Ultrasensitive Ag-coated TiO_2 nanotube arrays for flexible SERS-based optofluidic devices. Journal of Materials Chemistry C. 2015;**3**:6868-6875. DOI: 10.1039/c5tc01154j

[18] Kudelski A, Pisarek M, Roguska A, Hołdyński M, Janik-Czachor M. Surface-enhanced Raman scattering investigations on silver nanoparticles deposited on alumina and titania nanotubes: Influence of the substrate material on surface-enhanced Raman scattering activity of Ag nanoparticles. Journal of Raman Specroscopy. 2012;**43**:1360-1366. DOI: 10.1002/jrs.4075

[19] Pisarek M, Roguska A, Kudelski A, Andrzejczuk M, Janik-Czachor M, Kurzydlowski KJ. The role of Ag particles deposited on TiO_2 or Al_2O_3 self-organized nanoporous layers in their behavior as SERS-active and biomedical substrates. Materials Chemistry and Physics. 2013;**139**:55-65. DOI: 10.1016/j.matchemphys.2012.11.076

[20] Lee K, Mazare A, Schmuki P. One-dimensional titanium dioxide nanomaterials: Nanotubes. Chemical Reviews. 2014;**114**:9385-9454. DOI: 10.1021/cr500061m

[21] Roguska A, Pisarek M, Belcarz A, Marcon L, Hołdyński M, Andrzejczuk M, Janik-Czachor M. Improvement of the bio-functional properties of TiO_2 nanotubes. Applied Surface Science. 2016;**388**:775-785. DOI: 10.1016/j.apsusc.2016.03.128

[22] Roguska A, Pisarek M, Andrzejczuk M, Dolata M, Lewandowska M, Janik-Czachor M. Characterization of a calcium phosphate – TiO_2 nanotube composite layer for biomedical applications. Materials Science and Engineering: C. 2011;**31**:906-914. DOI: 10.1016/j.msec.2011.02.009

[23] Pisarek M, Nowakowski R, Kudelski A, Hołdyński M, Roguska A, Janik-Czachor M, Kurowska-Tabor E, Sulka GD. Surface modification of nanoporous alumina layers by deposition of Ag nanoparticles. Effect of alumina pore diameter on the morphology of silver deposit and its influence on SERS activity. Applied Surface Science. 2015;**357**: 1736-1742. DOI: 10.1016/j.apsusc.2015.10.011

[24] Regonini D, Bowen CR, Jaroenworaluck A, Stevens R. Effect of heat treatment on the properties and structure of TiO_2 nanotubes: Phase composition and chemical composition. Materials Science and Engineering R. 2013;**74**:377-406. DOI: 10.1016/j.mser.2013.10.001

[25] Wang AX, Kong X. Review of recent progress of Plasmonic materials and nano-structures for surface-enhanced Raman scattering. Materials. 2015;**8**:3024-3052. DOI: 10.3390/ma8063024

[26] Edited by Chastain J, King Jr. RC: Handbook of X-Ray Photoelectron Spectroscopy, a Reference Book of Standard Spectra for Identification and Interpretation of XPS Data. Physical Electronics, Inc. Eden Praire; 1995. ISBN 0-9648124-1-X

[27] Kumar R, Rashid J, Barakat MA. Zero valent Ag deposited TiO_2 for the efficient photocatalysis of methylene blue under UV-C light irradiation. Colloids and Interface Science Communications. 2015;**5**:1-4. DOI: 10.1016/j.colcom.2015.05.001

[28] Gross PA, Pronkin SN, Cottineau T, Keller N, Keller V, Savinova ER. Effect of deposition of Ag nanoparticles on photoelectrocatalytic activity of vertically aligned TiO_2 nanotubes. Catalysis Today. 2012;**189**:93-100. DOI: 10.1016/j.cattod.2012.03.054

[29] Zhang J, Li Y, Zhang Y, Chen M, Wang L, Zhang C, He H. Effect of support on the activity of Ag-based catalysts for formaldehyde oxidation. Scientific Reports. 2015;**5**:12950. DOI: 10.1038/srep12950

[30] Albiter E, Valenzuela MA, Alfaro S, Valverde-Aguilar G, Martinez-Pallares FM. Photocatalytic deposition of Ag nanoparticles on TiO_2: Metal precursor effect on the structural and photoactivity properties. Journal of Saudi Chemical Society. 2015;**19**:563-573. DOI: 10.1016/j.jscs.2015.05.009

[31] Goodman DW. "Catalytically active Au on Titania": Yet another example of a strong metal support interaction (SMSI)? Catalysis Letters. 2005;**99**(1-2):1-4. DOI: 1011-372X/05/0100-0001/0

[32] Lopez N, Nørskov JK. Theoretical study of the Au/TiO_2(110) interface. Surface Science. 2002;**515**:175-186 S00 39-6028 (02) 01873- 3

[33] Pisarek M, Krajczewski J, Wierzbicka E, Hołdyński M, Sulka GD, Nowakowski R, Kudelski A, Janik-Czachor M. Influence of the silver deposition method on the activity of platforms for chemometric surface-enhanced Raman scattering measurements: Silver films on ZrO_2 nanopore arrays. Spectrochimica Acta – Part A: Molecular and Biomolecular Spectroscopy. 2017;**182**:124-129. DOI: 10.1016/j.saa.2017.04.005

[34] Zuo C, Jagodzinski PW. Surface-enhanced Raman scattering of pyridine using different metals: Differences and explanation based on selective formation of α-Pyridyl on metal surfaces. The Journal of Physical Chemistry. B. 2005;**109**:1788-1793. DOI: 10.1021/jp0406363

[35] DY W, Li JF, Ren B, Tian ZQ. Electrochemical surface-enhanced Raman spectroscopy of nanostructures. Chemical Society Reviews. 2008;**37**:1025-1041. DOI: 10.1039/b707872m

[36] Michaels AM, Nirmal M, Brus LE. Surface enhanced Raman spectroscopy of individual Rhodamine 6G molecules on large Ag Nanocrystals. Journal of the American Chemical Society. 1999;**121**:9932-9939. DOI: 10.1021/ja992128q

[37] Liu YC, ChCh Y, Sheu SF. Low concentration rhodamine 6G observed by surface-enhanced Raman scattering on optimally electrochemically roughened silver substrates. Journal of Materials Chemistry. 2006;**16**:3546-3551. DOI: 10.1039/b609417a

[38] Dieringer JA, Wustholz KL, Masiello DJ, Camden JP, Kleinman SL, Schatz GC, Van Duyne RP. Surface-enhanced Raman excitation spectroscopy of a single Rhodamine 6G molecule. Journal of the American Chemical Society. 2009;**131**:849-854. DOI: 10.1021/ja8080154

[39] Pristinski D, Tan S, Erol M, Du HSS. In situ SERS study of Rhodamine 6G adsorbed on individually immobilized Ag nanoparticles. Journal of Raman Specroscopy. 2006;**37**:762-770. DOI: 10.1002/jrs.1496

Chapter 4

Precisely Controllable Synthesized Nanoparticles for Surface Enhanced Raman Spectroscopy

Meng Xu and Jiatao Zhang

Additional information is available at the end of the chapter

http://dx.doi.org/10.5772/intechopen.73086

Abstract

Surface-enhanced Raman scattering (SERS) is a powerful technique for trace molecular detection because of its ultrahigh molecular structure sensitivity and unique fingerprinting spectra. The morphology, size and structure of the plasmonic nanoparticles seriously influence the Raman scattering intensity of sample. In this chapter, we focus on the influence of nanoparticle morphology. By tailoring the plasmonic properties of anisotropic Au, Ag nanoparticles and generating electromagnetic "hot spots" of SERS active substrate, the SERS intensity can be seriously influenced. We also focus on providing a general introduction to understand the main parameters of anisotropic noble metal nanoparticles of SERS performance.

Keywords: surface enhanced Raman spectroscopy (SERS), localized surface plasmon resonances (LSPRs), noble metal nanoparticles, core/shell nanoparticles, controllable synthesis

1. Introduction

The utilization of the enhancement effect of Raman scattering by using the plasmonic effect for nanoparticles (NPs) is mainly determined by the physical and chemical properties of the surface-enhanced Raman spectroscopy (SERS) active substrates. SERS is a powerful technique for ultrasensitive detection through the enhancing of electromagnetic fields (E-field) generated by localized surface plasmon resonance (LSPR) effect [1–4]. It can provide specific fingerprint information about a wide range of target molecules [5]. Noble metal NPs (e.g., Au, Ag, Cu, etc.) display a variety of plasmonic behaviors. The extremely inefficient inelastic light scattering of Raman scattering of the molecules or materials is the most important factor restricting its applications [6]. However, the bottleneck can be overcome when the molecules are adsorbed onto a rough surface of noble metal NPs by enhancing the vibrational excitation of the adsorbed molecules or materials in a process commonly known as SERS. To develop size, shape and chemical composition, controllable anisotropic NPs to generate the "hot spot" can maximize the Raman signal to enhance the molecule detecting capacity. A wide variety of anisotropic noble metal NPs have been developed as a suitable candidate for SERS detecting and sensing. By controlling the size, shape, chemical component and structure of the NPs, the absorption range of noble metal NPs can be tuned form the visible to the near-infrared (NIR) range, and the resonances of noble metal NPs occur in the visible and NIR range of the electromagnetic spectrum [7]. Synthetizing the precisely controllable anisotropic noble metal NPs has proven to be an extremely powerful tool to tune the plasmon resonance for rendering them suitable for ultrasensitive detecting.

By varying the diameters of spherical noble metal NPs, the plasmon resonances of NPs can only be tuned in a relatively narrow wavelength range (always only tens of nanometers). However, the anisotropic shape NPs provide one or several additional degrees of freedom, which allows controlling the plasmon resonance wavelengths, ranging from the visible to the NIR range, by varying the aspect ratio (AR) of the nanorods (NRs) or other topography parameters of NPs.

This chapter focuses on recent researches on the most commonly used anisotropic Ag, Au, and so on noble metal NPs for SERS application. Herein, we focus on the intrinsic shape-dependent SERS property of the precisely controlled anisotropic NPs, synthesized mainly in solution.

2. Nanorods

In the early twenty-first century, the seed-mediated Au and Ag NRs growth method was developed, and this method has become the most popular approach for Au and Ag NRs synthetizing [8]. In general, this method comprises two steps: (1) the synthetizing of the small metal NPs, that is, so-called "seeds" and (2) the overgrowing of the seed to the larger, mature metal NPs. In addition, the shape of the mature metal NPs can be controlled by using the different surfactants, salt precursors and solvents. In this method, the separation of the nucleation and growth of the nanocrystals synthesis process lead to precise control of shape and size and the high uniformity of the metal NRs. In 2001, the precisely size controlled Ag NRs were successfully synthetized by using silver nitrate as the Ag precursor, ascorbic acid (A.A.) as the reductant and cetyltrimethylammonium bromide (CTAB) as the surfactant with the presence of Ag seeds and NaOH [9]. By reducing the amounts of NaOH, the ARs of the Ag NRs increased. As a pioneering Au NRs synthetizing research, the high-crystallinity colloidal Au NRs have been prepared with a high 90% shape yield by

Nikoobakht and El-Sayed [10]. But the Ag ions are always needed in the high shape yield Au NRs synthetic method [11].

The typical UV-vis-NIR spectra of Au and Ag NRs show two obvious plasmon bands, corresponding to the transverse (short wavelength) and longitudinal modes (long wavelength). The transverse mode at ca. 520 nm can be hardly tuned, but the longitudinal mode can be precisely controlled by tuning the AR of noble metal NRs, and the intensity is much stronger than the transverse mode as shown in **Figure 1** [12]. The simulation and experimental studies of the external electromagnetic fields (E-field) distribution indicated that the high localized E-field enhancement at the noble metal NRs tips, so-called hot spots, along the dipolar longitudinal mode [13, 14].

Noble metal NRs are the most commonly used anisotropic SERS active substrate materials. Tailoring the LSPR relative to the laser excitation wavelength is one of the most significant approaches to obtain a high SERS signals. The search has indicated that the Ag NRs can generate the higher E-field than that of Au NRs under the same wavelength of incident light. In addition, the E-field intensity increases with the increasing of AR for Ag NRs [15]. Han's group developed an efficient platform for investigating the kinetics of catalytic reactions with SERS by using Au@AuPd NRs as the SERS active materials as shown in **Figure 2** [16].

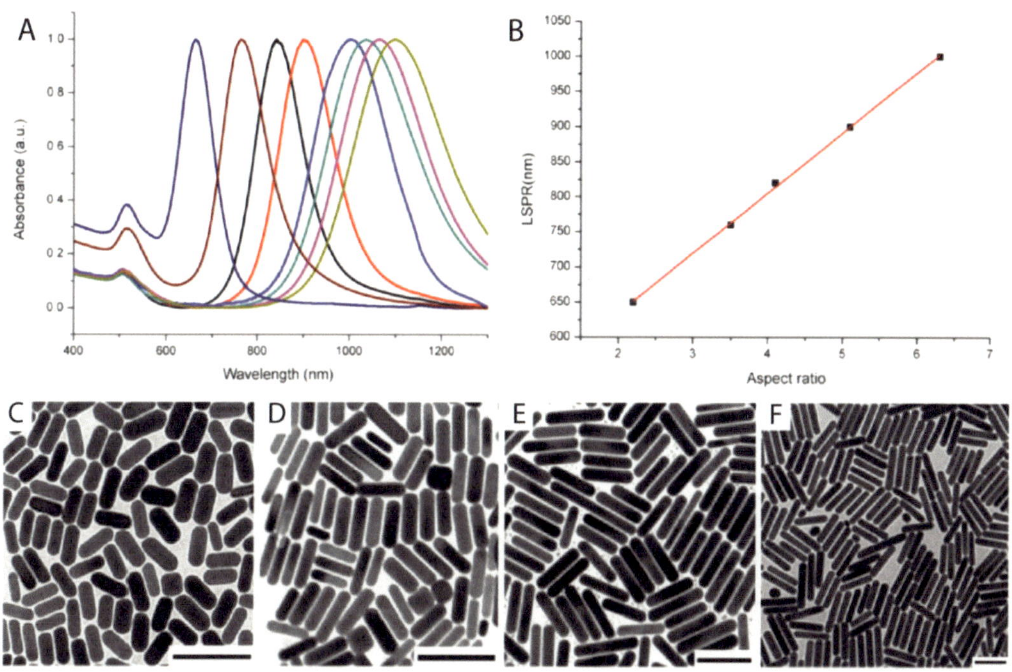

Figure 1. The optical properties and morphology of Au nanorods with different AR. (A) The UV-vis-NIR spectra of Au NRs with different ARs. (B) Relationship between longitudinal LSPR and AR. (C–F) Representative TEM images of Au NRs with different ARs. Scale bars: 100 nm. Adapted from Ref. [12] with permission, Copyright Royal Society of Chemistry.

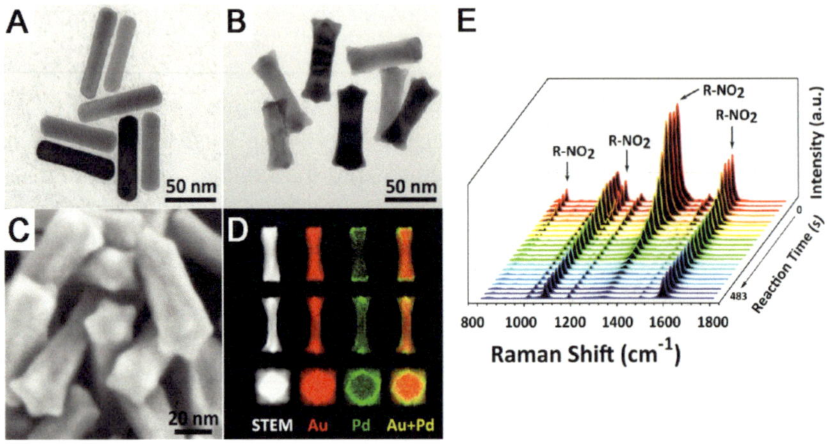

Figure 2. (A) TEM image of the Au NRs seeds. (B) TEM image and (C) SEM image of the Au @AuPd NRs. (D) STEM images of Au@AuPd NRs taken along the [100], [110], and [001] axes (the first column, from the top down), and the corresponding elemental mappings for Au (the second column) and Pd (the third column); the last column shows combined mappings, in which the simultaneous presence of Au and Pd appears yellowish. (E) Successive SERS spectra of the reduction of 4-nitrothiophenol by H_2 collected on a planar platform made of Au@AuPd NRs. Copyright American Chemical Society.

The results indicated that the Ag NRs show the high Raman enhancement than the Au NRs because of the stronger E-field of Ag NRs than that of Au NRs. This phenomenon can be attributed to higher plasmonic intensity at the tips of Ag NRs and higher Rayleigh scattering for Au NRs.

3. Nanoplates

Triangular nanoplates are the circular, triangular or hexagonal nanomaterials in which one of the dimensions is much smaller than the other two. The most commonly used SERS active substrate nanoplates are the Au and Ag nanotriangles and nanohexagons. Because of the sharp corners and edges, the nanotriangles exhibit the strong E-field enhancements. **Figure 3** shows the morphology and optical property of nanotriangles [17].

Au and Ag nanotriangles exhibit the excellent SERS signal because of the high generation of E-field, good assembling behavior and the tunability of the LSPR bands. Although the nanotriangles own the better optical property than that of NRs, the synthetic difficulties of nanotriangles restrict its application as the SERS active substrate materials. Study has shown that the Ag nanotriangles with sharp corners and edges exhibit the larger Raman enhancement than the rounded corner nanotriangles as shown in **Figure 4** [18]. In addition, the zigzag edges of nanotriangles can generate a further enhancement compared

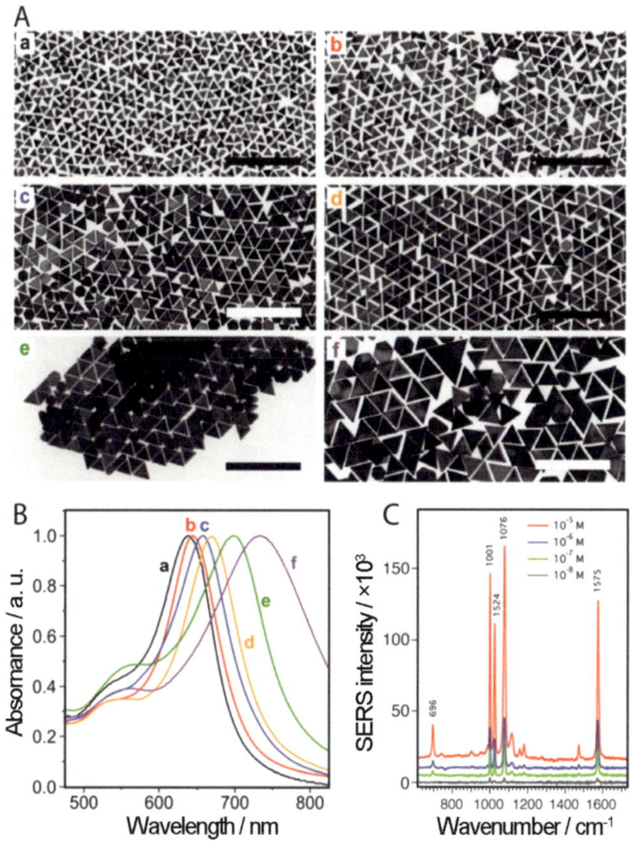

Figure 3. (A) TEM images of Au nanotriangles with different edge lengths (scale bar 500 nm). (B) UV-vis-NIR spectra of the Au nanotriangles depicted in TEM images (a–f). (C) SERS performance of Au NTs in solution for thiophenol (TP) excited at 785 nm and concentrations varying between 10^{-5} and 10^{-8} M. Adapted from Ref. [17] with permission, Copyright American Chemical Society.

to the straight edge nanotriangles. The most widely used Ag nanotriangles synthetizing methods include light-induced synthesis and wet chemistry methods. For the Ag nanotriangles synthetizing methods, polyvinylpyrrolidone (PVP), cetyltrimethylammonium chloride (CTAC) or CTAB acting as the surfactants can induce the Ag growth to the plate-like morphologies with sharp corners and edges [19]. For Au nanotriangles, Scarabelli et al. reported the controllable synthesis of Au nanotriangles with tunable size and high yield (95%) upon simple purification [17]. Tan et al. synthesized Ag nanotriangles with different edge lengths from 30 up to 210 nm with corresponding LSPR band between 485 nm and 1130 nm [20]. Ag nanotriangles SERS substrate was also used for biosensing and measured the E-field of adenosine triphosphate (ATP) with two different incident light wavelengths [20].

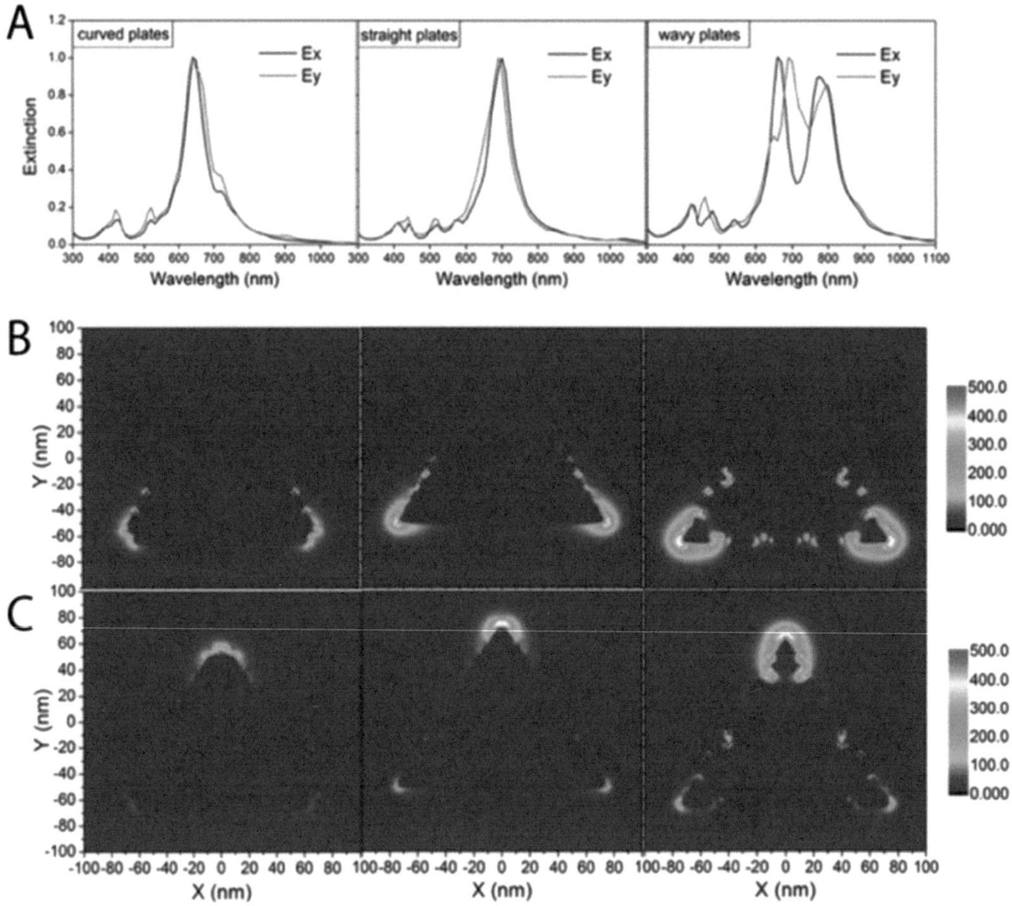

Figure 4. (A) The simulation results of Ag nanotriangles with curved, straight and zigzag edges, using finite-difference time-domain (FDTD) calculations. (B and C) E-field amplitude patterns of nanotriangles with E-field along the x-axis (Ex) and y-axis (Ey), for B and C, respectively. Adapted from Ref. [18] with permission, Copyright Wiley-VCH.

4. Nanocubes

From the synthetic angle, Ag nanocubes (NCs) can be easily obtained with high crystallinity, monodispersion and uniformity than that of Au NCs. In the most commonly used Ag NCs synthetizing method, the Ag ions of the $AgNO_3$ are reduced by ethylene glycol in the presence of HCl and PVP [21, 22]. These Ag NCs were single crystals and were characterized by a slightly truncated shape bounded by {100}, {110} and {111} facets (**Figures 5** and **6**).

The synthesis of monodisperse Au NCs is still a huge challenge. The key parameters such as reproducibility and fine size control still require further optimization. The most commonly used synthetizing method is also based on seed-mediated growth, in combination with CTAC, CTAB or CPC (cetylpyridinium chloride) acting as the surfactants. By precisely controlling the

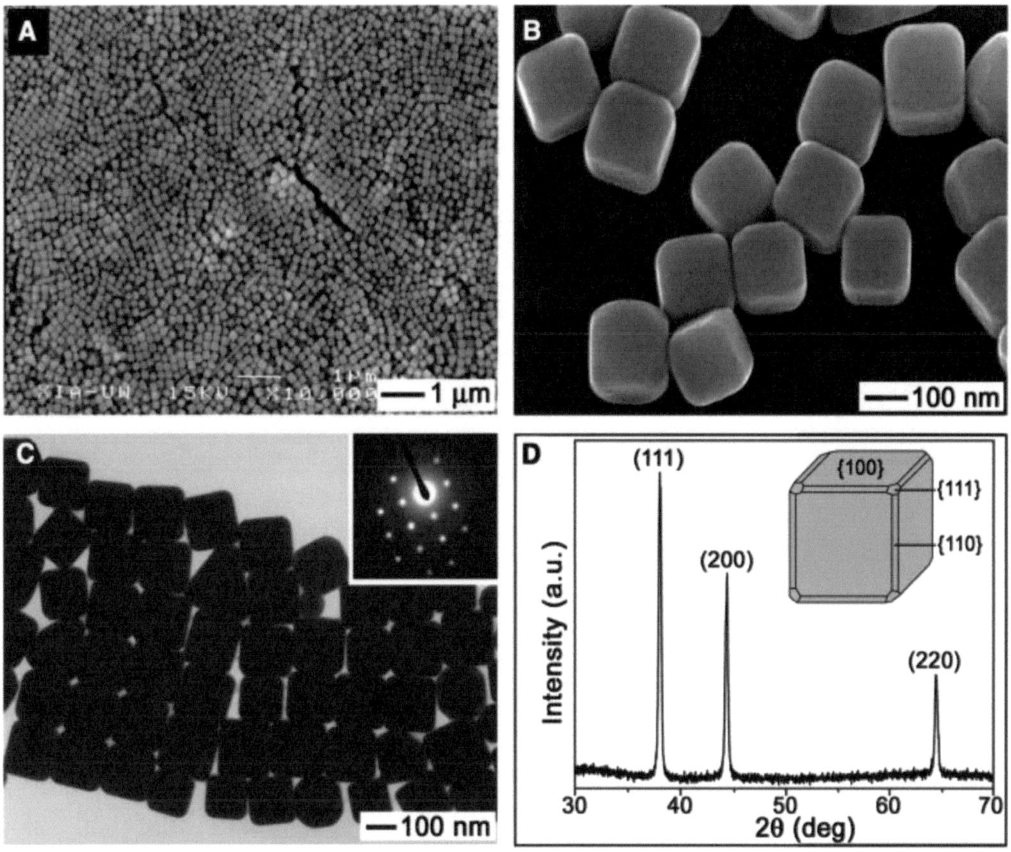

Figure 5. (A) Low- and (B) high-magnification SEM images of Ag NCs. (C) The TEM image of the same batch of Ag NCs. Inset: Diffraction pattern of an individual cube. (D) The XRD pattern of the Ag NCs. Adapted from Ref. [22] with permission, copyright science.

dynamic process, reductant concentration and the selective absorption facet of surfactants, the Au NCs can be obtained with requested shape and single crystal structure [23]. The simulation results indicated that the dipolar LSPR charges of Ag and Au NCs tend to accumulate at corner sites [24, 25]. Due to the strong LSPR and hot spots highly localized at corners, NCs are excellent candidates as SERS substrates. When the other symmetric Au NPs such as rhombic dodecahedra and octahedra were used as the SERS active substrate materials, the SERS signal of these Au NPs was still observed [26].

Because certain facets show higher chemical activities, concave NCs can also be prepared via modifications of the seed-mediated growth method, where CTAC provides control over the concave morphology of the final product. Thereby, the concave NCs can effectively enhance the performance as the SERS active substrates. The LSPR band of concave NCs red-shift obviously compared to the NCs with flat faces. Because of the sharp corner of concave NCs, the higher E-field enhancements can be expected [27].

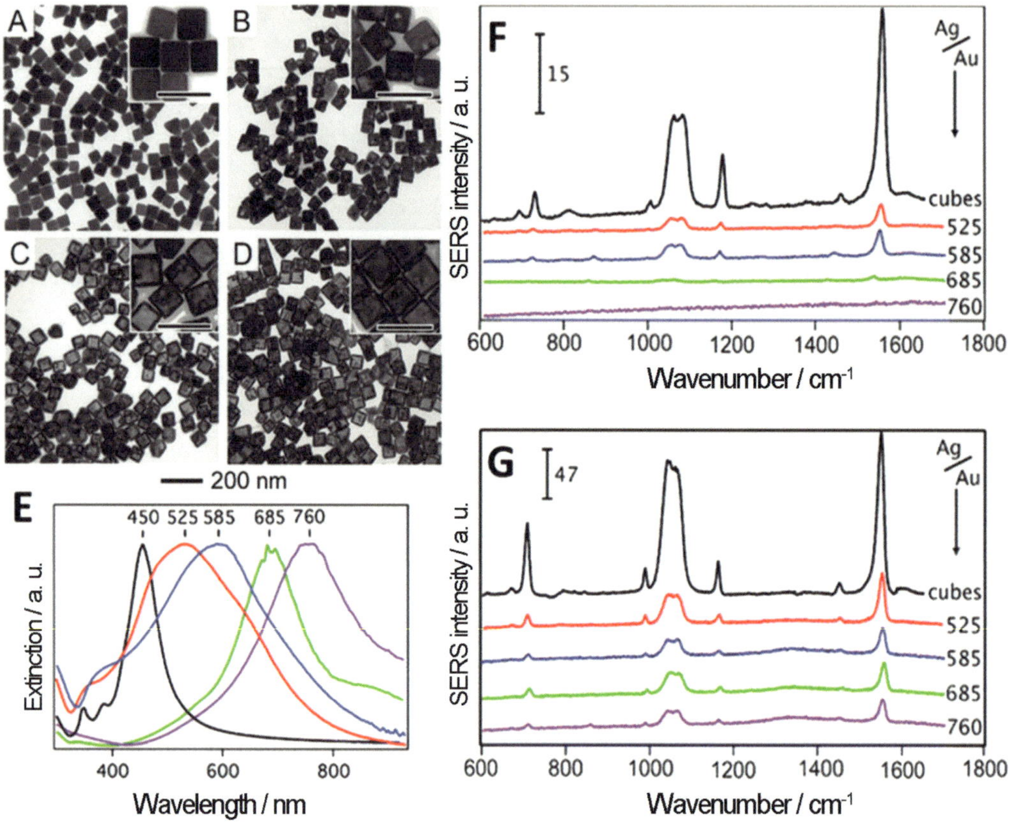

Figure 6. (A–D) TEM images of nanocages at degrees of galvanic replacement. (E) UV-vis-NIR spectra of nanocages with varying Au amounts. (F and G) SERS spectra of 1,4-BDT with nanocages used as the SERS substrate. Adapted from Ref. [28] with permission, copyright Royal Society of Chemistry.

When the inner atoms are etched, the nanocubes can be transformed to nanocages. Galvanic replacement is the most common method for synthetizing hollow nanocages. Au nanocages can be created by galvanic replacement by using Ag NCs as the templates with chloroauric acid in water phase [21]. A valuable property of Au nanocages is the red-shifting of LSPR bands into the NIR range, which is particularly useful for biological sensing and detecting applications. In addition, the galvanic replacement can also produce the bimetallic Au-Ag alloy nanocages, in which the SERS intensity shows a strong relationship between excitation wavelength and Au [28].

5. Nanostars

Nanostars and nanoflowers, so-called branched NPs, are formed by a central body and several arms or tips. These nanoarms and tips can effectively enhance the Raman scattering signal. A lot of synthetic methods based on wet chemistry, for example, the seed-mediated growth and other methods, have been reported for the preparation of nanostars for Au, Ag and other noble metals. For the seed-mediated growth method, in the presence of Ag ions,

Au ion can be reduced by ascorbic acid and CTAB or CTAC acting as the surfactants or using N,N-dimethyl-formamide (DMF) as solvent and reductant in the presence of PVP as surfactant. For template-based methods, using mesoporous silica as the template, Au nanotips can be grown on the surface of mesoporous silica. The seed-mediated growth method can also be used to synthetize Ag nanostars. In the presence of the sodium polyacrylate as the seeds, Ag ion can be reduced to Ag nanostars by ascorbic acid as the reductant.

The UV-vis-NIR spectra of Au nanostars indicate that it shows a plasmon band in the range from 600 nm to 1200 nm, corresponding to tailor the sharpness and/or AR of the tips as shown in **Figure 7** [29, 30]. EELS mapping showed an extremely high E-field intensity at the tips of the nanostars, which can enhance the Raman signal efficiently [31].

Nanostars show a higher E-field at the resonance wavelength than that of NRs or nanospheres with the similar dimers. The SERS detection limit can achieve to enhance the Raman signal via the plasmon coupling between the adjacent Au tips or between the Au tips and Au core. Au nanostars acting as the SERS active substrate material can achieve an ultra-sensitive 4-mercaptobenzoic acid (4-MBA) detection, with the detection limits as low as 10 fM [32]. Based on the SIE-MoM, by increasing the surface coverage, the relatively constant enhancement can be observed [33].

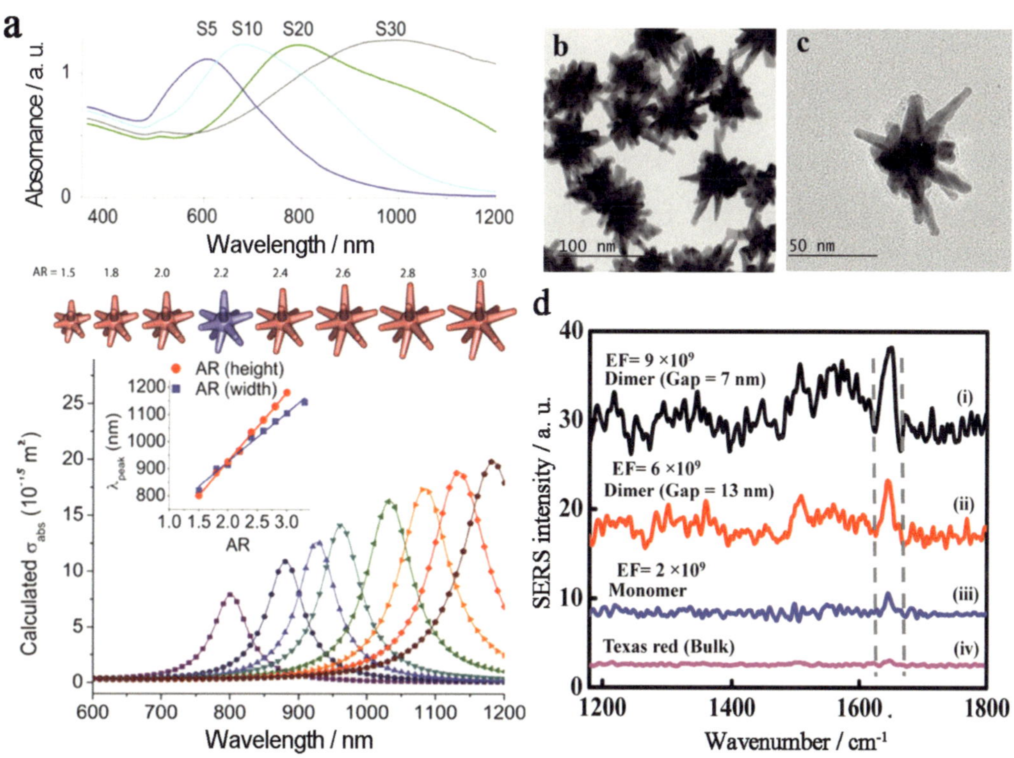

Figure 7. (a) Optical properties of Au nanostars with different branching degrees. An increase in the branching produces a red-shift in the corresponding spectra. Adapted from Ref. [30] with permission, Copyright IOP Publishing. (b and c) TEM images of Au nanostars. (d) SERS spectra of Texas red (TR) dye bound to Au nanostar dimers with average gaps of 7 nm (curve (i)) and 13 nm (curve (ii)), and TR dye bound to Au nanostar monomer on DNA origami (curve (iii)) and bulk TR dye (curve (iv)) recorded using 532 nm laser. Adapted from Ref. [34] with permission, Copyright American Chemical Society.

6. Conclusions and outlook

In summary, the synthesis of precisely controllable anisotropic NPs, with high uniformity and yield, plays the significant role in the SERS detection and imaging. This chapter has demonstrated that the recent research efforts to synthesize the anisotropic noble metal NPs with different morphologies, the E-field distribution, the key morphological parameters to achieve the strong E-field and ultra-sensitive SERS detection. The biggest influence of SERS performance is the effective near-field Raman scattering enhancement, which can be achieved by tuning the LSPR wavelength of the anisotropic noble metal NPs matching with the excitation light wavelength to induce the strong plasmon oscillation. In addition, the anisotropic NPs with high curvature give rise to the highest polarizability at the corners and edges. Many other factors can also powerfully impact the SERS signal output, for example, the binding affinities of analytes to the SERS active materials, the stability of the analyte-NPs system and so on. The interplay of these factors offers a huge potential for the NPs used as SERS active material, and the multifunctional design of the anisotropic noble metal NPs can be applied for the specific application and creates the next generation SERS substrate.

Author details

Meng Xu and Jiatao Zhang*

*Address all correspondence to: zhangjt@bit.edu.cn

Beijing Key Laboratory of Construction-Tailorable Advanced Functional Materials and Green Applications, School of Materials Science and Engineering, Beijing Institute of Technology, Beijing, China

References

[1] Lane LA, Qian X, Nie S. Chemical Reviews. 2015;**115**:10489

[2] Moskovits M. Journal Raman Spectroscopy. 2005;**36**:485

[3] Smith WE. Chemical Society Reviews. 2008;**37**:955

[4] Willets KA, Van Duyne RP. Annual Review of Physical Chemistry. 2007;**58**:267

[5] McCreery RL. Raman Spectroscopy for Chemical Analysis Measurement Science and Technology. 2001;**12**:653

[6] Fleischmann M, Hendra PJ, McQuillan AJ. Raman Spectra of Pyridine Adsorbed at a Silver Electrode Chemical Physics Letters. 1974;**26**:163

[7] Jiang RB, Li BX, Fang CH, Wang JF. Metal/Semiconductor Hybrid Nanostructures for Plasmon-Enhanced Applications Advanced Materials. 2014;**26**:5274

[8] Chen H, Shao L, Li Q, Wang J. Gold Nanorods and Their Plasmonic Properties Chemical Society Reviews. 2013;**42**:2679

[9] Jana NR, Gearheart L, Murphy CJ. Wet Chemical Synthesis of Silver Nanorods and Nanowires of Controllable Aspect Ratio Chemical Communications. 2001:617

[10] Nikoobakht B, El-Sayed MA. Preparation and Growth Mechanism of Gold Nanorods (NRs) Using Seed-Mediated Growth Method Chemistry of Materials. 2003;**5**:1957

[11] Ye X, Zheng C, Chen J, Gao Y, Murray CB. Preparation and Growth Mechanism of Gold Nanorods (NRs) Using Seed-Mediated Growth Method Nano Letters. 2013;**13**:765

[12] Reguera J, Langer J, de Aberasturiab DJ, Liz-Marzán LM. Anisotropic Metal Nanoparticles for Surface Enhanced Raman Scattering Chemical Society Reviews. 2017;**46**:3866

[13] Chu MW, Myroshnychenko V, Chen CH, Deng JP, Mou CY, García de Abajo FJ. Probing Bright and Dark Surface-Plasmon Modes in Individual and Coupled Noble Metal Nanoparticles Using an Electron Beam Nano Letters. 2009;**9**:399

[14] Guiton BS, beriV I, Li S, Leonard DN, Parish CM, Kotula PG, Varela M, Schatz GC, Pennycook SJ, Camden JP. Correlated Optical Measurements and Plasmon Mapping of Silver Nanorods Nano Letters. 2011;**11**:3482

[15] Orendorff CJ, Gearheart L, Jana NR, Murphy CJ. Aspect Ratio Dependence on Surface Enhanced Raman Scattering Using Silver and Gold Nanorod Substrates Physical Chemistry Chemical Physics. 2006;**8**:165

[16] Huang JF, Zhu YH, Lin M, Wang QX, Zhao L, Yang Y, Yao KX, Han Y. Site-specific growth of Au–Pd alloy horns on Au nanorods: A platform for highly sensitive monitoring of catalytic reactions by surface enhancement Raman spectroscopy. Journal of the American Chemical Society. 2013;**135**:8552

[17] Scarabelli L, Coronado-Puchau M, Giner-Casares JJ, Langer J, Liz-Marzán LM. Monodisperse Gold Nanotriangles: Size Control, Large-Scale Self-Assembly, and Performance in Surface-Enhanced Raman Scattering. ACS Nano. 2014;**8**:5833

[18] Yang Y, Zhong XL, Zhang Q, Blackstad LG, Fu ZW, Li ZY, Qin D. The Role of Etching in the Formation of Ag Nanoplates with Straight, Curved and Wavy Edges and Comparison of Their SERS PropertiesSmall. 2014;**10**:1430

[19] Qian H, Xu M, Li W, Ji M, Cheng L, Shoaib A, Liu J, Jiang L, Zhu H, Zhang JT. Surface Micro/nanostructure Evolution of Au–Ag Alloy Nanoplates: Synthesis, Simulation, Plasmonic Photothermal and Surface-enhanced Raman Scattering ApplicationsNano Research. 2016;**9**:876

[20] Tan T, Tian C, Ren Z, Yang J, Chen Y, Sun L, Li Z, Wu A, Yin J, Fu H. LSPR-dependent SERS Performance of silver Nanoplates with Highly Stable and Broad Tunable LSPRs Prepared through an Improved Seed-mediated Strategy Physical Chemistry Chemical Physics. 2013;**15**:21034

[21] Skrabalak SE, Au L, Li X, Xia Y, Facile synthesis of Ag nanocubes and Au nanocages Nature Protocols. 2007;**2**:2182

[22] Sun YG, Xia YN. Shape-Controlled Synthesis of Gold and Silver Nanoparticles Science. 2002;**298**:2176

[23] O'Brien MN, Jones MR, Brown KA, Mirkin CAJ. American Chemical Society. 2014;**136**: 7603

[24] Cortie MB, Liu F, Arnold M D, Niidome Y. Multimode Resonances in Silver Nanocuboids Langmuir. 2012;**28**:9103

[25] Wiley BJ, Im SH, Li ZY, McLellan J, Siekkinen A, Xia Y. Maneuvering the Surface Plasmon Resonance of Silver Nanostructures through Shape-Controlled Synthesis. J. Physical Chemistry B. 2006;**110**:15666

[26] Wu H-L, Tsai H-R, Hung Y-T, Lao K-U, Liao C-W, Chung P-J, Huang J-S, Chen I-C, Huang MH. A Comparative Study of Gold Nanocubes, Octahedra, and Rhombic Dodecahedra as Highly Sensitive SERS Substrates Inorganic Chemistry. 2011;**50**:8106

[27] Romo-Herrera JM, Gonzalez AL, Guerrini L, Castiello FR, Alonso-Nunez G, Contreras OE, Alvarez-Puebla RA. A Study of the Depth and Size of Concave Cube Au Nanoparticles as Highly Sensitive SERS Probes Nanoscale. 2016;**8**:7326

[28] Rycenga M, Hou KK, Cobley CM, Schwartz AG, Camargo PHC, Xia Y. Probing the Surface-Enhanced Raman Scattering Properties of Au–Ag Nanocages at Two Different Excitation Wavelengths Physical Chemistry Chemical Physics. 2009;**11**:5903

[29] Niu W, Chua YAA, Zhang W, Huang H, Lu X. Highly Symmetric Gold Nanostars: Crystallographic Control and Surface-Enhanced Raman Scattering Property J. American Chemical Society. 2015;**137**:10460

[30] Yuan H, Khoury CG, Hwang H, Wilson CM, Grant GA, Vo-Dinh T. Gold Nanostars: Surfactant-free Synthesis, 3D Modelling, and Two-photon Photoluminescence Imaging Nanotechnology. 2012;**23**:075102

[31] Kumar PS, Pastoriza-Santos I, Rodríguez-González B, Javier García de Abajo F, Liz-Marzan LM. High-yield Synthesis and Optical Response of Gold NanostarsNanotechnology. 2008;**9**:015606

[32] Indrasekara ASDS, Meyers S, Shubeita S, Feldman LC, Gustafsson T, Fabris L. Gold Nanostar Substrates for SERS-based Chemical Sensing in the Femtomolar Regime Nanoscale. 2014;**6**:8891

[33] Solís DM, Taboada JM, Obelleiro F, Liz-Marzán LM, García de Abajo FJ. Optimization of Nanoparticle-Based SERS Substrates through Large-Scale Realistic Simulations ACS Photonics. 2017;**4**:329

[34] Tanwar S, Haldar KK, Sen T. DNA Origami Directed Au Nanostar Dimers for Single-Molecule Surface-Enhanced Raman Scattering. Journal of the American Chemical Society. 2017;**139**:17639

Chapter 5

Nanoscale Insights into Enhanced Raman Spectroscopy

Izabela Rzeznicka and Hideyuki Horino

Additional information is available at the end of the chapter

http://dx.doi.org/10.5772/intechopen.72284

Abstract

Enhanced Raman spectroscopies, such as surface-enhanced Raman spectroscopy (SERS) and tip-enhanced Raman spectroscopy (TERS), are based on the amplification of intrinsically weak Raman signals of a molecule by metallic nanostructures. The main enhancement is attributed to electromagnetic enhancement. Chemical effects, such as formation of a surface complex, or a charge-transfer complex, co-adsorbed anion effect, also add to the enhancement of the signal. Using SERS, it has been difficult to study details of chemical enhancement and polarization effects due to limited optical resolution of the technique and usage of roughened metal surfaces. These obstacles were overcome with the development of the TERS technique. TERS has extended Raman spectroscopy into the nanoscale region. In this chapter, nanoscale insights into surface chemistry that lead to Raman signal enhancement are described. The effect of molecular binding and orientation as well as commonly used in SERS chloride activation of metal surfaces is discussed. Finally, we describe the future prospects of TERS and the challenges that keep us from harnessing the full potential of the technique.

Keywords: tip-enhanced Raman spectroscopy, surface-enhanced Raman spectroscopy, scanning tunneling microscopy, halogen overlayer, bipyridine, chemical enhancement, molecular orientation, Raman tensor

1. Introduction

Raman spectroscopy is an analytical technique based on inelastic scattering of light. The light scattered by a molecule shows a wavenumber shift with respect to the excitation line. This effect was first described by C. V. Raman in 1928 and quickly became used as a powerful method for identifying molecules through their vibrational motions [1]. However, until the 1970s, Raman spectroscopy did not receive much attention of scientists working in the field of surface analysis. This is because intrinsic Raman scattering cross-sections of molecules is small, on the order of 10^{-32} cm^2 sr^{-1} per molecule [2]. The observation of Raman signal enhancement of molecules adsorbed on roughened metal surfaces was a corner stone in the development of a family of enhanced vibrational spectroscopies with the surface-enhanced Raman spectroscopy (SERS) being the most widely practiced [3–5]. In SERS, Raman signals are amplified by placing a molecule in the vicinity of metal nanostructures. A large number of papers have been published over a period of 40 years on the origin of Raman signal enhancement on SERS active substrates [6–8]. Consequently, the overall enhancement of the signal has been attributed to two main mechanisms: electromagnetic enhancement (EME) and chemical enhancement (CE).

EME is explained by the enhancement of the electromagnetic field at the molecule's position by excitation of the localized surface plasmon resonances of metallic nanostructures [9, 10]. Basic EME is now well understood and believed to be the major contribution to the enhancement of Raman signals on SERS active substrates. In the process, both the incoming and outgoing field is amplified. For a single molecule, basic EME, excluding polarization effects and tensorial nature of the Raman polarizability, scales as E^4, where E is the intensity of the electromagnetic field. Polarization effects influence signal intensity but in a more complex way, depending on the symmetry of the vibrational mode. Moskovits, in his 1982 work, described Raman surface selection rules, that is, modification of band intensities for a molecule adsorbed on a flat metal surface [11]. Experimental justification of polarization effects requires Raman data obtained on flat metal surfaces and a nanoscale characterization of the environment in which a molecule is present. With the development of tip-enhanced Raman spectroscopy (TERS), we can now define the nanoscale environment of a molecule and correlate it directly with the Raman signal. It is expected that polarization studies in TERS settings, supported by calculations of the Raman polarizability tensor components of a molecule, will bring further insights into mechanism of Raman signal enhancement and surface selection rules.

CE describes various effects affecting the Raman polarizability α^R with respect to the α_0^R of a free molecule [12]. Among them is the formation of a metal-molecule complex, anion and field gradient effects [8, 13–16]. CE can be separated into a resonant and nonresonant mechanism. The resonant mechanism is discussed based on the strength of a metal-molecule coupling. In the weak coupling regime, the molecular electronic states may shift and broaden upon their interaction with the metal, and enhancement of the Raman scattering can be achieved by the resonance Raman effect, in which the incident beam is in resonance with the electronic excitation of a molecule. In the strong coupling regime (chemisorption), new electronic states appear, so-called charge-transfer (CT) states that act as intermediate states in Raman scattering [6, 17, 18]. The enhancement of the signal is due to resonance of the incident beam with an excitation from the metal to the molecule or *vice versa*. The nonresonant effect, called the static chemical enhancement, involves various effects, such as the binding geometry of the adsorbate, the influence of co-adsorbed anions and the presence of non-zero static electric fields [14–16, 19, 20]. It has been difficult to investigate how these effects influence the strength of the Raman signals as the signals measured in SERS are averaged over the diffraction-limited spot. A deeper understanding of CE requires Raman studies with nanoscale resolution aided by suitable electronic structure calculations of the molecule-metal system.

Combining Raman spectroscopy with scanning probe microscopy into so-called TERS has made it possible to investigate CE and EME in detail with nanoscale resolution [21, 22]. TERS is an offspring of SERS, in which a "hot spot" is created between a metallic tip and a flat surface.

http://dx.doi.org/10.5772/intechopen.72284

The TERS technique eliminates the need for rough or nanostructured surfaces, allowing for investigation of the details of molecular adsorption under well-defined conditions. The biggest advantage of TERS over SERS is its capability to acquire Raman signals and nanoscale images of the molecule adsorbed on a solid substrate simultaneously.

In this chapter, readers will be introduced to the experimental and theoretical aspects of TERS based on a scanning-tunneling microscopy (STM-TERS). Subsequently, the results of TERS studies of molecules adsorbed on flat metal surfaces are summarized with an emphasis on the molecular orientation and surface selection rules. Later, the effect of chlorine activation of metal surfaces on the adsorption of organic molecules and halogen overlayer-templated growth of surface-grown metal-organic layered structures are described. At the end of the chapter, future prospects and challenges of TERS in studies of molecular adsorption on metal surfaces are discussed.

2. Tip-enhanced Raman spectroscopy

2.1. TERS technique: a historical overview

Nanoscale vibrational spectroscopy has been a longstanding dream of scientists working in various fields. In the early 1980s, a scanning tunneling microscope was invented which brought us the capability to explore surfaces with unprecedented subnanometer scale resolution [23]. Later on, images of molecules adsorbed on metal and semiconductor surfaces were obtained [24]. STM is based on a tunneling current between a conductive surface and a sharp metallic tip. In the basic mode of operation, STM does not yield vital chemical information, and formation of surface complexes cannot be confirmed. Consequently, inelastic electron tunneling spectroscopy (IETS) was developed to obtain chemical information from single molecules adsorbed on metal surfaces [25]. Despite its ultimate resolution and sensitivity, IETS has not evolved into a common nanoscale vibrational spectroscopic method. This is because IETS is a very challenging technique requiring low temperature, ultra-high vacuum conditions and ultra-low noise electronics.

From the development of the field of plasmonics and the demonstration of large electromagnetic field enhancement in SERS experiments, the idea of nanoscale vibrational spectroscopy on surfaces using Raman scattering has been brought to life [26]. Raman spectroscopy has many advantages as it is based on the optical response of the system and can be applied under ambient conditions and in water. However, Raman scattering is intrinsically a very weak process with only one in every 10^6–10^8 incident photons being scattered [27]. Theoretical reports in the late 1990s showed that the electric field at the metallic tip end is dramatically enhanced under certain polarization direction [28]. This report inspired scientists working in the field of optical microscopy and surface science to utilize the enhancement effect to develop a high-resolution molecular spectroscopic technique. The Kawata group in Japan and the Zenobi group in Switzerland simultaneously reported TER spectra from multilayers of organic molecules deposited on a glass surface using a metalized cantilever probe of an atomic force microscope (AFM) [21, 22]. Later, Pettinger *et al.* reported TER spectra from the monolayer of malachite green adsorbed on CN^- modified gold surface, using STM tip [29]. Ultimate sensitivity and resolution of a single molecule were demonstrated in 2008 with TERS optics incorporated into ultra-high vacuum environment (UHV-TERS) [30]. Over the last decade, TERS has been adapted as an analytical tool in chemistry, biology and materials science [31].

2.2. Experimental aspects of STM-TERS

2.2.1. Tip fabrication and related issues

A sharp metallic tip is a central part of the TERS setup and its quality (size and shape) defines the spatial resolution of TERS and, to some extent, the magnitude of the enhanced field. Owing to its importance, several papers have discussed various tip fabrication methods [32–34]. Tips are usually made of gold or silver as these metals have plasmon resonances in the visible region of the electromagnetic spectrum. Electrochemical etching is a common method to prepare the tips. In a typical procedure, Au tips are prepared by direct current electrochemical etching in a 50:50 (v:v) mixture of concentrated HCl and ethanol. Au wire (a tip after etching) serves as the anode and a gold or platinum ring acts as the cathode. The end of the Au wire is submerged 1–2 mm into the solution at the center of the gold ring. A voltage in the range of 1.7–2.5 V is applied between the cathode and the anode. The etching reaction proceeds until the electrochemical current drops to zero. Typically tips with a tip-apex size in the range of 20–50 nm are obtained in this way.

2.2.2. Optical geometries

TERS requires integration of STM/AFM with the optical components used in Raman spectroscopy. Various optical geometries have been adapted in the past 15 years to work with a variety of different samples [34]. They include a bottom-, side-, top- and a parabolic mirror illumination depicted in **Figure 1**. Each geometry has its own advantages and disadvantages, which are summarized hereafter.

The bottom illumination geometry (**Figure 1a**), in which the incident laser light is fed through the microscope objective placed at the bottom and the rear side of the sample, is not practiced among STM-based TERS users as it requires transparent samples. This geometry is commonly used in AFM-based TERS, in which an inverted optical microscope platform is used. The advantage of this geometry is a very high collection efficiency of scattered light as oil-immersion objectives with a high numerical aperture (NA) can be used. Recently, the Deckert group has modified the bottom illumination geometry to work with opaque samples by using

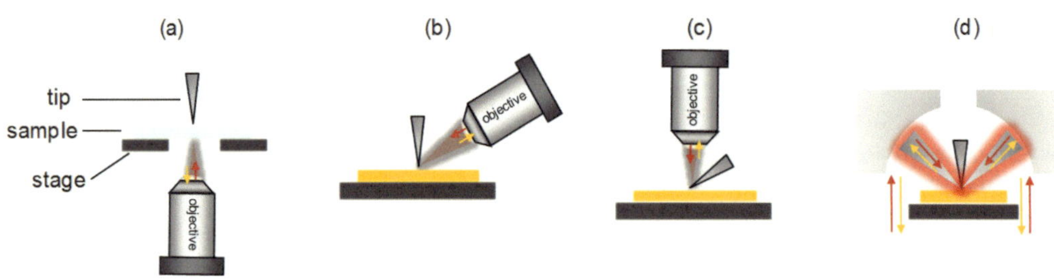

Figure 1. Four common optical geometries used in TERS (a) bottom illumination, (b) side illumination, (c) top illumination, (d) parabolic mirror illumination.

a dichroic mirror and passing the incident light through the transparent side of the opaque sample [35]. The bottom illumination is well suited for biological samples.

Side illumination (**Figure 1b**) is the most commonly used setup in TERS built on STM. In this geometry, the incoming laser light as well as the outgoing scatter signal is collected by a long working distance microscope objective placed at an angle in the range of 45–70° relative to the tip axis that maximizes the electric light component along the tip axis. The advantage of this geometry is an easy integration with commercial STM setups. It can be used both for transparent and nontransparent samples. However, only conductive samples can be studied. The disadvantage of the side illumination geometry is relatively low collection efficiency in comparison to the bottom light illumination geometry. This is because objective lenses with a low numerical aperture are employed due to space restriction around the STM tip. Another disadvantage of the side illumination geometry is an asymmetric illumination of the tip, which leads to larger spot size and higher far-field background. In addition to tip resonances, so-called gap mode resonances can be excited in the side illumination geometry [36, 37]. They play a big role in the enhancement of the signal when the distance of the tip to the surface is below 2 nm [38]. Excitation of the gap-mode resonances improves sensitivity and resolution of the TERS technique. The side illumination geometry is widely used in the field of surface science.

Top illumination (**Figure 1c**) is the second most commonly used geometry in the STM-TERS community. This geometry makes focusing of the incident beam on the STM tip easier. The tip is placed at an angle to the surface with the microscope objective placed normal to the surface. Although some of the scattered light is shadowed by the tip, this geometry offers still higher excitation and collection efficiency than the side-illumination geometry does. Another advantage is a reduction in the far-field background. Both opaque and transparent samples can be studied. This geometry is ideally suited for investigating nanoscale phenomena on silicon or graphene samples.

A geometry utilizing a parabolic mirror (**Figure 1d**) was first demonstrated by researchers working with an STM unit operating in UHV conditions [30]. It was developed in order to increase the collection efficiency by allowing collection of light from all directions. In this geometry, a hole is made in a parabolic mirror to accommodate the STM tip. STM-TERS setups with a parabolic mirror work in a reflection mode, allowing both opaque and transparent samples to be studied. Compared to the side and top illumination geometries, the parabolic mirror geometry offers high collection efficiency. Two major disadvantages of this geometry are: it is difficult to integrate into commercial STM units and optical alignment is difficult. A small angular mismatch results in defocusing of the laser spot and loss of the signal. Thus, this geometry has not been widely used.

2.3. Understanding TERS

The origin of Raman signal enhancement in TERS is similar to that in SERS, which is due to EME and CE. In contrast to SERS, in which Raman signals are obtained from multiple hot spots across the surface, the signal in TERS originates from one central hot spot, which is created between the apex of a tip and a flat metal substrate.

Various effects, such as electrostatic lightening rod effect, excitation of localized surface plasmon polaritons (SPPs) on the tip and antenna resonances, contribute to EM field enhancement [39]. The lightening rod effect is independent of the excitation wavelength. On the other hand, excitation of SPPs is wavelength dependent with maximum field enhancement achieved when the laser energy coincides with the localized surface plasmon resonances of the tip. Finite-difference time domain (FDTD) calculations show that the magnitude of the enhancement due to plasmon excitation depends on the laser light polarization, tip radius and dielectric properties of the surrounding medium [38]. In-plane light polarization (*p*-polarization, parallel to the tip axis) gives much higher enhancement than the out-of-plane polarization (*s*-polarization) does. The maximum enhancement is predicted for tips with the apex radius of 15–20 nm [38]. Tips made of silver provide higher enhancement when visible light excitation is used. **Figure 2** shows all possible effects contributing to the enhancement of the signal in TERS.

When the tip-metal surface distance is smaller than 2 nm, additional EM field enhancement is observed. At this distance, LSP of the tip and a metal interact with each other to form hybridized modes, called gap modes [37]. The enhancement due to excitation of the gap mode resonances depends strictly on the tip-metal surface separation [40, 41]. The gap modes are efficiently excited when $D/R < 1$, where D is the distance of the particle from the surface and R is the radii of the tip apex. The enhancement of the scattered light intensity is found to be as high as 10^{12} for a 20 nm radius gold tip and tip-substrate separation of 1 nm [38]. Such small tip-substrate separations are easily controlled by the tunneling feedback function of the STM.

CE due to chemisorption, formation of a surface-complex and anion surface modification can be studied with excellent resolution using STM-TERS. These studies began in the field of surface science. A resonance enhancement of 10^6 has been reported by Pettinger *et al.* for a malachite green molecule adsorbed on an Au(111) surface [42]. Ren *et al.* have shown that Raman signal can be obtained from monolayers of non-resonant molecules with weak Raman cross-sections [43]. Observed frequency shifts between Au and Pt surfaces indicate that TERS is sensitive enough to identify molecular orientation and revealed details of molecule-surface interaction.

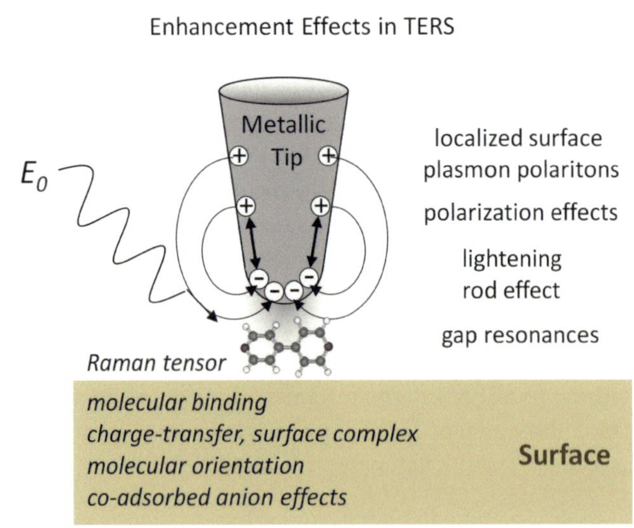

Figure 2. Possible CE and EME effects contributing to Raman signal enhancement in TERS.

Although simultaneous observation of Raman and infrared (IR) vibrational modes are exclusive to each other in the case of centrosymmetric molecules, some TER spectra published in the literature show the presence of IR active or silent modes [44]. Polbutko explained appearance of these lines due to strong quadrupole light-molecule interactions arising from strongly inhomogeneous electromagnetic fields, which exist near rough metal surface [45].

The Dong group has recently demonstrated sub-nanometer resolution in TERS experiments [46]. As subnanometer resolution is difficult to understand in terms of the classical electromagnetic theory, these results have inspired theoreticians to work on proposing new mechanisms that could explain the results. Duan and Luo have proposed involvement of nonlinear optical processes [47]. Creation of an "atomic-scale hot spot" has also been proposed [48]. In addition, multiple elastic scattering of light between molecular dipoles adsorbed on the surface has been proposed to explain the improved signal intensity and TERS spatial resolution [49].

The ultimate goal to understand TERS from molecules adsorbed on metal surfaces is to understand how relative intensities of Raman lines depend on the molecular orientation and polarization direction of the excitation light. These studies are still challenging as there are many parameters, such as molecular binding geometry, Raman tensors and direction of local field polarization, that have to be determined. It is still not possible to formulate rules similar to the IR metal surface-selection rule (SSR). The IR SSR states that, for a molecule adsorbed on a metal surface, vibrational modes having a dipole moment perpendicular to the surface are the most enhanced [50]. Moskovits described the concept of SSR in SERS experiments for flat metal surfaces [11]. The author's work in 1982 gave theoretical grounds for local field polarization. Recently, Ru *et al.* have experimentally validated Moskovits theory by studying polarization and incident angle dependences of the SERS signals [51]. Similar studies using TERS can offer more insights into the Raman SSR as the Raman signal can be directly related to the molecules present under the tip.

In order to comprehend enhancement mechanisms in detail, more sophisticated experiments and theoretical analysis are required. Emerging TERS studies under UHV and on well-defined systems can deliver more results, which should eventually bring us closer to understanding origin of signal enhancement and contribute to the development of TERS as a reliable analytical tool.

3. Enhancement of Raman signals due to chemisorption

This section summarizes our studies on molecular adsorption and orientation of 4,4′-bipyridine (4,4′-BiPy) and 4,4′-bipyridine *N,N′*-dioxide (4,4′-BiPyO_2) in monolayers formed on gold thin films deposited on muscovite mica substrates using STM-TERS supported by calculated Raman tensor polarizability components. The enhancement of the Raman signals is attributed to the formation of a chemisorbed overlayer with a standing up molecular configuration [52].

3.1. Experimental details

A TERS setup with the side-illumination geometry was used in the experiments described here. The setup consists of a commercial STM unit (Nanoscope E, Veeco Instruments Inc., USA), a spectrograph (SP-2150i, Roper Scientific, GmbH) and optical components. The STM has a modified piezo scanner head which allows to install a high numerical aperture objective lens (Mitutoyo, LWD 100×, NA = 0.7, WD = 6 mm) in front of the STM tip. The lens is placed at an angle of 60° to the surface normal with the light polarization parallel to the tip axis (*p*-polarization). This objective lens is used to deliver the excitation laser beam as well as to collect the backscattered light from a tip-surface junction.

The optical pathway adapted in the study is shown in **Figure 3**. A red, He-Ne laser beam (632.8 nm, max. Output 30 mW, CVI Melles Griot, USA) with circular polarization was used for the excitation. The laser light was allowed to pass through a band-pass filter (Sigma Koki, Japan, bandwidth = 3 nm) and a polarizer. A transmitted light was reflected by a mirror, passed through a 45° dichroic beam splitter (RazorEdge, type U, Semrock), and reflected by two other mirrors before being focused on the tip-surface junction by the objective lens.

The backscattered radiation is collected by the same objective lens and reflected by two mirrors before falling on the dichroic beam splitter. The scattered signal passes through an ultra-steep long-pass edge filter (RazorEdge, type E, Semrock), and is focused by a lens (diameter = 25 mm, focal length = 100 mm) onto the slit of the spectrograph. A back-illuminated, charge-coupled device (CCD) camera (Spec-10, Princeton Instruments) cooled by liquid nitrogen was used to acquire Raman spectra. The spectrograph was installed with 300 g/mm diffraction grating. The spectral resolution of the system was 10 cm^{-1}. All experiments were carried out in ambient conditions with the incident laser power of 0.4 mW, giving power density of 8×10^7 W/m^2 in the focal region.

3.2. Details of theoretical calculations used to derive molecular orientation

The geometry of the molecules and fundamental vibrational frequencies were calculated using the Gaussian 09 package. Molecular structures in the ground state were optimized by the B3LYP exchange-functional of the density functional theory and 6-31G++(d,p) basis set [53]. The optimized geometries of both molecules (4,4′-BiPy and 4,4′-BiPyO_2) are nonplanar

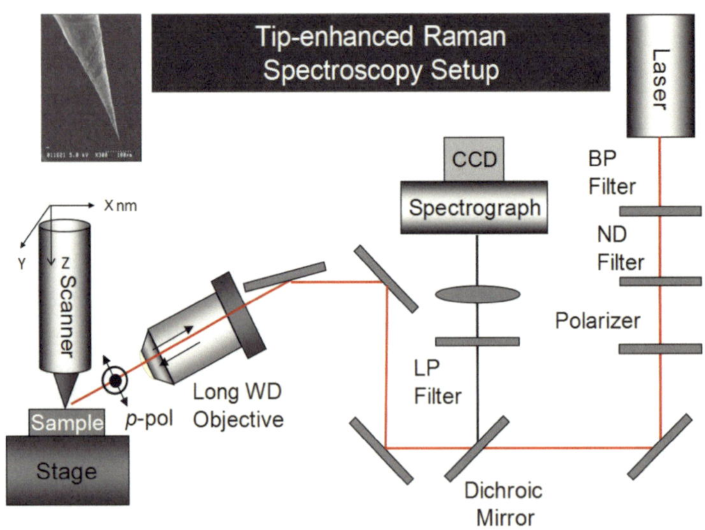

Figure 3. STM-TERS setup. The inset shows a SEM picture of the Au tip, etched at a bias of 2.4 V. Adapted from Rzeznicka *et al.* [52]. Copyright@Elsevier B.V.

with D_2 symmetry. Each vibrational mode has been ascribed to a given symmetry mode and Raman polarizability tensor components were calculated for each of the mode.

The intensities of the Raman scattering were evaluated with the matrix elements of the Raman tensor, $\langle k | \hat{\alpha}_{ij} | 0 \rangle$, where $\hat{\alpha}_{ij}$ $(i, j = x, y, z)$ denotes the polarizability tensor in usual (electronic off-resonant) conditions, and $|0\rangle$ and $|k\rangle$ are the vibrational ground state and the first excited state for the normal mode k (= 1, 2,…, 3n–6), respectively [54]. This matrix element is represented in the harmonic approximation with the polarizability derivative with respect to the normal mode coordinate Q_k as follows:

$$\langle k | \hat{\alpha}_{ij} | 0 \rangle = \sqrt{\frac{\hbar}{2 \mu_k \omega_k}} \frac{\partial \alpha_{ij}}{\partial Q_k} \tag{1}$$

where $\hbar$ is the Planck constant divided by 2π, ω_k and μ_k are the angular frequency and the reduced mass of the mode k, respectively. The molecule-fixed coordinates were defined with the principal axes of inertia, where the z axis is along the long molecular axis, and the x axis is nearly perpendicular to the rings. The principal axes of polarizability tensor coincide with the *x, y and z* axes, to give α_{xx}, α_{yy} and α_{zz}. The vibrational analysis was performed to obtain the frequency ω_k, normal mode coordinate Q_k and the reduced mass μ_k of each mode k. The polarizability tensor components, α_{ij}, were calculated by the facilities involved in Gaussian 09 with varying molecular geometries displaced along the normal coordinate Q_k, and the derivatives $\partial \alpha_{ij} / \partial Q_k$ were obtained by five-point numerical differentiation of the calculated polarizability tensor. The Raman tensor elements, $\langle k | \hat{\alpha}_{ij} | 0 \rangle$, in Eq. (1) were derived in this way. Subsequently, Raman scattering intensity was simulated. The intensity of the light scattered from the molecule is proportional to the square of the electric vector of the Raman scattered light, E_{sc}, which is related to the electric vector of the incident light, E_0 through the characteristic molecular Raman tensor, (α_{ij}) $(i, j = x, y, z)$

$$|E_{sc}| \propto |(\alpha_{ij}) E_0| \tag{2}$$

In order to determine orientation of the molecule, the experimental scattering intensities were compared with the scattering intensities calculated for three representative molecular orientations. The direction of the incident radiation was described in the surface-fixed coordinate system (*X, Y, Z*). The polarization of an incident laser beam in our TERS experiment was adjusted in a way that the electric vector, $E_0//Z$, that is, *Z* axis is perpendicular to the surface. The molecules can take various orientations having various molecular Euler angles, with respect to the Au substrate plane. Three representative configurations were considered, that is, when *x//Z*, *y//Z* and *z//Z*. The corresponding molecular orientation for each case is shown in **Figure 4**. For the molecule perpendicular to the surface ("end-on" configuration), the molecular *z* axis is parallel to the surface-fixed *Z* axis (*z//Z*), that is, perpendicular to the surface. The molecule with the "edge-on" configuration and the "face-on" configuration are denoted as *y//Z* and *x//Z*, respectively.

3.3. Adsorption and molecular orientation of 4,4′-BiPy on Au(111)

Adsorption of 4,4′-BiPy on the surface of an Au thin film proceeded in two stages. A first adsorption stage was observed after a short immersion time (3–5 h) of the Au film into a 1 mM ethanolic solution of 4,4′-BiPy. An image of the surface at this stage is shown in **Figure 5a**. No well-defined

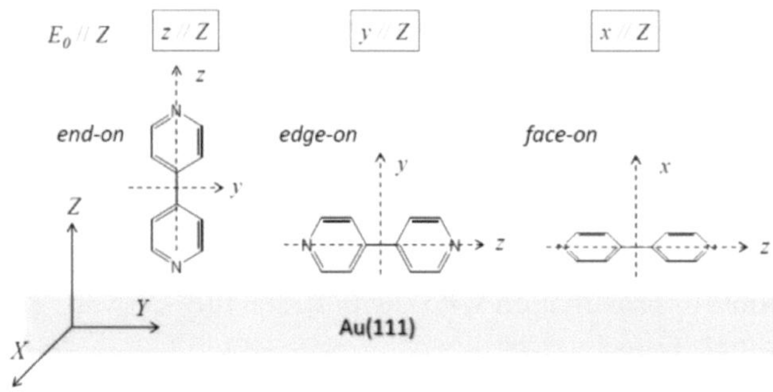

Figure 4. Molecular coordinate system. The molecular axes are x, y and z, and the surface axes are X, Y and Z. The three cases of adsorption configuration discussed in the text are shown. *E* defines the electric vector of an incident radiation. Adapted from Rzeznicka *et al.* [52]. Copyright@Elsevier B.V.

overlayers were observed. The surface of Au looked very rough and dynamic. Imaging was very unstable due to apparent adsorbate-induced surface reconstruction. Surface reconstruction is associated with the ejection of gold atoms and their diffusion over the surface. Low-coordinated gold atoms are highly reactive, and they may form a complex with molecules in the solution and diffuse over the surface to stable adsorption sites. These transient species are seen in the image as whitish spots. A second stage of adsorption was observed upon a prolonged immersion time. In this stage, a well-defined overlayer was formed. An STM image of the surface immersed into

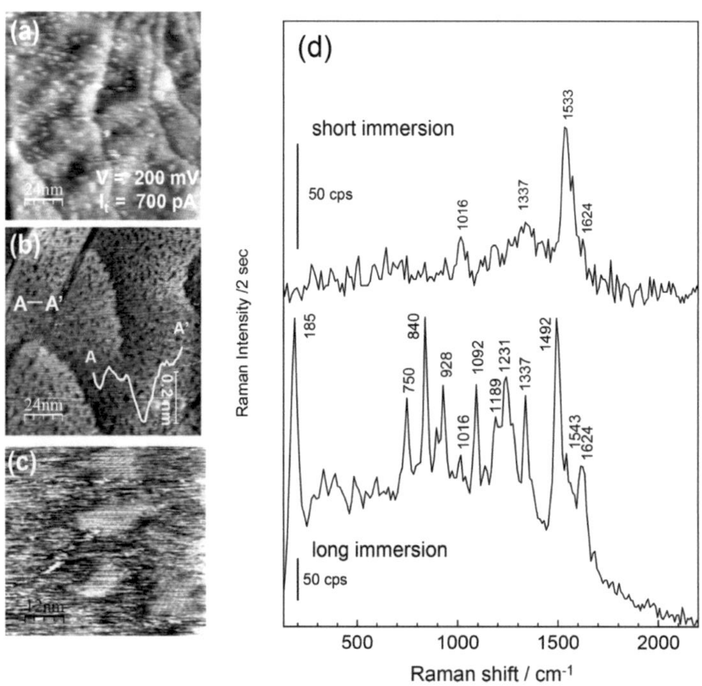

Figure 5. STM images of 4,4-BiPy adlayer formed on Au(111) after immersion of the film into a 1 mM ethanolic solution for (a) 3 hours, (b) 4 days and (c) zoom into (b). (d) TERS spectra corresponding to the layer shown in image a and b. A depth profile across the A-A′ line is shown in image (b). Adapted from Rzeznicka *et al.* [52]. Copyright@Elsevier B.V.

the solution for 4 days is shown in **Figure 5b, c**. A homogeneous monolayer with pits having a depth of a single-gold-atom was observed. It looks similar to monolayers formed by alkanethiols on Au surface, indicating involvement of Au adatoms in the process of self-assembly. TER spectra for the short immersion time and the long immersion time are shown in **Figure 5d**. The Raman spectrum for the short immersion time has only few bands with very low intensity. The absence of low-frequency vibrational signals, which could be assigned to Au-N stretching band, indicated that molecules were only weakly adsorbed (physisorbed) on the surface. In the case of the long immersion time, intensities of the Raman signals were higher, and many vibrational bands, which were not observed in the case of the short immersion time, appeared. An intense Au-N stretching signal was detected at 185 cm^{-1}, indicating that molecules were chemisorbed on the surface.

Vibrational frequencies for the two cases are summarized in **Table 1**. Each mode has been ascribed to a given symmetry mode, and Raman polarizability tensor components were calculated for each of the mode. The scattering intensities for the three possible molecular orientations were calculated and used to aid in determining the molecular orientation.

First molecular orientation of 4,4′-BiPy in the case of long immersion time is discussed.

Experimental TERS peak position/cm^{-1}		**Results of calculations**						**Modal assignment**
Figure 5	**Figure 5**				**$\|(\alpha_{ij})E_0\|(\|E_0\| = 1)/$ atomic unit**			
Short immersion	**Long immersion**	**Frequency /cm^{-1}**	**Symmetry class**	**Mode number *k***	***x*//*Z***	***y*//*Z***	***z*//*Z***	
	185 *s*	—	—	—	—	—	—	ν(Au-N)
	750 *m*	688	B_2	13	0.25	0	0.25	pyridyl ring deformation
	840 *s*	864	B_2	18	0.26	0	0.26	γ(C–H) + γ(C–C) + γ(C–C)int + γ(C-N)
	928 *m*	984	B_2	22	0.01	0	0.01	γ(C-H)
1016 *w*	1016 *w*	1014	*A*	26	0.44	0.73	1.94	γ(C-H) + δ(C-C) + δ(C-N) + ν(C-C) + ν(C-N)
	1092 *m*	1098	*A*	29	0.15	0.11	0.52	δ(C-H) + δ(C-C) + δ(C-N) + ν(C-N)
	1231 *m*	1276	B_2	34	0.16	0	0.16	ν(C-C) + ν(C-N)
1337 *m*	1337 *m*	1364	B_3	38	0	0.32	0.32	δ(C-H) + ν(C-C)
1533 *s*	1492 *s*	1540	*A*	42	0.14	0.47	0.88	δ(C-H) + ν(C-C) + ν(C-C)int + ν(C-N)
	1543 *m*	1583	B_2	43	0.07	0	0.07	δ(C-H) + ν(C-C) + ν(C-N)
1624 *w*	1624 *m*	1645	*A*	46	0.22	0.83	3.43	δ(C-H) + ν(C-C) + ν(C-C)int + ν(C-N)

s-strong, *m*-medium, *w*-weak intensity.

The symmetry index stands on D_2 class for a free molecule.

ν-stretching; δ-in-plane bending; γ-out-of-plane bending; ν(C-C)int denotes interring C-C vibration.

Adapted from Rzeznicka *et al.* [52]. Copyright@Elsevier B.V.

Table 1. Experimental and calculated Raman scattering data for 4,4′-BiPy.

As expected for the polarization direction perpendicular to the surface only vibrational modes with *A*, B_2 (*xz*) and B_3 (*yz*) symmetry are observed. For the observed vibrational modes, only the "end-on" orientation does not have null Raman intensity values suggesting that the "end-on" orientation is the most plausible. The values are equally distributed over all symmetry modes, which imply that the molecule is tilted in all three directions of Au(111) surface. A presence of the Au-N stretching peak is another strong evidence to support the "end-on" orientation 4,4′-BiPy. Henceforward, we concluded that the 4,4′-BiPy molecules, in the case of long immersion time are adsorbed in a standing-up but tilted orientation, with one of two nitrogen ends anchored to Au.

In the case of short immersion time, many of the vibrational peaks seen in the long immersion time spectrum were missing. There was no Au-N stretching signal, and the B_2-symmetry vibrational modes were not observed which rejects possibility of the "face-on" configuration. The peak intensities coincide with the y//Z-values of $|(\alpha_{ij})E_0|(|E_0| = 1)$ in **Table 1**. For the missing signals, the calculated y//Z-value of $|(\alpha_{ij})E_0|(|E_0| = 1)$ is zero or nearly zero. Henceforward, we concluded the 4,4′-BiPy has y//Z orientation, that is, the "edge-on" orientation, without the N atoms bonded to the Au substrate.

Our analysis is based on Raman polarizability tensor components calculated for a free molecule and on the assumption that a local electric field is perpendicular to the surface. As in the case of long immersion, chemisorption may change polarizability of the bonds, and Raman tensor elements may be different than the tensor elements calculated for a free molecule.

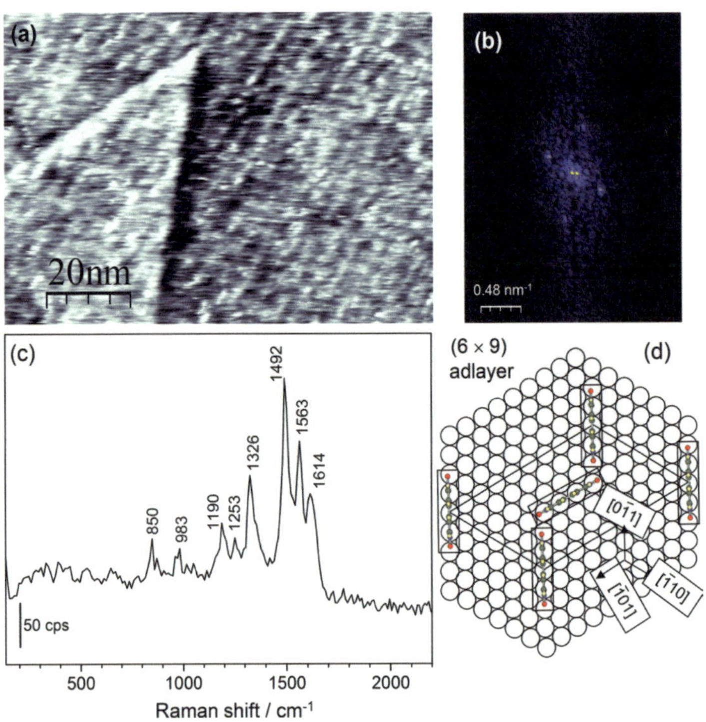

Figure 6. (a) 100 × 70 nm constant current STM image of 4,4-BiPyO_2 adlayer formed on Au(111) after immersion of the film in a neutral, 1 mM ethanolic solution for 6 h. (b) the two-dimensional Fourier-transform of the image. (c) TERS spectra of the overlayer. (d) Schematic representation of a (6 × 9) BiPyO_2 adlayer. Adapted from Rzeznicka *et al.* [52]. Copyright@Elsevier B.V.

3.4. Adsorption and molecular orientation of 4,4′-BiPyO$_2$ on Au(111)

Figure 6a shows an STM image of the Au surface upon 30 min immersion into a neutral 1 mM ethanolic solution of 4,4′-BiPyO$_2$. A two-dimensional overlayer consisting of parallel rows, extending over a triangular terrace of the Au(111) surface was observed. A two-dimensional Fourier transform (2D–FFT) of the image, shown in **Figure 6b**, revealed the spacing between parallel rows to be 1.5 and 2.2 nm^{-1}, respectively. The angle between stripes and the edges of the terrace was 30°. The overlayer is designated as (6 × 9) overlayer. **Figure 6c** shows TER signals from an Au thin film surface immersed for 6 h in a neutral 1 mM ethanolic solution of 4,4′-BiPyO$_2$. The spectrum contains a peak at 850 cm^{-1}, assigned to the in-plane ring vibrations and the N-O stretching vibrations, and a peak at 1190 cm^{-1}, which draws its intensity mainly from the in-plane C-H bending vibrations. The position of these bands falls into the frequency region of the uncoordinated 4,4′-BiPyO$_2$. The most intense band is at 1492 cm^{-1} followed by peaks at 1563 and 1614 cm^{-1}, similarly to 4,4′-BiPy. No Au-N or Au-O stretching bands were found in TER spectrum, which indicated rather weak interaction of 4,4′-BiPyO$_2$ with the Au substrate. Orientation of the 4,4′-BiPyO$_2$ is deduced in the same manner, as done for 4,4′-BiPy.

Vibrational frequencies, their symmetry modes and calculated Raman intensities are summarized in **Table 2**. TER spectrum for 4,4-BiPyO$_2$, shown in **Figure 6c** contains three bands in A symmetry. Since neither Au-O nor Au-N vibrational modes were observed, it is more likely that a molecule has its molecular long axis parallel to the Au surface.

In **Table 2**, x//Z-values of $|(\alpha_{ij})E_0|(|E_0| = 1)$ do not follow the real spectral intensity. The calculated x//Z-values are zero for the B_3 bands. On the other hand, the values for y//Z and z//Z

Experimental TERS peak position/cm^{-1}		Results of calculations						Modal assignment
Figure 6					$\|(\alpha_{ij})E_0\|(\|E_0\| = 1)$/ atomic unit			
TERS	Powder Raman	Frequency/ cm^{-1}	Symmetry class	Mode number *k*	*x//Z*	*y//Z*	*z//Z*	
850 *w*	852 *m*	858	B_1	24	0.22	0.22	0	δ(C-H) + δ(C-C) + ν(C-N) + ν(N-O)
1190 *m*	1202 *m*	1210	*A*	37	0.10	0.63	3.55	δ(C-H)
1253 *w*		1266	B_3	39	0	0.51	0.51	ν(C-C) + ν(C-N)
1326 *m*	1300 *m*	1319	*A*	40	0.23	0.51	7.16	δ(C-H) + δ(C-C) + δ(C-N) + ν(C-C) + ν(C-C)int
1492 *s*	1512 *m*	1499	B_3	46	0	0.23	0.23	δ(C-H) + ν(C-C)
1563 *s*		1572	B_3	50	0	0.39	0.39	δ(C-H) + ν(C-C) + ν(C-N)
1614 *m*	1617 *s*	1667	*A*	51	0.27	0.92	9.67	δ(C-H) + δ(C-C) + δ(C-N) + ν(C-C) + ν(C-C)int + ν(C-N) + ν(N-O)

s-strong, *m*-medium, *w*-weak intensity.
The symmetry index stands on D_2 class for a free molecule.
ν-stretching; δ-in-plane bending; γ-out-of-plane bending; ν(C-C)int denotes interring C-C vibration.
Adapted from Rzeznicka *et al.* [52].Copyright@Elsevier B.V.

Table 2. Experimental and calculated Raman scattering data for 4,4′-BiPyO$_2$.

follow the observed frequencies, except for the peak at 850 cm^{-1}. The value of $|(\alpha_{ij})E_0|(|E_0|=1)$ is zero for z//Z. The appearance of this 850 cm^{-1} band denies z//Z orientation. In conclusion, the Raman signal intensity supports the "edge-on" orientation.

4. Enhancement of Raman signals due to halogen overlayer-templated crystal growth

This section describes particular surface chemistry leading to the growth of metal-organic surface crystals in the presence of halogen overlayer. The crystals were grown on an Au surface from ethanolic solutions of 4,4′-BiPy, in the presence of HCl. STM-TERS and ordinary Raman spectroscopy were used to reveal details of a crystal growth [55].

4.1. Chlorine overlayer-templated growth of Au-4,4′-BiPy crystals on Au(111)

Figure 7a shows a large area STM image of the Au surface obtained after immersion of Au/mica film into a 4,4′-BiPy solution, adjusted with 0.1 M HCl to pH 3, for 2 days at room temperature. A zoom into the flat part of the image shows a periodic overlayer structure, shown in **Figure 7b**. The overlayer consists of bright stripes having a width of ~7.5 Å. The width is close to the length of 4,4-BiPy, which measures ~ 7.1 Å. Each stripe shows contrast modulation with periodicity of

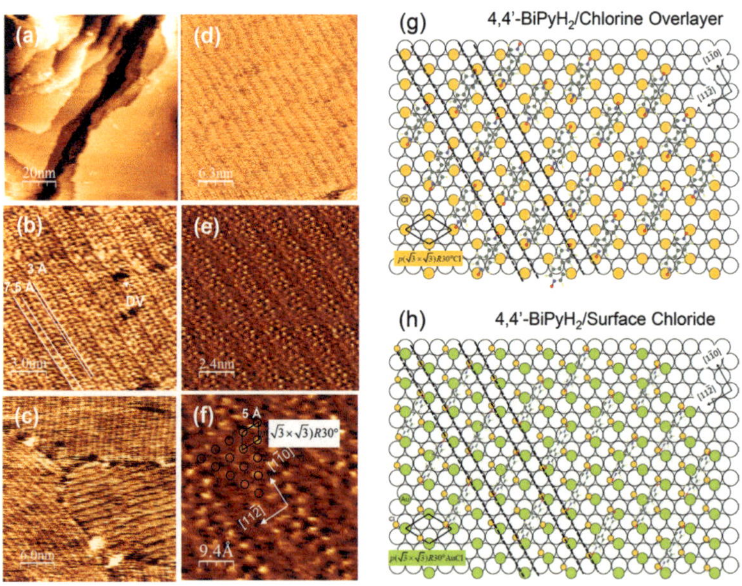

Figure 7. STM images after immersion of Au slide into 1 mM ethanolic solution of 4,4-BiPy, acidified to pH = 3 with HCl: (a) 100 × 100 nm image showing well-defined overlayer; (b) 15 × 15 nm zoomed image into (a) showing a striped structure; (c) 30 × 30 nm image showing rotational domains; (d) 31 × 31 nm image showing a p(√3 × √3)R30°-Cl overlayer structure; (e) zoomed 12 × 12 nm image of (d); (f) zoomed 4.7 × 4.7 nm image of (e) into the p(√3 × √3)R30°-Cl overlayer. Possible molecular models of the striped structure; (g) growing on top of chlorine overlayer and (h) growing on top of surface chloride. Adapted from Rzeznicka *et al.* Copyright@Elsevier B.V [55].

~ 3 Å, as indicated in the figure. The overlayer has few dark vacancies (DV). The depth of dark vacancies is in the range of 1.3–1.5 Å. The overlayer was observed to grow along the crystallographic directions of the underlying Au(111) surface as shown in **Figure 7c**. By careful alternation of the tunneling current and scanning speed, another structure originated from the underlying layer was detected, as shown in **Figure 7d** (notice the transition at the bottom of the image). In this underlying layer, individual atoms are found to be arranged in a rectangular lattice with a unit cell of $a = 5$ Å. **Figure 7(e, f)** shows large area and a zoom image of the lattice. This atomic arrangement is assigned to the $p(\sqrt{3} \times \sqrt{3})R\,30^{\circ}$ structure, which has been observed upon adsorption of 0.33 ML of Cl_2 onto the Au(111) surface at room temperature [56]. The density functional theory (DFT) calculations predicted that the fcc hollow site is the most stable adsorption site for chlorine in this overlayer [56]. Based on these facts, the most possible molecular model for the stripe phase observed after adsorption of 4,4′-BiPy onto the Au(111) surface in the presence of chlorine ions was proposed. The proposed overlayer structure is shown in **Figure 7g**. In this model, $[4,4'\text{-BiPyH}_2]^{2+}$ bipyridine cations are assumed to have a "flat-on" or "edge-on" orientation. Molecules are aligned along the $[10\bar{1}]$ direction. The self-assembly is mainly driven by an electrostatic interaction between protonated bipyridine cations and chlorine anions. **Figure 7h** shows a model for the overlayer growth on top of a surface chloride.

Figure 8a shows an STM image of Au surface after prolonged immersion of Au/mica film into the acidic solution of 4,4′-BiPy. A new overlayer with a long-range order was observed as shown in **Figure 8b**. The overlayer consists of bright stripes with a periodicity of ~ 10 Å. A growth of the next top layers can be seen at the left side of **Figure 8c**. The top layer, is rotated in respect to the bottom layer, at an angle of 120°, indicating a three-dimensional growth with the Au(111) surface registry. The stripes of the top layer consist of bright protrusions with a height of 1.5–1.8 Å. A TER spectrum taken on this surface is shown in **Figure 8f**. In-plane vibrational modes are observed above 1000 cm^{-1}. Six vibrational peaks are found in the spectrum: peaks at 1606, 1503, 1293, 1225, 1071 and 1017 cm^{-1}. The observed vibrational frequencies correspond to protonated form of 4,4′-BiPy [57]. No out-of-plane modes are observed, suggesting the "edge-on" molecular orientation. Below 1000 cm^{-1}, only a small peak at ~255 cm^{-1} is observed. Pettinger *et al.* assigned vibration at this frequency to the metal-halogen vibration of a surface complex containing metal adatom, halogen ions and pyridine [58].

An optical microscopic image of the sample after a prolonged immersion into the solution is shown in **Figure 8d**. Rectangular shaped, 3D islands of different sizes are found on the surface. Depression defects are always seen near the islands. We speculate that these defects act as a supply of Au adatoms that are further incorporated into the crystal. **Figure 8f** shows a Raman spectrum taken within the area of a large 3D island using a confocal Raman unit. The spectrum above 1000 cm^{-1} is consistent with the Raman spectrum of a solid $BiPyH_2Cl_2$ [59]. The spectrum is similar to the TER spectrum but bands are more intense.

In contrast to TER spectrum, out-of-plane modes are also observed suggesting that molecules with "flat-on" orientation are also present. A very weak Raman signals were also observed at 2460 and 3450 cm^{-1}. They were assigned to the $N\text{-}H^{+}\ldots Cl^{-}$–4,4′-BiPy stretching, and free N-H stretching vibrations, respectively [60, 61]. At very low frequencies, two strong peaks at ~88 and 116 cm^{-1} with the shoulder at 134 cm^{-1} were observed. Similarly, low-frequency Raman peaks are observed

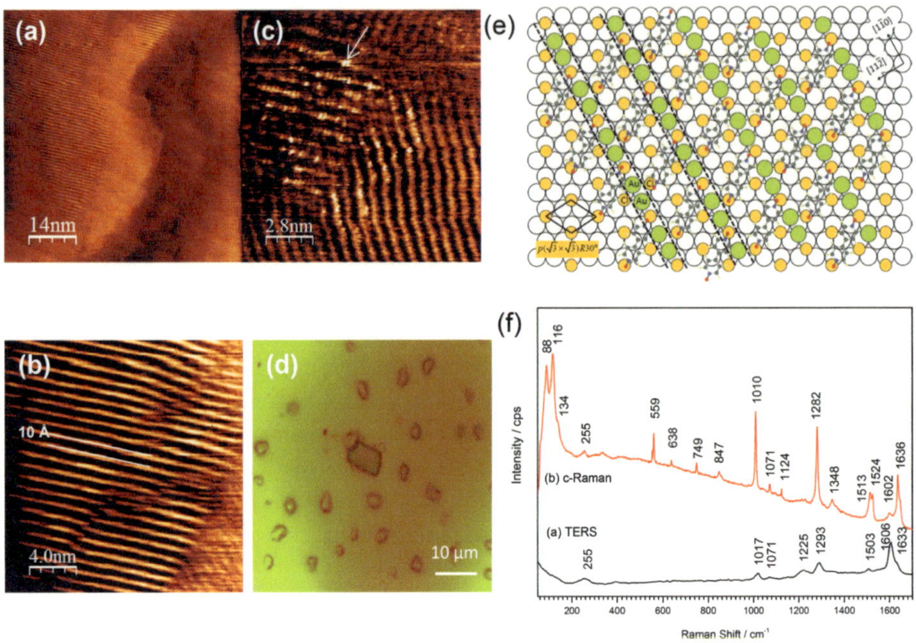

Figure 8. STM images after prolonged immersion of an Au slide into 1 mM ethanolic solution of 4,4-BiPy, acidified to pH = 3 with HCl: (a) a 70 × 70 nm image showing a chain structure; (b) a 20 × 20 nm zoomed image of the chain structure; (c) a 14 × 14 nm zoomed image of the chain structure showing development of the next top layers; (d) a bright-field microscope image of the Au slide showing surface-grown large 3D crystals; (e) a possible molecular model of the chain structure and (f) Raman spectra. Adapted from Rzeznicka *et al.* Copyright@Elsevier B.V [55].

in dinuclear Au complexes containing Cl, and in the case of pyridine adsorption on Ag electrodes. In dimethylgold halides, Au-Au vibrations are found at ~74 cm^{-1} [62]. Thus, these two peaks were assigned to Au-Au and Au-Cl stretching vibrations, respectively. The assignment was supported by the results of secondary mass ion spectrometry (SIMS) which yields information on the surface species. A highest intensity gain was observed for $m/z = 465$ corresponding to Au_2Cl_2 species [55].

5. Future prospects and challenges

In this chapter, we have reviewed the principles of STM-based TERS and discussed how molecular binding and halogen overlayer influence the intensity of the Raman signals. These two effects contribute to CE, which is system specific, that is, its magnitude depends on the metal-molecule system and experimental conditions of sample preparation. Further studies on the effects of molecular orientation on signal enhancement under more well-defined conditions, such as those provided by UHV environment, can bring us more knowledge on the mechanism of TERS.

Combined studies using cryogenic, polarized UHV-TERS and nanolithographically fabricated model nanostructures, supported by the state-of-the-art calculations to determine the Raman polarizability tensor components of a molecule-metal can lead to the formulation of TERS surface selection rules [46, 63, 64]. Home built STM-TERS systems in the Duyene and the Wang group are making first steps in this direction. We have collaborated with the company Unisoku in Japan in the development of a commercial UHV-TERS and have shown its capability to obtain relatively strong Raman signals from organic molecules adsorbed on a metal surface. Cryogenic cooling has been found to resolve issues of spectral fluctuations, as shown in **Figure 9**.

Fukumura *et al.* have proposed that single molecule sensitivity could be facilitated by employing vibrational excitation of molecules using inelastic scattering of tunneling electrons synchronized with the laser excitation to the excited states [65]. The technical challenge with this approach lies in the synchronization of the laser pulse with the electric pulse. The Duyene group has just started incorporating ultra-short laser pulses with UHV-TERS [66]. Apart from a purely academic interest, STM-TERS could contribute to understanding surface chemistry under ambient or solution conditions and aid in the development of large-scale metal protective organic layers. Moreover, metal leads are also important in electrical applications. It is a challenging task to minimize Ohmic losses for metal electrodes covered with thin organic films. As demonstrated in this chapter, halogen-modified surfaces could act as templates for the subsequent growth of metal-organic framework structures directly on the surfaces of metals.

Studies using electrochemical STM-TERS (EC STM-TERS) could assist in the fabrication of conductive metal/organic molecule thin films by utilizing anion-overlayers as templates for formation of well-defined organic thin films, as demonstrated here. Such organic thin films are increasingly important in the field of sensing, molecular electronics and optoelectronics. A challenge in Raman spectroscopy of organic molecules adsorbed on metal surfaces is detection of low frequency Raman signals, which give information on the chemical state of the molecule and possible metal–organic surface complex formation. Utilizing ultra-narrowband notch filters and a pinhole in front of the spectrograph slit, we recently observed signals down

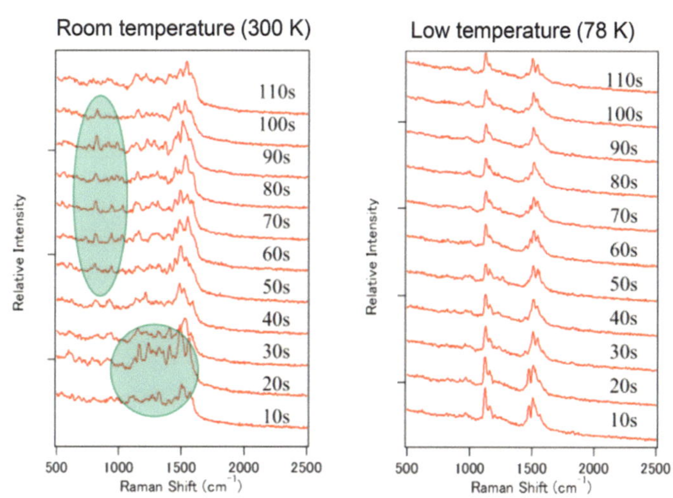

Figure 9. UHV-TER spectra of 1, 2-di-(4-pyridyl)-ethylene (BPE) at room (300 K) (left) and liquid nitrogen temperature (78 K)(right). Adapted from http://www.unisoku.com/products.

to 15 cm^{-1}. Improvement in the optical density of the filters would allow for detection of Raman signal from organic molecules that have weak Raman scattering cross-sections.

In the field of lithium-ion batteries, growth of conductive metal-organic interfaces with small contact resistances and catalytic functions is very attractive but remains very challenging. New experimental setups based on the TERS idea could allow for the study of interfacial processes during battery operation. A challenge in this case is a strong fluorescence signal from various battery components such as organic electrolytes, additives, binders, and so on. In this respect, systems based on near-infrared excitation would offer elimination of the fluorescence signal. Another advantage of this approach based on hyper Raman phenomena is that Raman signals originate from the small focal volume, which allow for distinguishing the interface signals from the signals originating from the bulk.

In summary, the demand for chemical analysis with nanoscale resolution makes SPM-based TERS attractive in many fields of science and engineering. We expect that, in the next 10 years, we will witness further developments in this technique and obtain more system-specific information, which will expand our knowledge of surface chemistry and the interactions of molecules with light in confined fields. Understanding system-specific chemical enhancement will advance the field of molecular plasmonics, which is an emerging field of science exploiting the molecule-plasmon interactions to harness light at the nanoscale for nanophotonic devices.

Author details

Izabela Rzeznicka[1]* and Hideyuki Horino[2]

*Address all correspondence to: izabela@shibaura-it.ac.jp

1 Graduate School of Science and Engineering, Shibaura Institute of Technology, Tokyo, Japan

2 Department of Chemistry for Materials, Mie University, Tsu, Japan

References

[1] Raman CV, Krishnan KS. A theory of the optical and electrical properties of liquids. Proceedings of the Royal Society of London. 1928;**117**:589-599. DOI: 10.1098/rspa.1928.0021

[2] Schrötter HW, Klöckner HW. Raman spectroscopy of gases and liquids. In: Weber A, editors. Topics in Current Physics. 1st ed. Berlin: Springer Berlin Heidelberg; 1979. pp. 123-166. DOI: 10.1007/978-3-642-81279-8_4

[3] Fleischmann M, Hendra PJ, McQuillan AJ. Raman spectra of pyridine adsorbed at a silver electrode. Chemical Physics Letters. 1974;**26**:163-166. DOI: 10.1016/0009-2614(74)85388-1

[4] Albrecht MG, Creighton JA. Anomalously intense Raman spectra of pyridine at a silver electrode. Journal of the American Chemical Society. 1977;**99**:5215-5217. DOI: 10.1021/ja00457a071

[5] Jeanmaire DL, Van Duyne RP. Surface Raman spectroelectrochemistry. Journal of Electroanalytical Chemistry. 1977;**84**:1-20. DOI: 10.1016/S0022-0728(77)80224-6

[6] Otto A. In: Cardona M, Guntherodt G, editors. Light Scattering in Solids IV. Berlin: Springer Berlin Heidelberg; 1984. pp. 289-418. DOI: 10.1007/3-540-11942-6_24

[7] Moskovits M. Surface-enhanced spectroscopy. Reviews of Modern Physics. 1985;**57**:783-826. DOI: 10.1103/RevModPhys.57.783

[8] Campion A, Kambhampati P. Surface-enhanced Raman scattering. Chemical Society Reviews. 1998;**27**:241-250. DOI: 10.1039/a827241z

[9] Schatz GC, Young MA, Van Duyne RP. Electromagnetic mechanism of SERS. In: Katrin K, Martin M, Harald K, editors. Topics in Applied Physics. 1st ed. Berlin: Springer Berlin Heidelberg; 2006. pp. 19-45. DOI: 10.1007/3-540-33567-6_2

[10] Le Ru EC, Etchegoin PG. Principles of Surface-Enhanced Raman Spectroscopy. 1st ed. Amsterdam: Elsevier; 2009. 663 p. DOI: 10.1016/B978-0-444-52779-0.00012-X

[11] Moskovits M. Surface selection rules. The Journal of Chemical Physics. 1982;**77**:4408-4416. DOI: 10.1063/1.444442

[12] Le Ru EC, Etchegoin PG. Quantifying SERS enhancements. MRS Bulletin. 2013;**38**:631-640. DOI: 10.1557/mrs.2013.158

[13] Morton SM, Jensen L. Understanding the molecule–surface chemical coupling in SERS. Journal of the American Chemical Society. 2009;**131**:4090-4098. DOI: 10.1021/ja809143c

[14] Vivoni A, Birke RL, Foucault R, Lombardi JR. Ab Initio frequency calculations of pyridine adsorbed on an adatom model of a SERS active site of a silver surface. The Journal of Physical Chemistry. B. 2003;**107**:5547-5557. DOI: 10.1021/jp027642o

[15] Cardini G, Muniz-Miranda M. Density functional study on the adsorption of pyrazole onto silver colloidal particles. The Journal of Physical Chemistry. B. 2002;**106**:6875-6880. DOI: 10.1021/jp014205l

[16] Johansson P. Illustrative direct ab initio calculations of surface Raman spectra. Physical Chemistry Chemical Physics. 2005;**7**:475-482. DOI: 10.1039/b415535a

[17] Lombardi JR, Birke RL, Lu T, Xu J. Charge transfer theory of surface enhanced Raman spectroscopy: Herzberg–Teller contributions. The Journal of Chemical Physics. 1986;**84**:4174-4180. DOI: 10.1063/1.450037

[18] Lombardi JR, Birke RL. A unified approach to surface-enhanced Raman spectroscopy. Journal of Physical Chemistry C. 2008;**112**:5605-5617. DOI: 10.1021/jp800167v

[19] Wu D-Y, Ren B, Jiang Y-X, Xu X, Tian Z-Q. Density functional study and normal-mode analysis of the bindings and vibrational frequency shifts of the pyridine–M (M = Cu, Ag, Au, Cu+, Ag+, Au+, and Pt) complexes. The Journal of Physical Chemistry. A. 2002;**106**:9042-9052. DOI: 10.1021/jp025970i

[20] Futamata M. Adsorbed state of 4,4′-BiPy and BiPyH22+ on Au(111) electrode. The Journal of Physical Chemistry B. 2001;**105**:6933-6942. DOI: 10.1021/jp010344r

[21] Stöckle RM, Suh YD, Deckert V, Zenobi R. Nanoscale chemical analysis by tip-enhanced Raman spectroscopy. Chemical Physics Letters. 2000;**318**:131-136. DOI: 10.1016/S0009-2614(99)01451-7

[22] Hayazawa N, Inouye Y, Sekkat Z, Kawata S. Metallized tip amplification of near-field Raman scattering. Optics Communication. 2000;**183**:333-336. DOI 10.1016/S0030-4018(00)00894-4

[23] Binnig G, Rohrer H, Gerber C, Weibel E. Surface studies by scanning tunneling microscopy. Physical Review Letters. 1982;**49**:57-61. DOI: 10.1103/PhysRevLett.49.57

[24] Scanning tunneling microscopy I. In: Wiesendanger R, Gntherodt HJ, editors. Surface Sciences. Berlin: Springer-Verlag Berlin Heidelberg; 1992

[25] Jaklevic RC, Lambe J. Molecular vibration spectra by electron tunneling. Physical Review Letters. 1966;**17**:1139-1140. DOI: 10.1103/PhysRevLett.17.1139

[26] Dionne JA, Atwater HA. Plasmonics: Metal-worthy methods and materials in nanophotonics. MRS Bulletin. 2012;**37**:717-724. DOI: 10.1557/mrs.2012.171

[27] Raman Microscopy: Development and Applications. London: Academic Press Limited; 1996. 463 p

[28] Novotny L, Bian RX, Xie XS. Theory of nanometric optical tweezers. Physical Review Letters. 1997;**79**:645-648. DOI: 10.1103/PhysRevLett.79.645

[29] Pettinger B, Ren B, Picardi G, Schuster R, Ertl G. Nanoscale probing of adsorbed species by Tip-enhanced Raman spectroscopy. Physical Review Letters. 2004;**92**:096101. DOI: 10.1103/PhysRevLett.92.096101

[30] Steidtner J, Pettinger B. Tip-enhanced Raman spectroscopy and microscopy on single dye molecules with 15A nm resolution. Physical Review Letters. 2008;**100**:23610101. DOI: 10.1103/PhysRevLett.100.236101

[31] Schmid T, Yeo BS, Zhang W, Zenobi R. Tip enhancement. In: Kawata S, Shalaev VM, editors. Amsterdam: Elsevier; 2007. p.115-155. DOI: 10.1016/B978-044452058-6/50005-0

[32] Ren B, Picardi G, Pettinger B. Preparation of gold tips suitable for tip-enhanced Raman spectroscopy and light emission by electrochemical etching. The Review of Scientific Instruments. 2004;**75**:837-841. DOI: 10.1063/1.1688442

[33] Williams C, Roy D. Fabrication of gold tips suitable for tip-enhanced Raman spectroscopy. Journal of vacuum science & technology. B, Microelectronics and nanometer structures. 2008;**26**:1761-1764. DOI: 10.1116/1.2981078

[34] Wang X, Huang S-C, Huang T-X, Su H-S, Zhong J-H, Zeng Z-C, Li M-H, Ren B. Tip-enhanced Raman spectroscopy for surfaces and interfaces. Chemical Society Reviews. 2017;**46**:4020-4041. DOI: 10.1039/C7CS00206H

[35] Deckert-Gaudig T, Richter M, Knebel D, Jähnke T, Jankowski T, Stock E, Deckert V. A modified transmission tip-enhanced Raman scattering (TERS) setup provides access to opaque samples. Applied Spectroscopy. 2014;**68**:916-919. 10.1366/13-07419

[36] Aravind P, Metiu H. The effects of the interaction between resonances in the electromagnetic response of a sphere-plane structure; applications to surface enhanced spectroscopy. Surface Science. 1983;**124**:506-528. DOI: 10.1016/0039-6028(83)90806-3

[37] Nordlander P, Oubre C, Prodan E, Li K, Stockman MI. Plasmon hybridization in nanoparticle dimers. Nano Letters. 2004;**4**:899-903. DOI: 10.1021/nl049681c

[38] Downes A, Salter D, Elfick A. Finite element simulations of tip-enhanced Raman and fluorescence spectroscopy. The Journal of Physical Chemistry B. 2006;**110**:6692-6698. DOI: 10.1021/jp060173w

[39] Hartschuh A. Tip-enhanced near-field optical microscopy. Angewandte Chemie, International Edition, 2008;**47**:8178-8191. DOI: 10.1002/anie.200801605

[40] Notingher I, Elfick A. Effect of sample and substrate electric properties on the electric field enhancement at the apex of SPM nanotips. The Journal of Physical Chemistry B. 2005;**109**:15699-15706. DOI: 10.1021/jp0523120

[41] Pettinger B, Domke KF, Zhang D, Schuster R, Ertl G. Direct monitoring of plasmon resonances in a tip-surface gap of varying width. Physical Review B. 2007;**76**:113409. DOI: 10.1103/PhysRevB.76.113409

[42] Pettinger B, Ren B, Picardi G, Schuster R, Ertl G. Tip-enhanced Raman spectroscopy (TERS) of malachite green isothiocyanate at Au(111): bleaching behavior under the influence of high electromagnetic fields. Journal of Raman Specroscopy. 2005;**36**:541-550. DOI: 10.1002/jrs.1332

[43] Ren B, Picardi G, Pettinger B, Schuster R, Ertl G. Tip-enhanced Raman spectroscopy of benzenethiol adsorbed on Au and Pt single-crystal surfaces. Angewandte Chemie, International Edition. 2005;**44**:139-142. DOI: 10.1002/anie.200460656

[44] Fang Y, Zhang Z, Chen L, Sun M. Near field plasmonic gradient effects on high vacuum tip-enhanced Raman spectroscopy. Physical Chemistry Chemical Physics. 2015;**17**:783-794. DOI: 10.1039/c4cp03871a

[45] Golovin A, Polubotko A. Analysis of peculiarities of the SEHRS and SERS spectra of 4, 4′-Bipyridine molecule on the base of the dipole-quadrupole theory. Chemical Physics Letters. 2016;**662**:208-213. DOI: 10.1016/j.cplett.2016.09.047

[46] Zhang R, Zhang Y, Dong ZC, Jiang S, Zhang C, Chen LG, Zhang L, Liao Y, Aizpurua J, Luo Y, Yang JL, Hou JG. Chemical mapping of a single molecule by plasmon-enhanced Raman scattering. Nature. 2013;**498**:82-86. DOI: 10.1038/nature12151

[47] Duan S, Tian G, Luo Y. Visualization of vibrational modes in real space by tip-enhanced non-resonant Raman spectroscopy. Angewandte Chemie, International Edition. 2016; **55**:1041-1045. DOI: 10.1002/anie.201508218

[48] Benz F, Schmidt MK, Dreismann A, Chikkaraddy R, Zhang Y, Demetriadou A, Carnegie C, Ohadi H, de Nijs B, Esteban R, Aizpurua J, Baumberg JJ. Single-molecule optomechanics in "picocavities". Science. 2016;**354**:726-729. DOI: 10.1126/science.aah5243

[49] Zhang C, Chen B-Q, Li Z-Y. Optical origin of subnanometer resolution in tip-enhanced Raman mapping. Journal of Physical Chemistry C. 2015;**119**:11858-11871. DOI: 10.1021/acs.jpcc.5b02653

[50] Hoffmann FM. Infrared reflection-absorption spectroscopy of adsorbed molecules. Surface Science Reports. 1983;**3**:107109-107192. DOI: 10.1016/0167-5729(83)90001-8

[51] Le Ru E, Meyer S, Artur C, Etchegoin P, Grand J, Lang P, Maurel F. Experimental demonstration of surface selection rules for SERS on flat metallic surfaces. Chemical Communications. 2011;**47**:3903-3905. DOI: 10.1039/C1CC10484E

[52] Rzeźnicka II, Horino H, Kikkawa N, Sakaguchi S, Morita A, Takahashi S, Komeda T, Fukumura H, Yamada T, Kawai M. Tip-enhanced Raman spectroscopy of 4,4′-bipyridine and 4,4′-bipyridine N,N′-dioxide adsorbed on gold thin films. Surface Science. 2013;**617**:1-9. DOI: 10.1016/j.susc.2013.08.010

[53] M. J. Frisch et al., Gaussian 09, Gaussian, Inc., Wallingford CT In 2010

[54] Atkins PW, Friedman RS. Molecular Quantum Mechanics. 5th ed. Oxford: Oxford Univ. Press; 2010. 573 p

[55] Rzeźnicka I, Horino H, Yagyu K, Suzuki T, Kajimoto S, Fukumura H. Chlorine adlayer-templated growth of a hybrid inorganic–organic layered structure on Au (111). Surface Science. 2016;**652**:46-50. DOI: 10.1016/j.susc.2016.03.007

[56] Gao W, Baker TA, Zhou L, Pinnaduwage DS, Kaxiras E, Friend CM. Chlorine adsorption on Au(111): chlorine overlayer or surface chloride? Journal of the American Chemical Society. 2008;**130**:3560-3565. DOI: 10.1021/ja077989a

[57] Pérez-Jiménez AJ, Sancho-García JC, Pérez-Jordá JM. Torsional potential of 4,4′-bipyridine. Ab initio analysis of dispersion and vibrational effects. Journal of Chemical Physics. 2005;**123**:134309. DOI: org/10.1063/1.2043107

[58] Pettinger B, Wtzel H. Surface enhanced raman spectroscopy of pyridine on Ag electrodes. Surface complex formation. Chemical Physics Letters. 1981;**78**:398-403. DOI: 10.1016/0009-2614(81)80041-3

[59] Diao Y-X, Han M-J, Wan L-J, Itaya K, Uchida T, Miyake H, Yamakata A, Osawa M. Adsorbed structures of 4,4′-bipyridine on Cu(111) in acid studied by STM and IR. Langmuir. 2006;**22**:3640-3646. DOI: 10.1021/la052765w

[60] Huong PV, Schlaak M. Raman and infrared band shapes of the N-H+...X-hydrogen bond in trimethylaminium halide crystals. Chemical Physics Letters. 1974;**27**:111-113. DOI: 10.1016/0009-2614(74)80456-2

[61] Foglizzo R, Novak A. Low frequency infrared and Raman spectra of hydrogen bonded pyridinium halides. The Journal of Chemical Physics. 1969;**50**:5366-5373. DOI: 10.1063/1.1671056

[62] Bessonov AA, Basova TV, Kiselev VG, Sheludyakova LA, Morozova NB, Igumenov IK. Vibrational interactions in dimethylgold(III) halides and carboxylates. Vibrational Spectroscopy. 2009;**51**:283-288. DOI: 10.1016/j.vibspec.2009.07.003

[63] Jiang N, Foley ET, Klingsporn JM, Sonntag MD, Valley NA, Dieringer JA, Seideman T, Schatz GC, Hersam MC, Van Duyne RP. Observation of multiple vibrational modes in ultrahigh vacuum tip-enhanced Raman spectroscopy combined with molecular-resolution scanning tunneling microscopy. Nano Letters. 2012;**12**:5061-5067. DOI: 10.1021/nl2039925

[64] Nagasawa F, Takase M, Nabika H, Murakoshi K. Polarization characteristics of surface-enhanced Raman scattering from a small number of molecules at the gap of a metal nano-dimer. Chemical Communications. 2011;**47**:4514-4516

[65] Fukumura H, Irie M, Iwasawa Y, Masuhara H, Uosaki K, editors. Molecular Nano Dynamics. 1st ed. Verlag: John Wiley & Sons; 2009. 314 p

[66] Pozzi EA, Sonntag MD, Jiang N, Chiang N, Seideman T, Hersam MC, Van Duyne RP. Ultrahigh vacuum tip-enhanced Raman spectroscopy with picosecond excitation. Journal of Physical Chemistry Letters. 2014;**5**:2657-2661. DOI: 10.1021/jz501239z

Chapter 6

SERS Research Applied to Polymer Based Nanocomposites

Sara Fateixa, Helena I.S. Nogueira and Tito Trindade

Additional information is available at the end of the chapter

http://dx.doi.org/10.5772/intechopen.72680

Abstract

Polymer based nanocomposites containing metal nanoparticles (e.g. Au, Ag) have gained increased attention as a new class of SERS (Surface Enhanced Raman Scattering) substrates for analytical platforms. On the other hand, the application of SERS using such platforms can also provide new insights on the properties of composite materials. In this chapter, we review recent research on the development of SERS substrates based on polymer nanocomposites and their applications in different fields. The fundamentals of SERS are briefly approached and subsequently there is a reference to the strategies of preparation of polymer based nanocomposites. Here the main focus is on SERS studies that have used a diversity of polymer based nanocomposites, highlighting certain properties of the materials that are relevant for the envisaged functionalities. A final section is devoted to the joint use of Raman imaging and SERS in nanocomposites development, a topic that presents a great potential still to be explored as shown by the recent research in this field.

Keywords: SERS, metal nanoparticles, nanocomposites, polymers, Raman imaging

1. Introduction

There has been a long-standing interest in the application of metallic nanoparticles (NPs), in particular of gold (Au) and silver (Ag), due to their unique optical properties [1–6]. The dispersion of very small (submicrometric to nanometric) metal particles in materials that act as host matrices (e.g. glass) has been empirically exploited to confer bright and colorful effects, resulting from scattering and absorption of visible light. In fact, there are several examples of earlier technological applications that include the use of fine divided gold in glass materials in order to obtain beautiful colors, such as in stained glass windows [7, 8]. Nowadays, metal NPs are considered crucial in a series of devices derived from nanotechnological approaches and which are expected to expand in the near future in several applications and techniques. Among these techniques, Surface Enhanced Raman Scattering (SERS) has emerged in the past decades with great importance. Although the discovery of the SERS effect dates back to the 70's, [9] in the last decades, this technique has been improved due to the huge progress observed in Raman instrumentation and also in recent research specific to nanoscale materials. Indeed, the design of SERS active and highly sensitive analytical platforms has been a major goal in several fields, due to the impact in trace chemical analysis, environmental monitoring, medical applications and food safety [10–16].

This chapter intends to present recent developments concerning polymer based nanocomposites containing metal nanofillers for SERS applications, in some situations based on studies of metal loaded polymers prepared in our laboratory. In order to contextualize the SERS application of these polymer based nanocomposites, the synthesis of the nanostructures and some fundamental aspects of SERS are briefly introduced [17–25]. Polymer based composites for SERS, containing metallic fillers, have been largely documented for the vestigial detection of several bio-analytes. Illustrative examples on the use of polymer based composites as new platforms for SERS are described in more detail, reviewing their use for chemical analysis. The importance of Raman imaging in SERS studies is also explored because is a recent development with a great potential for the research on new SERS substrates.

2. Fundamentals of surface enhanced Raman scattering

Chandrasekhara Venkata Raman (Nobel Prize in Physics, 1930) and his student Kariamanickam Srinivasa Krishnan reported in a scientific paper entitled "A new type of secondary radiation," the phenomenon that we now call the Raman effect [26]. This effect involves inelastic scattering of light when the photons strike a certain medium containing molecules. Although the number of photons resulting from the elastic dispersion (Rayleight scattering) of light coming from a laser source is greater than that of the inelastic dispersion, thus giving rise to a more intense signal, it is the spectral information collected from the inelastic collision that is of interest for Raman spectroscopy [27–30]. Indeed, the bands that are observed in a Raman spectrum (Stokes and anti-Stokes shifted in relation to the excitation line) are associated to certain vibrational modes existing in molecules and materials and can thus give information about their structure and other properties. The use of Raman spectroscopy in certain areas of application was, until recently, limited by two main factors. One of these factors can be considered extrinsic to the phenomenon and depends essentially on the development of more sensitive and affordable equipment. The other factor is intrinsic to the Raman effect, since it is a physical process that results in low-intensity spectral bands, because of the low scattering cross-section (10^{-29} cm^2 $molecule^{-1}$), limiting its use as a high sensitive analytical technique [30–32].

The SERS (surface enhanced Raman scattering) effect was discovered about 44 years ago during studies applied to a silver electrode and an aqueous solution of pyridine [9]. Unexpectedly, this experiment revealed an increase in the Raman signal of pyridine adsorbed on the metal surface. A few years later, other research groups advanced the currently accepted explanation for this observation [33, 34]. The interpretation given considered that the Raman signal intensification occurred due to adsorption of molecules on rough metallic surfaces. It is generally accepted today that the intensification occurring in SERS is due to two distinct mechanisms: the chemical interaction of the analyte with the metallic surface (signal intensification $10–10^2$) and the enhancement of the local electric field at the junction of the metallic NPs (signal intensification can reach up to 10^{11}) [32, 35, 36]. The chemical mechanism (CM) for the SERS signal enhancement is related to the adsorption of molecular species on the metal surface, namely the surface selection rules, the type of interaction between the metal and the molecule and the chemical nature of the adsorbed molecule itself [36–38]. The electromagnetic mechanism (EM) is the dominant contribution to SERS and does not depend necessarily on the establishment of a chemical interaction between the analyte and the metal surface, contrary to what happens in the chemical mechanism. According to the EM, the intensification of the local electromagnetic field is mainly due to the excitation of the surface plasmons of the metal by the incident light [39–41]. While the CM is a short-range effect, the EM is a long-range effect, in the sense that it does not require that the molecular species is in contact with the metal surface and can still be observed a few nanometers of distance from the metal surface. **Figure 1** presents a scheme of

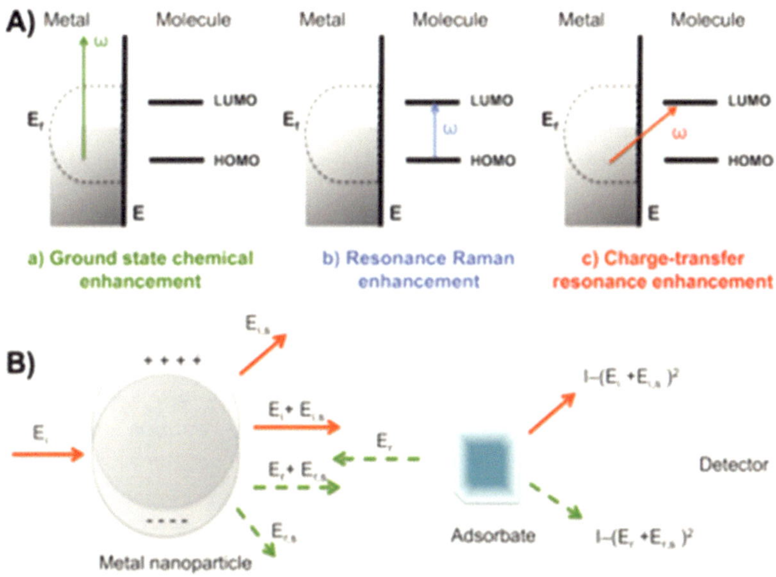

Figure 1. (A) Scheme illustrating the chemical mechanism (CM) that occurs in SERS. (a) Ground state chemical enhancement; (b) resonance Raman enhancement; (c) charge-transfer resonance enhancement, where E_f represents the local electromagnetic field; HOMO represents the highest occupied molecular orbital and LUMO represents the lowest unoccupied molecular orbital); (B) Scheme illustrating the electromagnetic mechanism (EF), in which the I represents the intensity of Raman; E_i represents the incident electric field; $E_{i,s}$ represents the field which is intensified by the metal and ($E_i + E_{i,s}$) represents the incident enhanced field; E_r represents the scattered Raman field, which can be intensified by the metal creating $E_{r,s}$; ($E_r + E_{r,s}$) represents the scattered enhanced field. ([10] - Reproduced by permission of The Royal Society of Chemistry.).

the different types of contributions to the chemical enhancement mechanism and the electromagnetic enhancement mechanism in SERS.

The contribution of nanotechnology, and in particular of the Chemistry of Nanomaterials, to the resurgence of studies and applications of Raman spectroscopy in materials characterization and the improvement of analytical techniques cannot be surprising. Indeed, the SERS effect allows the range of applications of Raman spectroscopy to be expanded but it is still very dependent on the quality of the substrates used to obtain signal intensification. The substrates most used for these purposes are based on Ag and Au nanostructures, although the SERS effect using other types of surfaces have been reported [42–45]. A variety of studies have demonstrated that the enhancement of the localized electromagnetic field occurs in close vicinity of metal NPs, metal nanotips or in metal surfaces with specific nanopatterns. The positions in which the strongest enhancement of the local electromagnetic field is observed correspond to the so called hotspots, due to the strongly enhanced Raman signals observed for certain molecular adsorbates [46–52]. Chemistry provides the synthetic tools to control the morphology and size of metal nanoparticles as well as other nanostructured materials and therefore has been widely used in the development of highly sensitive and reproducible SERS substrates. It should be noted that the sensitivity of some of the SERS substrates reported in the literature allows the analyte detection at the single-molecule level [53–55]. The fact that Raman spectroscopy is a non-destructive technique and can also be used for materials imaging, are additional factors that make this technique increasingly relevant in chemical analysis and materials characterization. In particular, the SERS technique is a valuable tool for the surface characterization of materials. For example, it is possible to obtain information about the surface of a metal using molecules that once adsorbed will function as molecular probes [56]. In addition, it is possible to obtain information about the orientation of these molecules, applying surface selection rules that were established in earlier studies [11, 39, 57].

3. Polymer based nanocomposites

By definition a composite is formed by at least two distinct materials whose chemical identity is preserved in the final material. Typically, a composite contains a material (filler) that is dispersed in a larger amount of a distinct material that acts as the host matrix. In the case of having a polymer as the host, such material is referred as a polymer based composite. For the particular situation, in which the fillers have at least one dimension at the nanoscale, the material is called a nanocomposite. Hence, polymers, either synthetic or of natural origin, that contain inorganic nanoparticles form an important class of nanocomposites. In particular, polymer based nanocomposites containing metallic nanoparticles (NPs) as fillers, are the object of this chapter due to their role as hybrid substrates for SERS analysis. Several approaches have been reported in order to produce polymer nanocomposites containing metal NPs as fillers. Briefly, these preparative methods can be divided in two main categories, depending if the metal nanofillers were generated in situ or previously prepared and then used for the composite fabrication. These approaches will be designated here as i) in situ and ii) ex situ. In the in situ method, the metallic nanofillers are prepared by chemical reduction methods in the presence of a polymer; in the ex situ approach, the metallic NPs are previously synthesized and then mixed with a polymer matrix, forming homogeneous blends composites (blending method) or, after surface modification procedures applied at the filler's surfaces [10, 58, 59]. These preparative methods will be briefly described in the next sections. **Table 1** lists some methods for preparing these composites based on polymers of natural and synthetic origin.

3.1. Chemical reduction (in situ method)

In this methodology, metallic nanofillers are produced by chemical reduction of a metal precursor using reducing agents such as sodium citrate or sodium borohydride, in the presence of a polymer. This strategy generates nanocomposites whose morphology can vary, such as in a polymeric shell and a metal core [60, 61, 77, 96, 102], a polymeric core and a metal shell [95] or a polymeric matrix having dispersed metallic fillers [68, 69, 97]. In particular cases, the polymer can act as reducing agent due to specific functional groups, avoiding the use of an external reducing agent [61, 73]. The advantages of this one-step approach relies on its

Polymer matrix	Preparative method	Metallic NPs
Natural polymers	Chemical reduction	Ag [60–72]
		Au [73–76]
		Cu [77–80]
	Blending or "grafting to" approach	Ag [66, 67, 81–87]
		Au [88–92]
		Cu [79]
		Au@Ag [93]
	UV light reduction of the metal ions	Ag [66, 94]
		Au [94]
		Au@Ag [94]
Synthetic polymers	Chemical reduction	Ag [95–102]
		Cu@Ag [103]
	Blending or "grafting to" approach	Ag [104, 105]
		Au [106–112]
	In situ polymerization or "grafting from" approach	Ag [113–118]
		Au [58, 108, 116, 119–124]
		Au@Ag [125]
		Fe_3O_4@Ag [48]
	Electrospinning method	Ag [126, 127]
		Au [126]
		Au@Ag [128]

Table 1. Preparative methods of polymer/metal nanocomposites.

simplicity, although the resulting nanocomposites may exhibit inhomogeneity in terms of morphology and fillers distribution in the polymer matrix [10, 129].

In situ preparation is commonly used to prepare biopolymers based nanocomposites because natural polymers might be used to induce control over NPs size and aggregation state [63, 70], promoting the required biocompatibility for specific applications namely medical diagnostic and target detection in SERS [60, 64]. Polysaccharides such as chitosan [64, 65, 71, 72, 74], agarose [60, 63, 70], glucose [102], hyaluronic acid potassium salt [75] and cellulose [67–69] have been used as polymeric matrices for the one-step preparation of such polymer based nanocomposites. The preparation of composites containing polysaccharides in which in situ chemical or UV light reduction of metal ions occurred has been subject of great interest. Indeed, these nanocomposites can be applied in a variety of domains such as antimicrobial agents [66, 80], platforms for chemical detection [68, 69], textile dyeing monitoring process [67] and electronic paper [79]. For example, bacterial cellulose has been investigated as an alternative host matrix to vegetable cellulose in the preparation of electronic paper, SERS substrates and antimicrobial agents by using metal NPs as fillers. More recently, we have demonstrated that Raman imaging is a useful technique to characterize and monitor the textile dyeing process of antimicrobial fabrics [67]. **Figure 2** provides examples of cellulose composites for diverse applications.

The use of polysaccharides in the form of hydrogels have also been reported as matrices for dispersing metallic NPs, in certain cases showing capacity to collapse or be lyophilized upon drying and recover their structure by rehydration and subsequent use in SERS analysis [60, 63, 65]. For

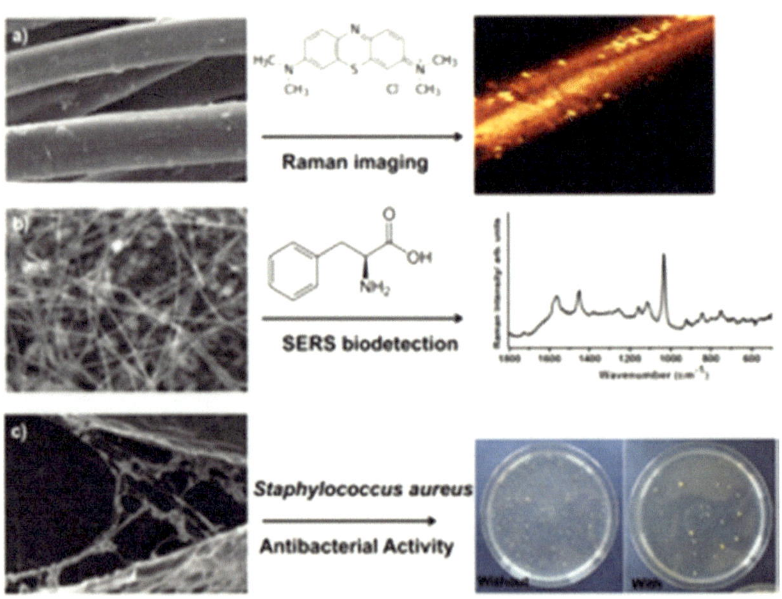

Figure 2. Examples of applications for cellulose based nanocomposites loaded with metal NPs: a) Raman image obtained using the integrated intensity of the Raman band at 1620 cm^{-1} in the SERS spectra of methylene blue 10^{-4} M using a linen based composite containing Ag NPs as substrate (excitation at 633 nm); [67] Ag/bacterial cellulose nanocomposites for SERS biodetection of phenylalanine; [69] c) Nanocomposites containing copper based NPs and cellulose to be used as antibacterial agents against *Staphylococcus aureus* [80].

example, Brust and co-workers have described the preparation of Ag loaded agarose hydrogels, in which the metallic nanofillers are trapped inside the polymer network due to the agarose capability to dry and rehydrate. In this particular case, the use of agarose as polymeric matrix provides the formation of a recyclable SERS substrate, in which the 1-naphthalenethiol used as analyte can be washed out by dialysis and the composite can be reused again [60]

Nanocomposites containing gelatin [61, 76, 77], natural rubber [73] and distinct synthetic polymers [98–101] have been also prepared using in situ methods. For instance, Wu and his team have reported the fabrication of Ag loaded poly(styrene) microspheres using the chemical reduction method for the SERS detection of organic molecules such as dyes. [98]. The authors have used the complex $[Ag(NH_3)_2]^+$ as precursor and the poly(vinylpyrrolidone) (PVP) and poly(dopamine) (PDA) as linkers to attach the Ag ions into the PS microspheres surface by hydroxylic and aminic groups.

3.2. Blending or "grafting to" approach

Blending metallic nanofillers with a polymeric matrix is a simple methodology to fabricate nanocomposites with efficient SERS activity [84, 104, 106, 107, 124]. Besides its simplicity, this is a cost-effective and easily scalable method over a large area [90, 130]. In the literature, there are some reports on the use of synthetic polymers such as poly(methyl metacrylate) [105, 109], poly(*t*-butyl acrylate) [107], polystyrene-block-poly(acrylic acid) [110] and poly(vinyl alcohol) [128] as matrices for the incorporation of metallic NPs and subsequent use as SERS platforms. Although not so common in SERS research context, natural polymers can also be used as polymeric matrices for the preparation of polymer based composites, in which water compatible metal NPs are normally used. Marsch *et al.* have reported the development of a surface enhanced resonance Raman scattering (SERRS) substrate containing Ag NPs with a positive surface charge, due to a poly-L-lysine coating, which was then used in the analysis of organic anions [86]. In their work, these polymer based composites were successfully used as SERS substrates for the detection of bilirubin, a organic molecule of clinical interest formed as a metabolic waste product of heme breakdown. More recently, Chang *et al.* have reported flexible SERS substrates based on common filter paper loaded with gold nanorods, which exhibited more than two orders of magnitude SERS enhancement compared to silicon-based SERS substrates [92]. The authors have demonstrated that these platforms are excellent candidates for trace chemical and biological detection due to their efficient uptake, and transport of the analytes from the dispersing liquid to the surface of metal nanostructures.

Other examples of bionanocomposites containing metallic NPs have been reported including those based on chitosan [81], pullulan [83], cellulose [67, 79, 82], gelatin [85, 88] and carrageenan [84, 89]. **Figure 3** illustrates the use of biopolymers and Ag NPs in the fabrication of composites for antimicrobial and SERS applications. In certain situations, these applications can be complementary, such as in the case of producing a gel that due to the presence of Ag not only has the ability for SERS detection but also lasts longer periods of time due to the antimicrobial characteristics.

In ex situ methods, the polymer can be used either as a continuous phase or as an aqueous emulsion. For example, Lee *et al.* have reported the preparation of several metallic nanofillers

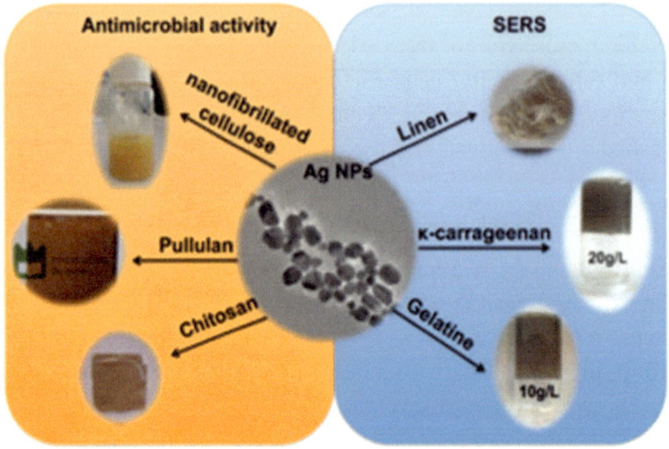

Figure 3. Blending of natural polymers and Ag NPs aiming at the fabrication of nanocomposites for SERS detection of analytes and antimicrobial applications.

with distinct sizes and shapes, which were used to decorated PS spheres stabilized in aqueous emulsion, producing active SERS platforms [104]. In addition, Fernández-López et al. have studied the thermoresponsive optical properties of poly(N-isopropylacrylamide) (pNIPAM) microgels doped with Au nanorods (Au NR) characterized by two different aspect ratios, observing a reversible behavior [112]. The thermoresponsive SERS sensitivity of the polymer composites was also analyzed using three different laser lines, demonstrating excitation wavelength-dependent efficiency, which can be controlled by either the aspect ratio (length/width) of the assembled Au NR or by the Au NR payload per microgel.

3.3. In situ polymerization and "grafting from" approach

Current polymerization procedures can be fine-tuned to fabricate metal load composites to be used as SERS active platforms. These include (mini)emulsion polymerization, suspension polymerization, atom transfer radical polymerization (ATRP) and reversible addition fragmentation chain transfer (RAFT). In in situ polymerization methods, the surface of the metallic nanofillers are modified with initiating species or chain transfer agents, allowing the growth of polymer chains from the metallic NPs surfaces [131]

By using the in situ emulsion polymerization, particles of the polymer based composite can be prepared with sizes ranging the micrometer and nanometer scale [117, 132]. In brief, this polymerization technique involves the formation of stable oil-in-water emulsions composed by small droplets of hydrophobic monomer dispersed in water that act as reactors after an initiator has been added. These micelles might contain surface modified inorganic NPs that at the end of the polymerization are coated or attached to the as prepared polymer.

Other polymerization strategies comprise the "grafting from" approach using reversible addition fragmentation chain transfer polymerization (RAFT) and atom-transfer radical polymerization (ATRP). There are a few reports on the use of ATRP and RAFT to produce metal/polymer SERS substrates. RAFT is a controlled radical polymerization approach, which involves multistep synthesis and sequential purification procedures. This polymeric approach is commonly used in the preparation of polymers with sulfur end groups, namely xanthates, dithioesters and thiocarbamates, that can be easily reduced to thiols [58, 133]. Besides the control of molecular weight and dispersity indexes of the polymers, several molecular structures can be achieved with the RAFT polymerization namely brush polymers, linear block copolymers, dendrimers and stars [58, 78, 116]. The polymer chains of the polymer based composites prepared by RAFT polymerization can be chemically functionalized with biomolecules that can be further used as SERS reports on the detection of specific analytes. In particular, metal loaded polymer composites functionalized with SERS reporters can be useful to conclude about the influence of the laser light source used in the target molecules or on the diffusion of the molecular probes through the polymer matrix. [58, 78, 134]. For example, Merican *et al.* have reported the surface modification of Au NPs with several SERS reporters, by using RAFT polymerization, in which a variety of polymers containing dithiocarbamate end groups were used. These polymer based composites were successfully used as SERS platforms for the detection of 2-naphthalenethiol and 2-quinolinethiol [58].

On the other hand, the ATRP approach controls the dispersity indexes and the molecular weight of the polymers by comprising an atom transfer step in the polymer chain growth phase in the polymerization [108, 120, 135, 136]. For example, core/shell structures prepared from surface functionalization of Au and Ag NPs by ATRP are often applied in a variety of fields such as catalysis, environmental monitoring, SERS detection and drug delivery [108, 120–122]. Mangeney and co-workers have fabricated Au–pNIPAM composites via ATRP method for the detection of methylene blue [121]. The authors have demonstrated that these hybrid materials are thermosensitive platforms, in which the Raman signal of the dye molecules increase with the increase of temperature. In addition, Yin *et al.* have reported the preparation of Au loaded polymer composites by ATRP approach to be used as SERS sensors with specific selectivity for Cd^{2+} [122].

Our own investigation in this field has the focus on the preparation of such polymer based composites by in situ polymerization using distinct synthetic strategies, such as emulsion [137], suspension [138, 139] and (mini)emulsion polymerization [140, 141], a variant of emulsion polymerization. In the latter, the emulsions are nanosized and organically capped NPs can be allocated in the interior of the hydrophobic monomer droplets [142, 143]. The final properties of the polymer based composites can be tuned by varying several parameters namely the amount of monomer, surfactant, the size and surface organic capping of the inorganic NPs [143]. We have reported several organically capped NPs successfully coated with a series of polymers, using the (mini)emulsion polymerization, namely metallic NPs [113–115, 119], quantum dots [141, 144–146], ferromagnetic NPs [147] and lanthanide compounds [148, 149] Esteves *et al.* have reported pioneer research in this field by describing the encapsulation of TOPO capped CdS and CdSe quantum dots (QDs) using poly(styrene) and poly(*t*-butylacrylate) as polymeric matrices [141]. They have demonstrated that by using the (mini) emulsion polymerization method, organically capped QDs NPs could be used as fillers leading to nanocomposites that still exhibit the typical photoluminescence of the dots at room temperature. Pereira and co-workers have synthesized oleylamine capped EuS nanocrystals from single-molecule precursors and investigated their use as fillers for polymer based composites [148]. In this work, it was demonstrated that the magnetic properties observed for the EuS/poly(styrene) nanocomposites have varied in comparison to the starting EuS nanocrystals, which result from surface effects due to dispersion of the EuS nanocrystals within the polymer beads. Martins *et al.* have obtained stable aqueous emulsions of distinct polymer based composites containing either Au NPs or $CoPt_3$ NPs *via* (mini)emulsion polymerization method. In the former, the optical properties of such composites depend not only on the Au NPs employed in their synthesis but also on the resulting morphology for the final composites [119]. In the later, they have demonstrated for the first time the preparation of a ferromagnetic polymer based composite composed by a magnetic core of $CoPt_3$ NPs encapsulated by poly(*t*-butylacrylate) [147]. As a proof of concept, the chemical binding of bovine IgG antibodies to the hydrolyzed surfaces of $CoPt_3$/PtBA nanocomposites was described. This strategy has established an interesting route for the development of nanocomposites materials for *in vitro* bioanalysis assays, which have been extended to other magnetic materials [150]. **Figure 4** presents TEM images of several polymer based composites prepared by (mini)emulsion polymerization. Although this method allows the preparation of diverse

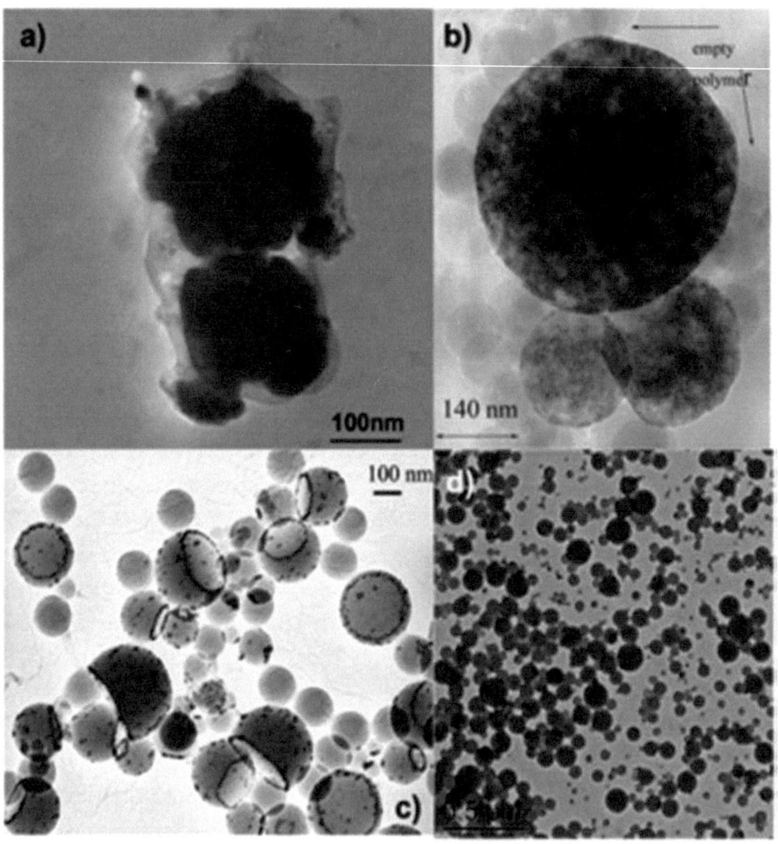

Figure 4. TEM images of nanocomposites prepared by (mini)emulsion polymerization: a) Ag/poly(methyl metacrylate); [115] b) EuS/poly(styrene) ([148]–Reproduced by permission of The Royal Society of Chemistry); c) Au/poly(styrene) (adapted with permission from [119]) and d) Fe_3O_4/poly(styrene) particles mixed with free PS beads (courtesy of P. C. Pinheiro).

metal loaded polymer nanocomposites, there is also the presence of free polymer particles in the final emulsion. For certain applications this is not necessarily a limitation but nevertheless other options should be considered. For example, the use of magnetic NPs as fillers, such as in the case shown in **Figure 4d** for magnetite in poly(styrene), allows the application of magnetic separation in order to separate the nanocomposite from the free polymer beads.

4. Metal loaded polymer nanocomposites as SERS substrates

In the last years, a series of polymer-based nanocomposites containing plasmonic metal NPs have been investigated as a new class of SERS substrates. The preparation and properties of such

Polymer matrix		**Metallic NPs**	**Applications**
Natural	Gelatin	Ag	Biomolecular detection [61] Environmental monitoring [85]
		Cu	Biomolecular detection [77]
	Gum Arabic	Ag	Biomolecular detection [62]
	Agarose	Ag	Biomolecular detection [63, 70] SERS mapping and imaging [60, 63] Ultradetection or single molecule detection [60]
	Cellulose based materials	Ag	Biomolecular detection [69] SERS mapping and imaging [67, 68]
		Au	Biomolecular detection [91] Medical diagnosis and target detection [157]
	Carrageenan	Ag	Biomolecular detection [84]
	Paraffin	Ag	Biomolecular detection [90]
	Alginate	Ag	Biomolecular detection [94]
		Au	Biomolecular detection [94]
		Au@Ag	Biomolecular detection [94]
	Chitosan	Ag	Medical diagnosis and target detection [64] In situ molecular changes monitoring [65]
		Au	Biomolecular detection [74]
		Cu	Substrate characterization [71]
		Au@Ag	Biomolecular detection [72]
	Poly-l-lysine	Ag	Medical diagnosis and target detection [86]
	Bovine serum albumin	Ag	Medical diagnosis and target detection [151]
	Natural rubber	Au	Ultradetection or single molecule detection [73] SERS mapping and imaging [73]
	Hyaluronic acid potassium salt	Au	Biomolecular detection [75]

Polymer matrix		Metallic NPs	Applications
Synthetic	Poly(amide)	Ag	Biomolecular detection [87]
	Poly(vinyl alcohol)	Ag	Biomolecular detection [95, 101] Substrate characterization [97]
		Au	Biomolecular detection [126] Medical diagnosis and target detection [126]
		Au@Ag	Biomolecular detection [128]
		Cu@Ag	Biomolecular detection [103]
	Poly(styrene)	Ag	Biomolecular detection [98] Substrate characterization [104] Medical diagnosis and target detection [98, 117]
	Poly(vinylpyrrolidone)	Ag	Biomolecular detection [99, 102, 105]
	Poly(aniline)	Ag	Biomolecular detection [100]
	Poly(*t*-butylacrylate)	Ag	Biomolecular detection [113, 114]
		Au	Biomolecular detection [107]
	Poly(methyl methacrylate)	Ag	Biomolecular detection [115]
	Poly(acrylamide)	Ag	Biomolecular detection [118]
	Poly(acryloyl) Hydrazine	Ag	Biomolecular detection [162]
	Poly(ethylene glycol)	Au	Medical diagnosis and target detection [11, 163] Heavy metal detection [122]
		Au@Ag	SERS mapping and imaging [93]
	Poly(pyrrole)	Ag	Substrate characterization [96]
	poly(ethylene glycol) diacrylate	Ag	Medical diagnosis and target detection [164]
	Poly(ethylene glycol dimethacrylate)	Au	Medical diagnosis and target detection [109]
	poly(ethylene glycol dimethacrylate-*co*-acrylonitrile)	Au	Medical diagnosis and target detection [124] SERS mapping and imaging [124]
	Poly(styrene)-*block*-poly(acrylic acid)	Au	Biomolecular detection [110]
	Poly(N-isopropylacrylamide)	Au	Biomolecular detection [121] Ultradetection or single molecule detection [123]
		Au@Ag	Biomolecular detection [125]
		Fe_3O_4@Ag	Environmental monitoring [48]
	Poly(vinylidene fluoride)	Ag	Biomolecular detection [87]
		Au	Biomolecular detection [165]
	Poly(hexamethylene adipamide)	Au	Ultradetection or single molecule detection [161]
	Poly(acrylonitrile)	Ag	Biomolecular detection [127]
	Poly(sodium 4-styrenesulfonate)	Au@Ag	Biomolecular detection [155]

Table 2. SERS applications of metal/polymer nanocomposites.

composites depends on the type of coating employed to stabilize the metal NPs, which were used as fillers. As discussed in the previous chapter, metal loaded polymer composites can be prepared by using different matrices such as synthetic [107, 113, 114, 123, 132] or natural polymers [61–63, 84, 87, 90, 151]. Polymer based composites containing metal NPs are of great interest due to their multifunctionality and potential for large-scale fabrication at low cost [23, 107, 152–154]

In the context of SERS, these composites appear as a promising alternative for the development of efficient and scalable substrates for the detection of molecular species, such as dyes [68, 73, 95, 102, 103, 105, 121], biomolecules [64, 74, 75, 101, 113, 123, 155], and environmental pollutants [48, 60, 85]. For example, the application of polymer-based composites as stable and active smart devices or SERS chips has increased in the last years [126, 156–159]. Gao *et al.* have described the preparation of chips composed by metal NPs embedded in polymer nanofibers mats. [126]. The authors have demonstrated that such composites doped with distinct metal NPs are facile to store and to transport, and can be easily fixed in slides or in microfluidic channels for SERS detection of a variety of analytes. Long and co-workers have reported SERS assays loaded with Ag NPs that result from a low-cost production process by using screen printing techniques [156]. Chen *et al.* have decorated commercial tape with Au NPs to be used as SERS platforms for the directly extraction and trace detection of pesticide in vegetables and fruits by a "paste and peel off" procedure [158]

Polymer nanocomposites containing metal NPs offer some advantages in SERS applications, namely by considering a judicious choice of the polymer used as the matrix. The polymer can be selected, for example, to provide a stimuli responsive platform or a porous polymeric matrix that facilitates the diffusion and entrapment of the biomolecules under analysis. In this context, an important objective in the application of SERS substrates has been the analysis of vestigial amounts of the target analyte for which minimal specimen preparation is required [69, 70, 72, 73, 107, 114, 115]. This is relevant to implement analytical protocols as many SERS applications are foreseen, such as environmental monitoring [48, 74], SERS mapping and imaging [67, 68, 93], medical diagnosis [109, 124, 126, 151, 160, 161] and substrate characterization [71, 96, 107]. **Table 2** lists polymer based composites mentioned in this chapter with the corresponding SERS applications by taking into account the type of polymer and the metal nanofillers employed.

In these nanocomposites, the polymer can also act as an active coating with influence in the interparticle distances of the dispersed plasmonic metal NPs and their close surroundings. These changes can endorse the formation of SERS active sites, the so called *hot spots*, that can be further accessed by molecular probes through diffusion in the polymeric matrix [113, 114, 152, 164]. Braun *et al.* have reported a general approach to create *hot spots* in SERS substrates via NPs linking, polymer coating and molecular permeation [152]. The polymer coating holds the analyte within the nanojunctions created by several Ag NPs trapped within the polymer. However, this approach requires the use of a linking agent prior substrate preparation and has been reported only for analyte detection in solution. More recently, Kim and co-workers have reported the preparation of microgels containing Ag nanocubes, providing molecular size-selective permeability and high SERS sensitivity for acetylsalicylic acid (aspirin) [164]. The size of the Ag aggregates in the microgel matrix allows selective diffusion of small molecules and also promotes the formation *hot spots* in the close vicinity of the Ag nanocubes. As such, these nanocomposite enable the Raman analysis of small molecules dissolved in complex mixtures of proteins and cells without sample pre-treatment.

Metal loaded polymer composites are also suitable for applying analyte trapping strategies [22, 118, 123, 152, 166]. For instance, the diffusion of the analyte molecules within the polymer towards the metal surface can be facilitated not only by existent porosity but also by external stimuli such as temperature, pH, ionic strength and dehydration [22, 84, 88, 123, 162, 163, 165]. Fateixa *el al.* have demonstrated that the gel strength of carrageenan hydrogel composites loaded with Ag NPs can be correlated with the signal enhancement observed in SERS studies [84]. This research has focused on the effect of hydrogel strength and temperature on the SERS behavior of these bionanocomposites for the detection of 2,2′- dithiodipyridine. In order to vary the gel strength of the biocomposite, several procedures were employed, such as the increase of the polysaccharide content in the gel, the addition of KCl as cross-linker, and varying the type of carrageenan (κ, ι, λ) gel. The authors have reported an increase in the SERS signal as the gel strength increased, which was attributed to the presence of sites with strong local electromagnetic field, that result from the formation of Ag particles nanojunctions as the carrageenan macromolecules tended to rearrange into stronger gels (**Figure 5**). Also, Contreras-Cáceres *et al.* have prepared core/shell microgel particles composed by a poly-(N-isopropylacrylamide) (pNIPAM) shell and a metallic core such as Au and Au@Ag nanospheres or nanorods [41, 123, 125]. In these systems, the molecules under analysis, which were trapped in the thermoresponsive polymer shells and at the vicinity of the metal surface, experience an enhancement of the Raman signal of about 10^5 times [41, 123, 125]. More recently, multifunctional substrates have been investigated by conferring magnetic properties to SERS substrates, thus allowing extraction of the molecular probes from liquid phase prior SERS analysis [48]. In this case, by encapsulating Fe_3O_4@Ag NPs in thermoresponsive pNIPAM shells, an increase of temperature promoted the formation of *hot pots* close to the Ag NPs. These new platforms were applied to monitor trace amounts of pentachlorophenol, a chlorinated environmental pollutant.

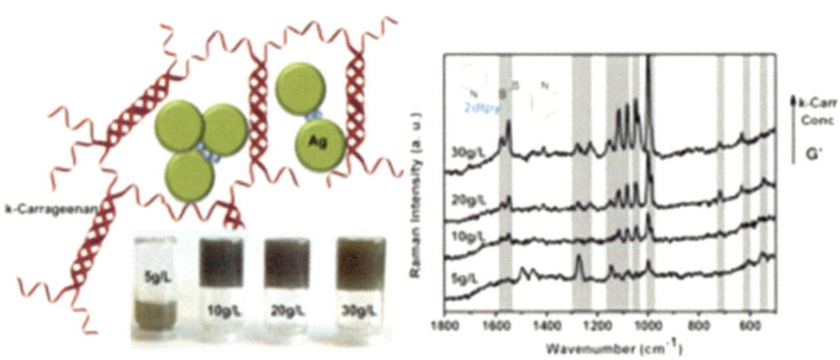

Figure 5. Illustration of biopolymer helices aggregation and consequently the creation of highly SERS sensitive Ag/k-carrageenan hydrogels and digital photographs of Ag/carrageenan hydrogels with variable carrageenan amount (left); SERS spectra of 2′2-dithiopyridine using Ag/k-carrageenan hydrogels as substrates with variable amount of biopolymer: (a) 5 g/dm³; (b) 10 g/dm³; (c) 20 g/dm³; (d) 30 g/dm³ (right) [84] (Reprinted with permission from ([84]). Copyright (2017) American Chemical Society.)

An interesting feature offered by certain polymer nanocomposites employed as SERS substrates is their chemical functionalization envisaging molecular recognition [106, 152, 167]. Nie *et al.* have reported a strategy to detect tumors in living animals, using SERS substrates based on *pegylated* Au NPs [111]. In this research, the surface of Au NPs were functionalized with organic dyes, namely diethylthiatricarbocyanine and malachite green, which acts as Raman reporters. The Au NPs were coated with a polyethylene glycol with a thiol group and then functionalized with antibody fragments, which would target the tumor. The tumor detection was successfully accomplished using the SERS technique, monitoring the signal of the corresponding Raman reporters. Additionally, Batt and co-workers have successfully prepared an apta-sensing SERS substrate composed by polymer-Au NP-aptamer composite microspheres that allows the detection of a target molecule such as malathion [109]. They have demonstrated that these nanocomposites when attached to an aptamer have extraction capabilities for a pesticide, whose Raman signal is strongly enhanced due to the presence of the Au NPs.

Our research group has reported a series of polymer-based nanocomposites that enable the SERS detection of several analytes (**Figure 6**). In this context, synthetic polymers as well as polymers of natural origin have been investigated for these purposes. Examples include nanocomposites based on poly(*t*-butylacrylate), poly(methyl metacrylate), cellulose, linen,

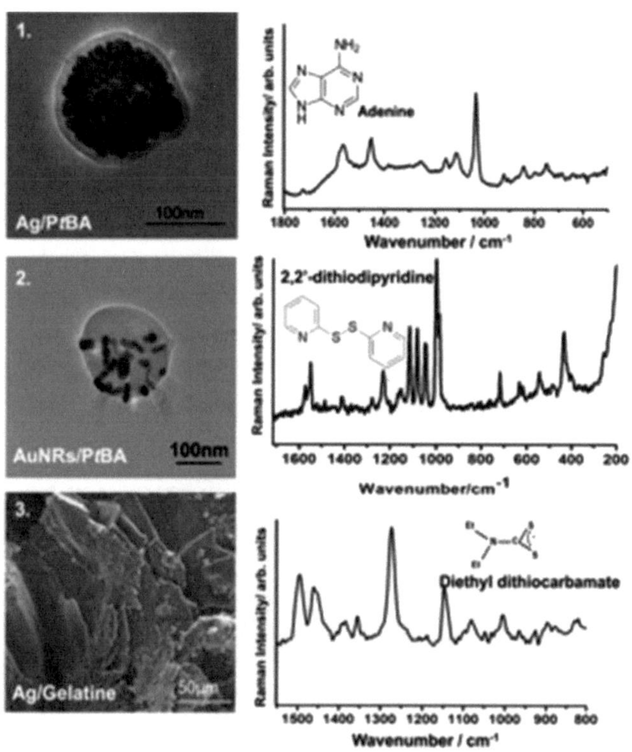

Figure 6. SERS analysis of biomolecules using polymer based composites as substrates. 1. detection of a DNA nucleobase using organically capped silver nanoparticles encapsulated with poly(*t*-butylacrylate) as SERS substrates; [113] 2. Au nanorods with poly(*t*-butylacrylate) for the detection of 2,2′-dithiodipyridine; [107] 3. Ag/gelatin nanocomposites as active SERS platforms for pesticide detection and release [85].

carrageenan and gelatin, loaded with colloidal Ag NPs as SERS substrates. Ultimately, these nanocomposites might find use to fabricate analytical platforms for distinct end uses, such as paper products, smart textiles, thermosensitive materials and drug delivery [67, 69, 84, 85]. Fateixa *et al.* have described Ag/gelatin A hydrogel samples with distinct gel strength as platforms for SERS detection and release of EtDTC, a pesticide model [85]. In this work, SERS was investigated as a spectroscopic method to detect the presence of low amounts of EtDTC in gelatin hydrogels, following the gradual release of the pesticide into water used as the dispersing medium. Noteworthy, this methodology can be used as an alternative to monitor the performance of hydrogel vehicles in the controlled release of pesticides, namely during the formulation and optimization stages of fabrication. On the other hand, P*t*BA matrices coating organically capped Ag NPs can be used either as aqueous emulsions or as cast films for active SERS substrates [114]. In this research, Trindade and co-workers have reported metal loaded polymer based composites with sensitivity for the SERS detection of thiosalicylic acid, even after the nanocomposite has been submitted to a temperature cycle (−60 to 65°C). The observations were interpreted by considering that the thiosalicylic acid molecules were entrapped within the polymer network and close to the Ag NPs. Similarly, DNA nucleobases, such as adenine, have been reported and detected by SERS using metal loaded polymer nanocomposites obtained by (mini)emulsion polymerization, namely Ag/PtBA and Ag/poly(metametylacrylate) (PMMA) [113, 115]. In addition, other plasmonic NPs can be used for the preparation of such polymer composites via (mini)emulsion polymerization namely gold nanorods (AuNRs). Fateixa *et al.* have reported SERS substrates obtained by a blending method of colloidal Au nanorods and PtBA aqueous emulsions [107]. The use of the blending method in this case limits morphological modifications of the rods that otherwise could occur in more drastic conditions. The composite blends were evaluated as SERS substrates, showing stronger signal enhancement when compared to the original Au NRs colloid, and using 2,2-dithiodipyridine as the analytical probe. This strategy involves a low-cost process with potential for the up-scale fabrication of SERS substrates, namely by using other types of polymers.

5. Raman imaging of functional polymer nanocomposites

In the last two decades, the rapid development of SERS has been in line with the scientific advances in nanofabrication and Raman instrumentation such as confocal Raman microscopy. Confocal Raman microscopy combines digital imaging technology with Raman spectroscopy in order to evaluate the chemical composition, molecular structure and spatial distribution of molecular components in a certain material, giving information about its homogeneity at the microscale level [168–171]. Delhaye and Dhamelincour have demonstrated for the first time the possibility to combine Raman spectroscopy and mapping microscopy in 1975 on a paper entitled "Raman microprobe and microscope with laser excitation" [172]. The authors have described the technique in detail, giving applications of the system such as the study of various materials such as rocks, plastics, composite materials, phases and inclusions and defects in solids. They emphasized that this new technique could became a valuable tool for the study of chemical reactions in micro-samples and also extended to biological samples.

In Raman imaging, thousands of linearly independent and spatially resolved spectra of the compounds existent in the specimen, are collected and analyzed. Among these spectra, the intensities of diagnosis bands for each species can be analyzed to generate images that are true maps for the spatial distribution of the compounds without the use of strains, dyes or contrast agents. This is a great advantage for materials characterization because little or no sample preparation is needed to characterize heterogeneous matrices [168–171, 173]. In fact, this technique is so versatile that has been applied in several fields, including pharmaceuticals' analysis [174–179], biology [180–182], biomedicine [183–186], label-free cell imaging [187–191], food industry [192–194], threat detection [195–197] and more fundamental research [198–202].

For pharmaceutical industry, the Raman imaging technique became an important analytical tool to trace the active pharmaceutical ingredient (API) heterogeneity in tablet or granulates, controlled release systems, and orally inhaled and nasal drug products [174–176]. Šašić has reported the use of Raman imaging for the spatial distribution of the API and major excipient (mannitol) on common pharmaceutical tablets and granulates [175, 176]. In the same year, Widjaja *et al.* have reported the combination of Raman imaging with advanced multivariate data analysis method, namely band-target entropy minimization (BTEM) for the identification of minor components of pharmaceutical drug tablets [177]. They have detected, in model pharmaceutical tablets, minor components level as low as 0.2% by weight. In addition, the identification and quantification of polymorph forms, unique crystal packing lattice forms of molecules, of API is another important issue for pharmaceutical analysis and nowadays is routinely performed using Raman imaging. For example, Henson and Zhang have reported the detection and spatial distribution of a polymorphic impurity (0.05% w/w) of active pharmaceutical ingredient in a tablet, using Raman imaging [178]. Lin *et al.* have demonstrated the use of Raman mapping for microscopic characterization of the surface of tablets containing chloramphenicol palmitate polymorphs [179]

Besides the chemical specificity, Raman spectroscopy coupled with microscopy maps may hold several other desirable properties for imaging applications, such as high spatial resolution, multiplexing capability, low background signal, and excellent photostability [170, 171]. Over the past years, Raman imaging techniques have been developed in step with the latest Raman fields, such as coherent anti-Stokes Raman spectroscopy (CARS) [203–207], surface enhanced Raman scattering (SERS) [67, 169, 173, 208–210] and tip-enhanced Raman scattering (TERS) [170, 211–215].

The current developments in Raman imaging have brought a new overview on SERS platforms. There are few reports on the use of Raman imaging together with SERS methods, which makes SERS imaging an unexploited resource to answer unsolved questions about the materials functionalities and NPs synthesis, and complement substrates characterization in SERS platforms [67, 169, 173, 208–210]. SERS imaging has been successfully applied in different areas such as cellular imaging [173, 208, 216, 217], in vivo biosensing [218–221], pharmaceutics and cosmetics [209, 210] and textile industries [67, 68]. Some examples can be cited, Chao et al. have reported the use of diamond nanoparticles with two different sizes (5 and 100 nm) as SERS probes to bio-label human lung epithelial cells [217]. The interaction of the nanoparticles and the cells were probed by SERS imaging, demonstrating that the nanoparticles are not toxic to the cells. Gambhir and co-workers have injected Au@SiO_2 nanoparticles into a live mouse and have studied the mouse's liver through the skin using SERS mapped images [221]. In addition, Firkala el al. have reported, for the first time, the application of surface enhanced Raman imaging on pharmaceutical tablets containing an API in very low concentrations, using Ag colloids as SERS probes [209].

It is of special interest to mention in the context of this chapter that SERS imaging has been a useful technique to identify the SERS active sites in polymer based nanocomposite substrates [60, 63, 67, 68, 73, 93, 160]. SERS imaging has been applied to Ag/agarose beads films, giving an idea of the distribution of the *hot spots* in the polymer matrix [63]. Highly sensitive biological imaging of HEK293 cells expressing PLCγ1 cancer markers were obtained, using Au/Ag core-shell NPs, conjugated with monoclonal antibodies [93]. A porphyrin–phospholipid conjugate with quenched fluorescence have been reported as a Raman reporter molecule for SERS based

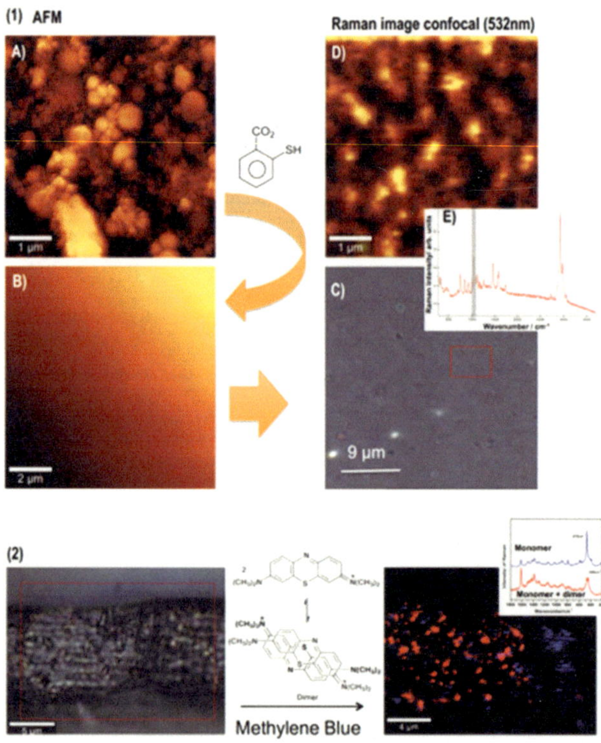

Figure 7. (1) AFM topography image of Ag/P*t*BA composites prepared by (mini)emulsion polymerization (A) before and (B) after the addition of 10 μL of an ethanolic solution of thiosalicylic acid 10^{-3} M; C) optical photograph of the sample with the scanned area marked in red; D) Raman images obtained using the integrated intensity of the Raman band at 1035 cm^{-1} in the SERS spectra of thiosalicylic acid 10^{-3} M using the Ag/P*t*BA composite as substrates; E) SERS spectrum of thiosalicylic acid (10^{-3} M) using Ag/P*t*BA composites as substrate (laser source: 532 nm); (2) Optical photograph (left) and combined Raman image (right), using two different Raman spectra of methylene blue (100 μM) adsorbed on Ag/linen composite Inset: Raman spectra of monomer form and a mixture of monomer and dimer of MB used to create the combined Raman image [68].

imaging [160]. The spatial distribution of thiophenol and an organic dye were achieved by SERS mapping using smart films composed by natural rubber containing Au NPs [73]

We have been particularly interested in exploring Raman imaging in the characterization of SERS substrates. An illustrative example comprises the development of SERS substrates based on P*t*BA polymer beads coated with organically capped Ag NPs, accomplishing information at the level of both the substrate surface and the molecular adsorbate distribution (**Figure 7-1**) [114]. By coupling SERS images with AFM, we observed a rearrange of the polymeric chains of the PtBA after the addition of the analyte dispersed in ethanol, allowing a better diffusion of the molecular probe through the matrix and closer to the Ag NPs. This proximity enhanced the Raman signal of the thiosalicylic acid and their spacial distribution can be observed in the SERS images. In addition, linen fibers loaded with Ag NPs and then stained with methylene blue (MB) were investigated using SERS imaging. MB was selected as the molecular probe not only because it is a common organic dye but also due to the formation of dimer or monomer species, each one with characteristic visible absorption and Raman spectra. We have demonstrated that the SERRS effect together with confocal Raman microscopy offer a new tool to map the local distribution of the MB dye in the fibers and consequently the distribution of Ag NPs over the fabrics, using Raman imaging [67]. In addition, it is also possible to assess the preferred adsorbate form of MB on distinct types of nanocomposite fibers and their local distribution (**Figure 7-2**) [68]. This investigation allows to foresee the use of this approach in terms of quality control of antimicrobial Ag containing fabrics, which is a market in great expansion.

6. Conclusions and outlook

The development of polymer based composites as active SERS substrates has contributed considerably to the rise of this methodology as an important and significantly tool in several analytical contexts. This improvement could only be achieved due to the development of more versatile and powerful instruments, including portable Raman equipment, and a deeper knowledge about the underlying mechanisms in the Raman effect that occur in molecules adsorbed at metal surfaces. It should be also emphasized the unprecedented progress observed in the past decades on the synthesis of nanostructures having controlled morphology. Interestingly, it has become clear that improvements in SERS have also impact on the knowledge that we have about the materials required as substrates for applying such spectroscopic technique. In this chapter, the most recent developments in metal loaded polymer nanocomposites for SERS studies were reviewed, showing their applicability into diverse areas due to their multifunctional properties. The vestigial SERS detection of specific molecules using Raman reporters, SERS tags and specific external stimuli are examples of further developments in SERS technologies dependent on materials development. This research has also shown that Raman imaging combined with other techniques such as SERS are valuable assets that complement or eventually provide unique characterization data, with particular relevance in the use of polymer based composites as SERS platforms.

Acknowledgements

This work was developed within the scope of the project CICECO-Aveiro Institute of Materials, POCI-01-0145-FEDER- 007679 (FCT Ref. UID/CTM/50011/2013), financed by national funds through the FCT/MEC and when appropriate co-financed by Fundo Europeu de Desenvolvimento Regional (FEDER) under the PT2020 Partnership Agreement. S. Fateixa thanks Fundação para a Ciência e Tecnologia (FCT) for the Grant SFRH/BPD/93547/2013.

Author details

Sara Fateixa*, Helena I.S. Nogueira and Tito Trindade

*Address all correspondence to: sarafateixa@ua.pt

Chemistry Department, CICECO-Aveiro Institute of Materials, University of Aveiro, Portugal

References

[1] Trindade T, Thomas PJ. Defining and using very small crystals. Comprehensive Inorganic Chemistry II: From Elements to Applications. 2013;**4**:343-369

[2] Grzelczak M, Perez-Juste J, Mulvaney P, Liz-Marzan LM. Shape control in gold nanoparticle synthesis. Chemical Society Reviews. 2008;**37**:1783-1791

[3] Liu S, Tang Z. Nanoparticle assemblies for biological and chemical sensing. Journal of Materials Chemistry. 2010;**20**:24-35

[4] Link S, El-Sayed MA. Spectral properties and relaxation dynamics of surface plasmon electronic oscillations in gold and silver nanodots and nanorods. The Journal of Physical Chemistry B. 1999;**103**:8410-8426

[5] Mock JJ, Barbic M, Smith DR, Schultz DA, Schultz S. Shape effects in plasmon resonance of individual colloidal silver nanoparticles. The Journal of Chemical Physics. 2002;**116**:6755-6759

[6] Zhao J, Pinchuk AO, McMahon JM, Li S, Ausman LK, Atkinson AL, Schatz GC. Methods for describing the electromagnetic properties of silver and gold nanoparticles. Accounts of Chemical Research. 2008;**41**:1710-1720

[7] Freestone I, Meeks N, Sax M, Higgitt C. The Lycurgus Cup - A Roman nanotechnology. Gold Bulletin. 2007;**40**:270-277

[8] Sciau P. Nanoparticles in Ancient Materials: The Metallic Lustre Decorations of Medieval Ceramics. In: Dr. Abbass A. Hashim editors. The Delivery of Nanoparticles. InTech, ch 25, 2012. DOI: 10.5772/34080

[9] Fleischmann M, Hendra PJ, McQuillan AJ. Raman spectra of pyridine adsorbed at a silver electrode. Chemical Physics Letters. 1974;**26**:163-166

[10] Fateixa S, Nogueira HIS, Trindade T. Hybrid nanostructures for SERS: Materials development and chemical detection. Physical Chemistry Chemical Physics. 2015;**17**:21046-21071

[11] Nogueira HIS, Teixeira-Dias JC, Trindade T. Nanostructured metals in surface enhanced Raman spectroscopy, Encyclopedia of Nanoscience and Nanotechnology, American Scientific Publishers ch. 2; 2004

[12] Moskovits M. Surface-enhanced Raman spectroscopy: A brief retrospective. Journal of Raman Specroscopy. 2005;**36**:485-496

[13] Abalde-Cela S, Aldeanueva-Potel P, Mateo-Mateo C, Rodríguez-Lorenzo L, Alvarez-Puebla RA, Liz-Marzán LM. Surface-enhanced Raman scattering biomedical applications of plasmonic colloidal particles. Journal of the Royal Society Interface. 2010;**7**:S435-S450

[14] Cotton TM, Kim J-H, Chumanov GD. Application of surface-enhanced Raman spectroscopy to biological systems. Journal of Raman Specroscopy. 1991;**22**:729-742

[15] Wei H, Hossein Abtahi SM, Vikesland PJ. Plasmonic colorimetric and SERS sensors for environmental analysis. Environmental Science: Nano. 2015;**2**:120-135

[16] Aoki PHB, Furini LN, Alessio P, Aliaga AE, Constantino CJL. Surface-enhanced Raman scattering (SERS) applied to cancer diagnosis and detection of pesticides, explosives, and drugs. Reviews in Analytical Chemistry. 2013;**32**:55-76

[17] Vo-Dinh T. Surface-enhanced Raman spectroscopy using metallic nanostructures. TrAC Trends in Analytical Chemistry. 1998;**17**:557-582

[18] Dieringer JA, McFarland AD, Shah NC, Stuart DA, Whitney AV, Yonzon CR, Young MA, Zhang X, Van Duyne RP. Surface enhanced Raman spectroscopy: New materials, concepts, characterization tools, and applications. Faraday Discussions. 2006;**132**:9-26

[19] Baker GA, Moore DS. Progress in plasmonic engineering of surface-enhanced Raman-scattering substrates toward ultra-trace analysis. Analytical and Bioanalytical Chemistry. 2005;**382**:1751-1770

[20] Sharma B, Frontiera RR, Henry A-I, Ringe E, Van Duyne RP. SERS: Materials, applications, and the future. Materials Today. 2012;**15**:16-25

[21] Fan MK, Andrade GFS, Brolo AG. A review on the fabrication of substrates for surface enhanced Raman spectroscopy and their applications in analytical chemistry. Analytica Chimica Acta. 2011;**693**:7-25

[22] Álvarez-Puebla RA, Liz-Marzán LM. Traps and cages for universal SERS detection. Chemical Society Reviews. 2012;**41**:43-51

[23] Cialla D, März A, Böhme R, Theil F, Weber K, Schmitt M, Popp J. Surface-enhanced Raman spectroscopy (SERS): Progress and trends. Analytical and Bioanalytical Chemistry. 2012;**403**:27-54

[24] Halas NJ, Moskovits M. Surface-enhanced Raman spectroscopy: Substrates and materials for research and applications. MRS Bulletin. 2013;**38**:607-611

[25] Cao Y, Li D, Jiang F, Yang Y, Huang Z. Engineering Metal Nanostructure for SERS Application. Journal of Nanomaterials. 2013;**2013**:12

[26] Raman CV, Krishnan KS. A new type of secundary radiation. Nature. 1928;**121**:501-502

[27] Ferraro RL, Nakamoto K. Introductory Raman spectroscopy. Boston: Academic Press; 1994

[28] Mulvaney SP, Keating CD. Raman spectroscopy. Analytical Chemistry. 2000;**72**:145R-157R

[29] Lin C-C, Kuo M-T, Chang H-C. Review: Raman spectroscopy – A novel tool for noninvasive analysis of ocular surface fluid. Journal of Medical and Biological Engineering. 2010;**30**:343-354

[30] Petry R, Schmitt M, Popp J. Raman spectroscopy—A prospective tool in the life sciences. Chemical Physical Chemistry. 2003;**4**:14-30

[31] Smith E, Dent G. Modern Raman spectroscopy: A practical approach. John Wiley and Sons; 2005

[32] Kumar GVP. Plasmonic nano-architectures for surface enhanced Raman scattering: A review. Journal of Nanophotonics. 2012;**6**:064503-064501

[33] Jeanmaire DL, Van Duyne RP. Surface raman spectroelectrochemistry: Part I. Heterocyclic, aromatic, and aliphatic amines adsorbed on the anodized silver electrode. Journal of Electroanalytical Chemistry. 1977;**84**:1-20

[34] Albrecht MG, Creighton JA. Anomalously intense Raman spectra of pyridine at a silver electrode. Journal of the American Chemical Society. 1977;**99**:5215-5217

[35] Hering K, Cialla D, Ackermann K, Dörfer T, Möller R, Schneidewind H, Mattheis R, Fritzsche W, Rösch P, Popp J. SERS: A versatile tool in chemical and biochemical diagnostics. Analytical and Bioanalytical Chemistry. 2008;**390**:113-124

[36] Jensen L, Aikens CM, Schatz GC. Electronic structure methods for studying surface-enhanced Raman scattering. Chemical Society Reviews. 2008;**37**:1061-1073

[37] Kambhampati P, Child CM, Foster MC, Campion A. On the chemical mechanism of surface enhanced Raman scattering: Experiment and theory. The Journal of Chemical Physics. 1998;**108**:5013-5027

[38] Arenas JF, Soto J, Tocón IL, Fernández DJ, Otero JC, Marcos JI. The role of charge-transfer states of the metal-adsorbate complex in surface-enhanced Raman scattering. The Journal of Chemical Physics. 2002;**116**:7207-7217

[39] Moskovits M. Surface-enhanced spectroscopy. Reviews of Modern Physics. 1985;**57**:783-826

[40] Stiles PL, Dieringer JA, Shah NC, PVDR. Surface-enhanced Raman spectroscopy. Annual Review of Analytical Chemistry. 2008;**1**:601-626

[41] Contreras-Cáceres R, Sierra-Martín B, Fernández-Barbero A. Surface-Enhanced Raman Scattering Sensors based on Hybrid Nanoparticles, Microsensors. In: Prof. Igor Minin editor. InTech, ch. 7, 2011. DOI: 10.5772/18735

[42] Wang Y, Asefa T. Poly(allylamine)-stabilized colloidal copper nanoparticles: Synthesis, morphology, and their surface-enhanced Raman scattering properties. Langmuir. 2010; **26**:7469-7474

[43] Xiong Y, McLellan JM, Chen J, Yin Y, Li Z-Y, Xia Y. Kinetically controlled synthesis of triangular and hexagonal nanoplates of palladium and their SPR/SERS properties. Journal of the American Chemical Society. 2005;**127**:17118-17127

[44] Liu Y, Hu Y, Zhang J. Few-layer graphene-encapsulated metal nanoparticles for surface-enhanced Raman spectroscopy. Journal of Physical Chemistry C. 2014;**118**:8993-8998

[45] Wang X, Shi W, She G, Mu L. Surface-enhanced Raman scattering (SERS) on transition metal and semiconductor nanostructures. Physical Chemistry Chemical Physics. 2012;**14**:5891-5901

[46] Camargo PHC, Cobley CM, Rycenga M, Xia Y. Measuring the surface-enhanced Raman scattering enhancement factors of hot spots formed between an individual Ag nanowire and a single Ag nanocube. Nanotechnology. 2009;**20**:434020

[47] Rycenga M, Camargo PHC, Li W, Moran CH, Xia Y. Understanding the SERS effects of single silver nanoparticles and their dimers, one at a time. Journal of Physical Chemistry Letters. 2010;**1**:696-703

[48] Contreras-Cáceres R, Abalde-Cela S, Guardia-Girós P, Fernández-Barbero A, Pérez-Juste J, Alvarez-Puebla RA, Liz-Marzán LM. Multifunctional microgel magnetic/optical traps for SERS ultradetection. Langmuir. 2011;**27**:4520-4525

[49] Camden JP, Dieringer JA, Wang Y, Masiello DJ, Marks LD, Schatz GC, Van Duyne RP. Probing the structure of single-molecule surface-enhanced Raman scattering hot spots. Journal of the American Chemical Society. 2008;**130**:12616-12617

[50] Zeng Z, Dan T, Liu L, Wang Y, Zhou Q, Su S, Hu D, Han B, Jin M, Ao X, Zhan R, Gao X, Lu X, Zhou G, Senz S, Zhang Z, Liu J. Highly reproducible surface-enhanced Raman scattering substrate for detection of phenolic pollutants. Nanotechnology. 2016;**27**:455301

[51] Aiming Y, Yi H, Vinayak PD. Locally enhanced surface plasmons and modulated "hot-spots" in nanoporous gold patterns on atomically thin MoS2 with a comparison to SiO2 substrate. Applied Physics Letters. 2016;**108**:091901

[52] Ingram W, He Y, Stone K, Dennis WM, Ye D, Zhao Y. Tuning the plasmonic properties of silver nanopatterns fabricated by shadow nanosphere lithography. Nanotechnology. 2016;**27**:385301

[53] Le Ru EC, Etchegoin PG. Single-molecule surface-enhanced Raman spectroscopy. Annual Review of Physical Chemistry. 2012;**63**:65-87

[54] Kneipp K, Wang Y, Kneipp H, Perelman LT, Itzkan I, Dasari RR, Feld MS. Single molecule detection using surface-enhanced Raman scattering (SERS). Physical Review Letters. 1997;**78**:1667-1670

[55] Nie S, Emory SR. Probing single molecules and single nanoparticles by surface-enhanced Raman scattering. Science. 1997;**275**:1102-1106

[56] Ding SY, Yi J, Li JF, Ren B, Wu DY, Panneerselvam R, Tian Z-Q. Nanostructure-based plasmon-enhanced Raman spectroscopy for surface analysis of materials. Nature Reviews Materials. 2006;**1**:16021

[57] Creighton JA. Surface Raman electromagnetic enhancement factors for molecules at the surface of small isolated metal spheres: The determination of adsorbate orientation from sers relative intensities. Surface Science. 1983;**124**:209-219

[58] Merican Z, Schiller TL, Hawker CJ, Fredericks PM, Blakey I. Self-assembly and encoding of polymer-stabilized gold nanoparticles with surface-enhanced Raman reporter molecules. Langmuir. 2007;**23**:10539-10545

[59] Esteves ACC, Barros-Timmons A, Trindade T. Nanocompósitos de matriz polimérica: estratégias de síntese de materiais híbridos. Quimica Nova. 2004;**27**:798-806

[60] Aldeanueva-Potel, P Faoucher E, Alvarez-Puebla RNA, Liz-Marzán LM, Brust M. Recyclable molecular trapping and SERS detection in silver-loaded agarose gels with dynamic hot spots. Analytical Chemistry. 2009;**81**:9233-9238

[61] Lee C, Zhang P. Facile synthesis of gelatin-protected silver nanoparticles for SERS applications. Journal of Raman Specroscopy. 2013;**44**:823-826

[62] Fan J, Shi Z, Ge Y, Wang J, Wanga Y, Yin J. Gum arabic assisted exfoliation and fabrication of Ag-graphene-based hybrids. Journal of Materials Chemistry. 2012;**22**:13764-13772

[63] Abalde-Cela S, Auguié B, Fischlechner M, Huck WTS, Alvarez-Puebla RA, Liz-Marzán LM, Abell C. Microdroplet fabrication of silver-agarose nanocomposite beads for SERS optical accumulation. Soft Matter. 2011;**7**:1321-1325

[64] Jung G-B, Kim J-H, Burm JS, Park H-K. Fabrication of chitosan-silver nanoparticle hybrid 3D porous structure as a SERS substrate for biomedical applications. Applied Surface Science. 2013;**273**:179-183

[65] Efeoglu E, Culha M. In situ monitoring of biofilm formation by using surface-enhanced Raman scattering. Appl. Spectrosc. 2013;**67**:498-505

[66] Pinto RJB, Marques PAAP, Neto CP, Trindade T, Daina S, Sadocco P. Antibacterial activity of nanocomposites of silver and bacterial or vegetable cellulosic fibers. Acta Biomaterialia. 2009;**5**:2279-2289

[67] Fateixa S, Wilhelm M, Nogueira HIS, Trindade T. SERS and Raman imaging as a new tool to monitor dyeing on textile fibres. Journal of Raman Spectroscopy. 2016;**47**:1239-1246

[68] Fateixa S, Wilhelm M, Jorge AM, Nogueira HIS, Trindade T. Raman imaging studies on the adsorption of methylene blue species onto silver modified linen fibers. Journal of Raman Specroscopy. 2017;**48**:795-802

[69] Marques PAAP, Nogueira HIS, Pinto RJB, Neto CP, Trindade T. Silver-bacterial cellulosic sponges as active SERS substrates. Journal of Raman Specroscopy. 2008;**39**:439-443

[70] Keating M, Chen Y, Larmour IA, Faulds K, Graham D. Growth and surface-enhanced Raman scattering of Ag nanoparticle assembly in agarose gel. Measurement Science and Technology. 2012;**23**:084006

[71] Tiwari AD, Mishra AK, Mishra SB, Kuvarega AT, Mamba BB. Stabilisation of silver and copper nanoparticles in a chemically modified chitosan matrix. Carbohydrate Polymers. 2013;**92**:1402-1407

[72] Huo D, He J, Yang S, Zhou Z, Hu Y, Epple M. Facile synthesis of polymer core@silver shell hybrid nanoparticles with super surface enhanced Raman scattering capability. Journal of Colloid and Interface Science. 2013;**393**:119-125

[73] Cabrera FC et al. Portable smart films for ultrasensitive detection and chemical analysis using SERS and SERRS. Journal of Raman Spectroscopy. 2012;**43**:474-477. DOI: 10.1002/jrs.3074

[74] Nhung TT, Bu Y, Lee S-W. Facile synthesis of chitosan-mediated gold nanoflowers as surface-enhanced Raman scattering (SERS) substrates. Journal of Crystal Growth. 2013;**373**:132-137

[75] Gao S, Zhang H, Wang X, Yang J, Zhou L, Peng C, Sun D, Li M. Unique gold sponges: Biopolymer-assisted hydrothermal synthesis and potential application as surface-enhanced Raman scattering substrates. Nanotechnology. 2005;**16**:2530

[76] Alsawafta M, Badilescu S, Truong V-V, Packirisamy M. The effect of hydrogen nano-bubbles on the morphology of gold–gelatin bionanocomposite films and their optical properties. Nanotechnology. 2012;**23**:065305

[77] Zhang D, Yang H. Gelatin-stabilized copper nanoparticles: Synthesis, morphology, and their surface-enhanced Raman scattering properties. Physica B: Condensed Matter. 2013;**415**:44-48

[78] Perrier S, Takolpuckdee P, Mars CA. Reversible Addition–Fragmentation Chain Transfer Polymerization: End Group Modification for Functionalized Polymers and Chain Transfer Agent Recovery. Macromolecules. 2005;**38**:2033-2036

[79] Pinto RJB, Neves MC, Neto CP, Trindade T. Growth and chemical stability of copper nanostructures on cellulosic fibers. European Journal of Inorganic Chemistry. 2012;**2012**:5043-5049

[80] Pinto RJB, Daina S, Sadocco P, Neto CP, Trindade T. Antibacterial activity of nanocomposites of copper and cellulose. BioMed Research International. 2013;**2013**:6

[81] Pinto RJB, Fernandes SCM, Freire CSR, Sadocco P, Causio J, Neto CP, Trindade T. Antibacterial activity of optically transparent nanocomposite films based on chitosan or its derivatives and silver nanoparticles. Carbohydrate Research. 2012;**348**:77-83

[82] Martins NCT, Freire CSR, Pinto RJB, Fernandes SCM, Neto CP, Silvestre AJD, Causio J, Baldi G, Sadocco P, Trindade T. Electrostatic assembly of Ag nanoparticles onto nanofibrillated cellulose for antibacterial paper products. Cellulose. 2012;**19**:1425-1436

[83] Pinto RJB, Almeida A, Fernandes SC, Freire CS, Silvestre AJ, Neto CP, Trindade T. Antifungal activity of transparent nanocomposite thin films of pullulan and silver against Aspergillus niger. Colloids and Surfaces, B: Biointerfaces. 2013;**103**:143-148

[84] Fateixa S, Daniel-da-Silva AL, Nogueira HIS, Trindade T. Raman signal enhancement dependence on the gel strength of Ag/hydrogels used as SERS substrates. The Journal of Physical Chemistry C. 2014;**118**:10384-10392

[85] Fateixa S, Soares SF, Daniel-da-Silva AL, Nogueira HIS, Trindade T. Silver-gelatine bionanocomposites for qualitative detection of a pesticide by SERS. Analyst. 2015;**140**: 1693-1701

[86] Marsich L, Bonifacio A, Mandal S, Krol S, Beleites C, Sergo V. Poly-l-lysine-coated silver nanoparticles as positively charged substrates for surface-enhanced Raman scattering. Langmuir. 2012;**28**:13166-13171

[87] Yu WW, White IM. A simple filter-based approach to surface enhanced Raman spectroscopy for trace chemical detection. Analyst. 2012;**137**:1168-1173

[88] Daniel-da-Silva AL, Salgueiro AM, Trindade T. 1. Effects of Au nanoparticles on thermoresponsive genipin-crosslinked gelatin hydrogels. Gold Bulletin. 2013;**46**:25-33

[89] Salgueiro AM, Daniel-da-Silva AL, Fateixa S, Trindade T. κ-Carrageenan hydrogel nanocomposites with release behavior mediated by morphological distinct Au nanofillers. Carbohydrate Polymers. 2013;**91**:100-109

[90] Kim N-J, Lin M, Hu Z, Li H. Evaporation-controlled chemical enhancement of SERS using a soft polymer substrate. Chemical Communications. 2009;**6246-6248**

[91] Abbas A, Brimer A, Slocik JM, Tian L, Naik RR, Singamaneni S. Multifunctional analytical platform on a paper strip: Separation, preconcentration, and subattomolar detection. Analytical Chemistry. 2013;**85**:3977-3983

[92] Lee CH, Tian L, Singamaneni S. Paper-Based SERS swab for rapid trace detection on real-world surfaces. ACS Applied Materials and Interfaces. 2010;**2**:3429-3435

[93] Lee S, Kim S, Choo J, Shin SY, Lee YH, Choi HY, Ha S, Kang K, Oh CH. Biological imaging of HEK293 cells expressing PLCγ1 using surface-enhanced Raman microscopy. Analytical Chemistry. 2007;**79**:916-922

[94] Saha S, Pal A, Pande S, Sarkar S, Panigrahi S, Pal T. Alginate gel-mediated photochemical growth of mono- and bimetallic gold and silver nanoclusters and their application to surface-enhanced Raman scattering. Journal of Physical Chemistry C. 2009;**113**:7553-7560

[95] Homan KA, Chen J, Schiano A, Mohamed M, Willets K. A, Murugesan S, Stevenson KJ, Emelianov S. Silver–polymer composite stars: synthesis and applications. Advanced Functional Materials. 2011;**21**:1673-1680

[96] Ye S, Fang L, Qing X, Lu Y. Surface-enhanced Raman scattering study of Ag@PPy nanoparticles. Journal of Raman Specroscopy. 2010;**41**:1119-1123

[97] Celebioglu A, Aytac Z, Umu OC, Dana A, Tekinay T, Uyar T. One-step synthesis of size-tunable Ag nanoparticles incorporated in electrospun PVA/cyclodextrin nanofibers. Carbohydrate Polymers. 2014;**99**:808-816

[98] Zhu W, Wu Y, Yan C, Wang C, Zhang M, Wu Z. Facile synthesis of mono-dispersed polystyrene (PS)/Ag composite microspheres via modified chemical reduction. Materials. 2013;**6**:5625-5638

[99] Shen XS, Wang GZ, Hong X, Zhu W. Nanospheres of silver nanoparticles: Agglomeration, surface morphology control and application as SERS substrates. Physical Chemistry Chemical Physics. 2009;**11**:7450-7454

[100] Xu P, Jeon SH, Mack NH, Doorn SK, Williams DJ, Han X, Wang HL. Field-assisted synthesis of SERS-active silver nanoparticles using conducting polymers. Nanoscale. 2010;**2**:1436-1440

[101] Karabiçak S, Kaya M, Vo-Dinh T, Volkan M. Silver nanoparticle-doped polyvinyl alcohol coating as a mediumfor surface-enhanced Raman scattering analysis. Journal of Nanoscience and Nanotechnology. 2008;**8**:955-960

[102] Mao A, Jin X, Gu X, Wei X, Yang G. Rapid, green synthesis and surface-enhanced Raman scattering effect of single-crystal silver nanocubes. Journal of Molecular Structure. 2012;**1021**:158-161

[103] Rao VK, Ghildiyal P, Radhakrishnan TP. In Situ fabricated cu–ag nanoparticle-embedded polymer thin film as an efficient broad spectrum sers substrate. The Journal of Physical Chemistry C. 2017;**121**:1339-1348

[104] Lee J-H, Mahmoud MA, Sitterle VB, Sitterle JJ, Meredith J, Highly C. Scattering, surface-enhanced Raman scattering-active, metal nanoparticle-coated polymers prepared via combined swelling–heteroaggregation. Chemistry of Materials. 2009;**21**:5654-5663

[105] Lin W. A durable plastic substrate for surface-enhanced Raman spectroscopy. Applied Physics A: Materials Science & Processing. 2011;**102**:121-125

[106] Bompart M, De Wilde Y, Haupt K. Chemical nanosensors based on composite molecularly imprinted polymer particles and surface-enhanced Raman scattering. Advanced Materials. 2010;**22**:2343-2348

[107] Fateixa S, Pinheiro PC, Nogueira HIS, Trindade T. Composite blends of gold nanorods and poly(t-butylacrylate) beads as new substrates for SERS. Spectrochimica Acta A. 2013;**113**:100-106

[108] Li D, Cui Y, Wang K, He Q, Yan X, Li J. Thermosensitive nanostructures comprising gold nanoparticles grafted with block copolymers. Advanced Functional Materials. 2007;**17**:3134-3140

[109] Barahona F, Bardliving CL, Phifer A, Bruno JG, Batt CA. An aptasensor based on polymer-gold nanoparticle composite microspheres for the detection of malathion using surface-enhanced Raman spectroscopy. Industrial Biotechnology. 2013;**9**:42-50

[110] Chen T, Wang H, Chen G, Wang Y, Feng Y, Teo WS, Wu T, Chen H. Hotspot-induced transformation of surface-enhanced Raman scattering fingerprints. ACS Nano. 2010; **4**:3087-3094

[111] Qian X, Peng XH, Ansari DO, Yin-Goen Q, Chen GZ, Shin DM, Yang L, Young AN, Wang MD, Nie S. In vivo tumor targeting and spectroscopic detection with surface-enhanced Raman nanoparticle tags. Nature Biotechnology. 2008;**26**:83-90

[112] Fernández-López C, Polavarapu L, Solís DM, Taboada JM, Obelleiro F, Contreras-Cáceres R, Pastoriza-Santos I, Pérez-Juste J. Gold nanorod–pNIPAM Hybrids with reversible plasmon coupling: Synthesis, modeling, and SERS properties. ACS Applied Materials and Interfaces. 2015;**7**:12530-12538

[113] Pinheiro PC, Fateixa S, Nogueira HIS, Trindade T. SERS study on adenine using a Ag/poly(t-butylacrylate) nanocomposite. Spectrochimica Acta A. 2013;**101**:36-39

[114] Fateixa S, Girao AV, Nogueira HIS, Trindade T. Polymer based silver nanocomposites as versatile solid film and aqueous emulsion SERS substrates. Journal of Materials Chemistry. 2011;**21**:15629-15636

[115] Pinheiro PC, Fateixa S, Nogueira HIS, Trindade T. SERS studies of DNA nucleobases using new silver poly(methyl methacrylate) nanocomposites as analytical platforms. Journal of Raman Specroscopy. 2014;**46**:47-53

[116] Liu J, Zhang L, Shi S, Chen S, Zhou N, Zhang Z, Cheng Z, Zhu X. A novel and universal route to SiO2-supported organic/inorganic hybrid noble metal nanomaterials via surface RAFT polymerization. Langmuir. 2010;**26**:14806-14813

[117] Quaroni L, Chumanov G. Preparation of polymer-coated functionalized silver nanoparticles. Journal of the American Chemical Society. 1999;**121**:10642-10643

[118] Guo Y, Kang L, Chen S, Li X. High performance surface-enhanced Raman scattering from molecular imprinting polymer capsulated silver spheres. Physical Chemistry Chemical Physics. 2015;**17**:21343-21347

[119] Martins MA, Fateixa S, Girão AV, Pereira SS, Trindade T. Shaping gold nanocomposites with tunable optical properties. Langmuir. 2010;**26**:11407-11412

[120] Li D, He Q, Cui Y, Wang K, Zhang X, Li J. Thermosensitive copolymer networks modify gold nanoparticles for nanocomposite entrapment. Chemistry--A European Journal. 2007;**13**:2224-2229

[121] Gehan H, Fillaud L, Chehimi MM, Aubard J, Hohenau A, Felidj N, Mangeney C. Thermo-induced electromagnetic coupling in gold/polymer hybrid plasmonic structures probed by surface-enhanced Raman scattering. ACS Nano. 2010;**4**:6491-6500

[122] Yin J, Wu T, Song J, Zhang Q, Liu S, Rong Xu R, Duan H. SERS-active nanoparticles for sensitive and selective detection of cadmium ion (Cd2+). Chemistry of Materials. 2011;**23**:4756-4764

[123] Álvarez-Puebla RA, Contreras-Cáceres R, Pastoriza-Santos I, Pérez-Juste J, Liz-Marzán LM. Au@pNIPAM colloids as molecular traps for surface-enhanced, spectroscopic, ultrasensitive analysis. Angewandte Chemie, International Edition. 2009;**48**:138-143

[124] Lee A, Dubinsky S, Tumarkin E, Moulin M, Beharry AA, Kumacheva E. Multifunctional hybrid polymer-based porous materials. Advanced Functional Materials. 2011;**21**: 1959-1969

[125] Contreras-Cáceres R, Pastoriza-Santos I, Alvarez-Puebla RA, Pérez-Juste J, Fernández-Barbero A, Liz-Marzán LM. Growing Au/Ag nanoparticles within microgel colloids for improved surface-enhanced Raman scattering detection. Chemistry—A European Journal. 2010;**16**:9462-9467

[126] Gao W, Chen G, Xu W, Yang C, Xu S. Surface-enhanced Raman scattering (SERS) chips made from metal nanoparticle-doped polymer fibers. RSC Advances. 2014;**4**:23838-23845

[127] Cruz SMA, Nogueira HIS, Marques PAAP. Potentialities of polymeric electrospun membranes decorated with silver nanoparticles and graphene oxide for biodetection by SERS. Ciência & Tecnologia dos Materiais. 2014;**26**:102-107

[128] Zhang C-L, Lv K-P, Huang H-T, Cong H-P, Yu S-H. Co-assembly of Au nanorods with Ag nanowires within polymer nanofiber matrix for enhanced SERS property by electrospinning. Nanoscale. 2012;**4**:5348-5355

[129] Ferhan AR, Kim D-H. Nanoparticle polymer composites on solid substrates for plasmonic sensing applications. Nano Today. 2016;**11**:415-434

[130] Balazs AC, Emrick T, Russell TP. Nanoparticle Polymer Composites: Where Two Small Worlds Meet. Science. 2006;**314**:1107-1110

[131] Gupta S, Agrawal M, Conrad M, Hutter NA, Olk P, Simon F, Eng LM, Stamm M, Jordan R. Poly(2-(dimethylamino)ethyl methacrylate) brushes with incorporated nanoparticles as a SERS active sensing layer. Advanced Functional Materials. 2010;**20**:1756-1761

[132] Chern CS. Emulsion polymerization mechanisms and kinetics. Progress in Polymer Science. 2006;**31**:443-486

[133] Chiefari J, Chong YK (Bill), Ercole F, Krstina J, Jeffery J, Le TPT, Mayadunne RTA, Meijs GF, Moad CL, Moad G, Rizzardo E, Thang SH. Living free-radical polymerization by reversible addition–fragmentation chain transfer: The RAFT Process. Macromol. 1998;**31**:5559-5562

[134] Zengin A, Tamer U, Caykara T. A SERS-based sandwich assay for ultrasensitive and selective detection of Alzheimer's Tau protein. Biomacromolecules. 2013;**14**:3001-3009

[135] Kato M, Kamigaito M, Sawamoto M, Higashimura T. Polymerization of methyl methacrylate with the carbon tetrachloride/dichlorotris- (triphenylphosphine)ruthenium(II)/methylaluminum Bis(2,6-di-tert-butylphenoxide) initiating system: Possibility of living radical polymerization. Macromolecules. 1995;**28**:1721-1723

[136] Wang J-S, Matyjaszewski K. Controlled/"living" radical polymerization. atom transfer radical polymerization in the presence of transition-metal complexes. Journal of the American Chemical Society. 1995;**117**:5614-5615

[137] Esteves AC, Neves MC, Barros-Timmons A, Bourgeat-Lami E, Liz-Marzán LM, Trindade T. Synthesis of SiO_2-coated Bi_2S_3/Poly(styrene) nanocomposites by in-situ polymerization. Journal of Nanoscience and Nanotechnology. 2006;**6**:414-420

[138] Esteves A. C. C. Barros-Timmons A. M. Martins J. A, Zhang W, Cruz-Pinto J, Trindade T. Crystallization behaviour of new poly(tetramethyleneterephthalamide) nanocomposites containing SiO_2 fillers with distinct morphologies. Composites Part B: Engineering. 2005;**36**:51-59

[139] Segala K, Dutra R. L, Franco C. V, Pereira A. S, Trindadeb T. In situ and ex situ preparations of ZnO/poly-{trans-[$RuCl_2(vpy)_4$]/styrene} nanocomposites. Journal of the Brazilian Chemical Society. 2010;**21**:1986-1991

[140] Landfester K. The generation of nanoparticles in miniemulsions. Advanced Materials. 2001;**13**:765-768

[141] Esteves ACC, Barros-Timmons A, Monteiro T, Trindade T. Polymer encapsulation of CdE (E = S, Se) quantum dot ensembles via radical polymerization in miniemulsion. Journal of Nanoscience and Nanotechnology. 2005;**5**:766-771

[142] Crespy D, Landfester K. Miniemulsion polymerization as a versatile tool for the synthesis of functionalized polymers. Beilstein. The Journal of Organic Chemistry. 2010;**6**:1132-1148

[143] Landfester K. Miniemulsion polymerization and the structure of polymer and hybrid nanoparticles. Angewandte Chemie, International Edition. 2009;**48**:4488-4507

[144] Esteves ACC, Bombalski L, Trindade T, Matyjaszewski K, Barros-Timmons A. Polymer grafting from CdS quantum dots via AGET ATRP in miniemulsion. Small. 2007; **3**:1230-1236

[145] Esteves ACC, Hodge P, Trindade T, Barros-Timmons AMMV. Preparation of nanocomposites by reversible addition-fragmentation chain transfer polymerization from the surface of quantum dots in miniemulsion. Journal of Polymer Science Part A: Polymer Chemistry. 2009;**47**:5367-5377

[146] Peres M, Costa LC, Neves A, Soares MJ, Monteiro T, Esteves ACC, Barros-Timmons AM, Trindade T, Kholkin A, Alves E. A green-emitting CdSe/poly(butyl acrylate) nanocomposite. Nanotechnology. 2005, 1969;**16**

[147] Martins MA, Neves MC, Esteves ACC, Girginova PI, Guiomar AJ, Amaral VS, Trindade T. Biofunctionalized ferromagnetic CoPt 3/polymer nanocomposites. Nanotechnology. 2007;**18**:215609

[148] Pereira AS, Rauwel P, Reis MS, Silva NJO, Barros-Timmons AM, Trindade T. Polymer encapsulation effects on the magnetism of EuS nanocrystals. Journal of Materials Chemistry. 2008;**18**:4572-4578

[149] Macedo AG, Martins MA, Fernandes SEM, Barros-Timmons AM, Trindade T, Carlos LD, Rocha J. Luminescent SiO_2-coated Gd_2O_3:Eu^{3+} nanorods/poly(styrene) nanocomposites by in situ polymerization. Optical Materials. 2010;**32**:1622-1628

[150] Daniel-da-Silva AL, Fateixa S, Guiomar AJ, Costa BFO, Silva NJO, Trindade T, Goodfellow BJ, Gil AM. Biofunctionalized magnetic hydrogel nanospheres of magnetite and κ -carrageenan. Nanotechnology. 2009;**20**:355602

[151] Sun L, Sung K-B, Dentinger C, Lutz B, Nguyen L, Zhang J, Qin H, Yamakawa M, Cao M, Lu Y, Chmura AJ, Zhu J, Su X, Berlin AA, Chan S, Knudsen B. Composite organic–inorganic nanoparticles as Raman labels for tissue analysis. Nano Letters. 2007;**7**:351-356

[152] Braun GB, Lee SJ, Laurence T, Fera N, Fabris L, Bazan GC, Moskovits M, Reich NO. Generalized approach to SERS-active nanomaterials via controlled nanoparticle linking, polymer encapsulation, and small-molecule infusion. Journal of Physical Chemistry C. 2009;**113**:13622-13629

[153] Folarin OM, Sadiku RE, Maity A. Polymer-noble metal nanocomposites: Review. International Journal of Physical Sciences. 2011;**6**:4869-4882

[154] He D, Hu B, Yao QF, Wang K, Yu SH. Large-scale synthesis of flexible free-standing SERS substrates with high sensitivity: Electrospun PVA nanofibers embedded with controlled alignment of silver nanoparticles. ACS Nano . 2009;**3**:3993-4002

[155] Oh JH, Shin H, Choi JY, Jung HW, Choi Y, Lee JS. Plate and on-plate structural control of ultra-stable gold/silver bimetallic nanoplates as redox catalysts, nanobuilding blocks, and single-nanoparticle surface-enhanced raman scattering probes. ACS Applied Materials & Interfaces. 2016;**8**:27140-27150

[156] Qu L-L, Li DW, Xue JQ, Zhai WL, Fossey JS, Long YT. Batch fabrication of disposable screen printed SERS arrays. Lab on a Chip. 2012;**12**:876-881

[157] Lee WWY, McCoy CP, Donnelly RF, Bell SEJ. Swellable polymer films containing Au nanoparticles for point-of-care therapeutic drug monitoring using surface-enhanced Raman spectroscopy. Analytica Chimica Acta. 2016;**912**:111-116

[158] Chen J, Huang Y, Kannan P, Zhang L, Lin Z, Zhang J, Chen T, Guo L. Flexible and adhesive surface enhance raman scattering active tape for rapid detection of pesticide residues in fruits and vegetables. Analytical Chemistry. 2016;**88**:2149-2155

[159] Kantarovich K, Tsarfati-BarAd I, Gheber LA, Haupt K, Bar I. Reading biochips by raman and surface-enhanced Raman spectroscopies. Plasmonics. 2013;**8**:3-12

[160] Tam NCM, McVeigh PZ, MacDonald TD, Farhadi A, Wilson BC, Zheng G. Porphyrin–lipid stabilized gold nanoparticles for surface enhanced Raman scattering based imaging. Bioconjugate Chemistry. 2012;**23**:1726-1730

[161] Chen L, Yan H, Xue X, Jiang D, Cai Y, Liang D, Jung YM, Han XX, Zhao B. Surface-enhanced Raman scattering (SERS) active gold nanoparticles decorated on a porous polymer filter. Applied Spectroscopy. 2017;**71**:1543-1550

[162] Yuan S, Ge F, Zhou M, Cai Z, Guang S. A new smart surface-enhanced Raman scattering sensor based on pH-responsive polyacryloyl hydrazine capped Ag nanoparticles. Nanoscale Research Letters. 2017;**12**:490

[163] Qian X, Li J, Nie S. stimuli-responsive sers nanoparticles: conformational control of plasmonic coupling and surface Raman enhancement. Journal of the American Chemical Society. 2009;**131**:7540-7541

[164] Kim DJ, Jeon TY, Park S-G, Han HJ, Im SH, Kim D-H, Kim S-H. Uniform microgels containing agglomerates of silver nanocubes for molecular size-selectivity and high SERS activity. Small. 2017;**13**:1604048

[165] Kim J-H, Twaddle KM, Cermak LM, Jang W, Yun J, Byun H. Photothermal heating property of gold nanoparticle loaded substrates and their SERS response. Colloids and Surfaces, A: Physicochemical and Engineering Aspects. 2016;**498**:20-29

[166] Anema JR, Li J-F, Yang Z-L, Ren B, Tian Z-Q. Shell-isolated nanoparticle-enhanced Raman spectroscopy: Expanding the versatility of surface-enhanced Raman scattering. Annual Review of Analytical Chemistry. 2011;**4**:129-150

[167] Chen HY, Abraham S, Mendenhall J, Delamarre SC, Smith K, Kim I, Batt CA. Encapsulation of single small gold nanoparticles by diblock copolymers. Chemical Physical Chemistry. 2008;**9**:388-392

[168] Treado PJ, Nelson MP. In Handbook of Raman spectroscopy: From the research laboratory to the process line (ed Marcel Dekker. Inc) Ch. 5, 2001;191-250

[169] Palonpon AF, Ando J, Yamakoshi H, Dodo K, Sodeoka M, Kawata S, Fujita K. Raman and SERS microscopy for molecular imaging of live cells. Nature Protocols. 2013;**8**:677-692

[170] Opilik L, Schmid T, Zenobi R. Modern Raman imaging: Vibrational spectroscopy on the micrometer and nanometer scales. Annual Review of Analytical Chemistry. 2013;**6**:379-398

[171] Stewart S, Priore RJ, Nelson MP, Treado PJ. Raman Imaging. Annual Review of Analytical Chemistry. 2012;**5**:337-360

[172] Delhaye M, Dhamelincourt P. Raman microprobe and microscope with laser excitation. Journal of Raman Specroscopy. 1975;**3**:33-43

[173] Zheng X, Zong C, Xu M, Wang X, Ren B. Raman imaging from microscopy to nanoscopy, and to macroscopy. Small. 2015;**11**:3395-3406

[174] Gendrin C, Roggo Y, Collet C. Pharmaceutical applications of vibrational chemical imaging and chemometrics: A review. Journal of Pharmaceutical and Biomedical Analysis. 2008;**48**:533-553

[175] Šašić S. An in-depth analysis of raman and near-infrared chemical images of common pharmaceutical tablets. Applied Spectroscopy. 2007;**61**:239-250

[176] Šašić S. Chemical imaging of pharmaceutical granules by Raman global illumination and near-infrared mapping platforms. Analytica Chimica Acta. 2008;**611**:73-79

[177] Widjaja E, Seah RKH. Application of Raman microscopy and band-target entropy minimization to identify minor components in model pharmaceutical tablets. Journal of Pharmaceutical and Biomedical Analysis. 2008;**46**:274-281

[178] Henson MJ, Zhang L. Drug characterization in low dosage pharmaceutical tablets using raman microscopic mapping. Applied Spectroscopy. 2006;**60**:1247-1255

[179] Lin W-Q, Jiang JH, Yang HF, Ozaki Y, Shen GL, Yu RQ. Characterization of chloramphenicol palmitate drug polymorphs by raman mapping with multivariate image segmentation using a spatial directed agglomeration clustering method. Analytical Chemistry. 2006;**78**:6003-6011

[180] Gantzsch SP, Kann B, Ofer-Glaessgen M, Loos P, Berchtold H, Balbach S, Eichinger T, Lehr CM, Schaefer UF, Windbergs M. Characterization and evaluation of a modified PVPA barrier in comparison to CaCO-2 cell monolayers for combined dissolution and permeation testing. Journal of Controlled Release. 2014;**175**:79-86

[181] Bennet M, Akiva A, Faivre D, Malkinson G, Yaniv K, Abdelilah-Seyfried S, Fratzl P, Masic A. Simultaneous Raman microspectroscopy and fluorescence imaging of bone mineralization in living Zebrafish Larvae. Biophysical Journal. 2014;**106**:L17-L19

[182] Noothalapati H, Sasaki T, Kaino T, Kawamukai M, Ando M, Hamaguchi HO, Yamamoto T. Label-free chemical imaging of fungal spore walls by Raman microscopy and multivariate curve resolution analysis. 2016;**6**:27789

[183] Otto C, de Grauw CJ, Duindam JJ, Sijtsema NM, Greve J. Applications of micro-Raman imaging in biomedical research. Journal of Raman Specroscopy 1997;**28**:143-150

[184] Timlin JA, Carden A, Morris MD, Bonadio JF, Hoffler CE, Kozloff K, Goldstein SA. In Spatial distribution of phosphate species in mature and newly generated mammalian bone by hyperspectral Raman imaging. Journal of Biomedical Optics—SPIE Digital Library. 1999;**4**:28-34

[185] Timlin JA, Carden A, Morris MD, Rajachar RM, Kohn DH. Raman spectroscopic imaging markers for fatigue-related microdamage in bovine bone. Analytical Chemistry. 2000;**72**:2229-2236

[186] Graef F, Vukosavljevic B, Michel JP, Wirth M, Ries O, De Rossi C, Windbergs M, Rosilio V, Ducho C, Gordon S, Lehr CM. The bacterial cell envelope as delimiter of anti-infective bioavailability—An in vitro permeation model of the Gram-negative bacterial inner membrane. Journal of Controlled Release. 2016;**243**:214-224

[187] El-Mashtoly SF, Petersen D, Yosef HK, Mosig A, Reinacher-Schick A, Kötting C, Gerwert K. Label-free imaging of drug distribution and metabolism in colon cancer cells by Raman microscopy. Analyst. 2014;**139**:1155-1161

[188] Klein K, Gigler AM, Aschenbrenner T, Monetti R, Bunk W, Jamitzky F, Morfill G, Stark RW, Schlegel J. Label-free live-cell imaging with confocal Raman microscopy. Biophysical Journal. 2012;**102**:360-368

[189] Mavarani L, Petersen D, El-Mashtoly SF, Mosig A, Tannapfel A, Kötting C, Gerwert K. Spectral histopathology of colon cancer tissue sections by Raman imaging with 532 nm excitation provides label free annotation of lymphocytes, erythrocytes and proliferating nuclei of cancer cells. Analyst. 2013;**138**:4035-4039

[190] Chan JW, Taylor DS, Zwerdling T, Lane SM, Ihara K, Huser T. Micro-Raman spectroscopy detects individual neoplastic and normal hematopoietic cells. Biophysical Journal. 2006;**90**:648-656

[191] Matthäus C, Chernenko T, Newmark JA, Warner CM, Diem M. Label-free detection of mitochondrial distribution in cells by nonresonant Raman microspectroscopy. Biophysical Journal. 2007;**93**:668-673

[192] Baranska M, Schulz H, Baranski R, Nothnagel T, Christensen LP. In Situ simultaneous analysis of polyacetylenes, carotenoids and polysaccharides in carrot roots. Journal of Agricultural and Food Chemistry. 2005;**53**:6565-6571

[193] Larmour IA, Faulds K, Graham D. Rapid Raman mapping for chocolate analysis. Analytical Methods. 2010;**2**:1230-1232

[194] Roeffaers MB, Zhang X, Freudiger CW, Saar BG, van Ruijven M, van Dalen G, Xiao C, Xie XS. Label-free imaging of biomolecules in food products using stimulated Raman microscopy. Journal of Biomedical Optics 2010;**15**:066016

[195] Nordberg M, Åkeson M, Östmark H, Carlsson TE. In development of particle standards for testing detection systems: Mass of RDX and particle size distribution of composition 4 residues, Proc. SPIE. 1994;**8017**:i80171B

[196] Priore RJ, Olkhovyk O, Drauch A, Treado P, Kim M, Chao K. Recent advances in chemical imaging technology for the detection of contaminants for food safety and security. Proceedings of SPIE. 1994;**7315**:731507

[197] Gresham GL, Davies JP, Goodrich LD, Blackwood LG, Liu BYH, Thimsen D, Yoo SH, Hallowell SF. Development of particle standards for testing detection systems: Mass of RDX and particle size distribution of composition 4 residues. Proceedings of SPIE. 1994;**2276**

[198] Stampfer C, Wirtz L, Jungen A, Graf D, Molitor F, Hierold C, Ensslin K. Raman imaging of doping domains in graphene on SiO_2. Applied Physics Letters. 2007;**91**:241907

[199] Sharma DKE, Ramana V, Fateixa S, Hortigüela MJ, Otero-Irurueta G, Nogueira HIS, Kholkin A. Pressure-dependent large area synthesis and electronic structure of MoS_2. Materials Research Bulletin. 2018;**97**:265-271

[200] Lopes J, Estrada A, Fateixa S, Ferro M, Trindade T. A general route for growing metal sulfides onto graphene oxide and exfoliated graphite oxide. Nanomaterials. 2017;**7**:245

[201] Sharma DK, Fateixa S, Hortigüela MJ, Vidyasagar R, Otero-Irurueta G, Nogueira HIS, KumarSingh M, Kholkin A. Defect concentration in nitrogen-doped graphene grown on Cu substrate: A thickness effect. Physica B: Condensed Matter. 2017;**513**:62-68

[202] Graf D, Molitor F, Ensslin K, Stampfer C, Jungen A, Hierold C, Wirtz L. Spatially resolved Raman spectroscopy of single- and few-layer graphene. Nano Letters. 2007;**7**:238-242

[203] El-Mashtoly SF, Niedieker D, Petersen D, Krauss SD, Freier E, Maghnouj A, Mosig A, Hahn S, Kötting C, Gerwert K. Automated identification of subcellular organelles by coherent anti-stokes Raman scattering. Biophysical Journal. 2014;**106**:1910-1920

[204] Zumbusch A, Holtom GR, Xie XS. Three-dimensional vibrational imaging by coherent anti-Stokes Raman scattering. Physical Review Letters. 1999;**82**:4142-4145

[205] Conor LE, Xie XS. Coherent anti-Stokes Raman scattering microscopy: Chemical imaging for biology and medicine. Annual Review of Analytical Chemistry. 1999;**1**:883-909

[206] Nan X, Cheng J-X, Xie XS. Vibrational imaging of lipid droplets in live fibroblast cells with coherent anti-Stokes Raman scattering microscopy. Journal of Lipid Research. 2003;**44**:2202-2208

[207] Evans CL, Xu X, Kesari S, Xie XS, Wong ST, Young GS. Chemically-selective imaging of brain structures with CARS microscopy. Optics Express. 2007;**15**:12076-12087

[208] Lahr RH, Vikesland PJ. Surface-enhanced Raman spectroscopy (SERS) cellular imaging of intracellulary biosynthesized gold nanoparticles. ACS Sustainable Chemical Engineer. 2014;**2**:1599-1608

[209] Firkala T, Farkas A, Vajna B, Farkas I, Marosi G. Investigation of drug distribution in tablets using surface enhanced Raman chemical imaging. Journal of Pharmaceutical and Biomedical Analysis. 2013;**76**:145-151

[210] De Bleye C, Sacré PY, Dumont E, Netchacovitch L, Chavez PF, Piel G, Lebrun P, Hubert P, Ziemons E. Development of a quantitative approach using surface-enhanced Raman chemical imaging: First step for the determination of an impurity in a pharmaceutical model. Journal of Pharmaceutical and Biomedical Analysis. 2014;**90**:111-118

[211] Kumar N, Mignuzzi S, Su W, Roy D. Tip-enhanced Raman spectroscopy: principles and applications. EPJ Techniques and Instrumentation. 2015;**2**:9

[212] Saito Y, Verma P, Masui K, Inouye Y, Kawata S. Nano-scale analysis of graphene layers by tip-enhanced near-field Raman spectroscopy. Journal of Raman Specroscopy. 2009;**40**:1434-1440

[213] Snitka V, Rodrigues RD, Lendraitis V. Novel gold cantilever for nano-Raman spectroscopy of graphene. Microelectronic Engineering. 2011;**88**:2759-2762

[214] Blum C, Schmid T, Opilik L, Weidmann S, Fagerer SR, Zenobi R. Understanding tip-enhanced Raman spectra of biological molecules: A combined Raman SERS and TERS study. Journal of Raman Specroscopy. 2012;**43**:1895-1904

[215] Opilik L, Bauer T, Schmid T, Stadler J, Zenobi R. Nanoscale chemical imaging of segregated lipid domains using tip-enhanced Raman spectroscopy. Physical Chemistry Chemical Physics. 2011;**13**:9978-9981

[216] Kneipp J, Kneipp H, Rajadurai A, Redmond RW, Kneipp K. Optical probing and imaging of live cells using SERS labels. Journal of Raman Specroscopy. 2009;**40**:1-5

[217] Chao J-I, Perevedentseva E, Chung PH, Liu KK, Cheng CY, Chang CC, Cheng CL. Nanometer-sized diamond particle as a probe for biolabeling. Biophysical Journal. 2007;**93**:2199-2208

[218] Henry A-I, Sharma B, Cardinal MF, Kurouski D, Van Duyne RP. Surface-enhanced raman spectroscopy biosensing: In vivo diagnostics and multimodal imaging. Analytical Chemistry. 2016;**88**:6638-6647

[219] Bodelón G, Montes-García V, Fernández-López C, Pastoriza-Santos I, Pérez-Juste J, Liz-Marzán LM. Bioimaging: Au@pNIPAM SERRS ags for multiplex immunophenotyping cellular receptors and imaging tumor cells. Small. 2015;**11**:4220-4220

[220] Mallia R, Veilleux I, Wilson BC, McVeigh P. Filter-based method for background removal in high-sensitivity wide-field-surface-enhanced Raman scattering imaging in vivo. Journal of Biomedical Optics. 2012;**17**:076017

[221] Keren S, Zavaleta C, Cheng Z, de la Zerda A, Gheysens O, Gambhir SS. Noninvasive molecular imaging of small living subjects using Raman spectroscopy. Proceedings of the National Academy of Sciences. 2008;**105**:5844-5849

Chapter 7

SERS-Based Sensitive Detection of Organophosphorus Nerve Agents

Qian Zhao, Guangqiang Liu and Weiping Cai

Additional information is available at the end of the chapter

http://dx.doi.org/10.5772/intechopen.72630

Abstract

Organophosphorus nerve agents, such as sarin, tabun, cyclosarin and soman, belong to the most toxic substances. So, it is very important to quickly detect it in trace-level and on-site or portable way. But, both fast and trace detections have been expected because current techniques are of low sensitivity or of poor selectivity and are time-consuming. The surface-enhanced Raman scattering (SERS)-based detection could be a suitable and effective method. However, the organophosphorus nerve agents only very weakly interact with highly SERS-activated noble metal substrates and are hardly adsorbed on them. In this case, it is difficult to detect such molecules, with reproducible or quantitative measurements and trace level, by the normal SERS technique. Recently, there have been some works on the SERS-based detection of the organophosphorus molecules. In this chapter, we introduce the main progresses in this field, including (1) the thin water film confinement and evaporation concentrating strategy and (2) the surface modification and amidation reaction. These works provide new ways for highly efficient SERS-based detection of the organophosphorus nerve agents and some other target molecules that weakly interact with the coin metal substrates.

Keywords: SERS-based sensitive detection, organophosphorus nerve agents, thin water film confinement, concentrating-enhanced Raman scattering effect, surface modification, amidation reaction

1. Introduction

Organophosphorus nerve agents (such as sarin, tabun, cyclosarin and soman, etc.) belong to the high-risk chemicals with strong poison [1]. When a person is exposed to such nerve agents, sarin for example, with 1.43 ppm, death would occur in few minutes if the agent is inhaled through his/her respiratory system. Even if the nerve agent enters the body through the skin or through consumption, poisoning would occur in few hours [2]. It is thus vital to fast detect them in highly sensitive and portable way. There have existed some methods for the detection of these organophosphorus molecules, such as gas chromatography coupled to a mass spectrometer, ion mobility time-of-flight mass spectrometry, gas sensing and microcantilever, and so on [2–5]. But, these current techniques are of low sensitivity or of poor selectivity and are time-consuming. Both quick and trace detection of them have been expected and are most challenging.

The detection based on surface-enhanced Raman scattering (SERS) effect, which has been extensively reported since its discovery in the 1970s [6–14], would be an appropriate and effective method for fast and ultrasensitive detection. The SERS effect originates from a significant enhancement in the effective Raman cross-section of the target molecules situated at or very near to the roughened noble metal surfaces or colloidal particles [15–25]. The detection based on the SERS effect has the characteristics of high sensitivity, fast response and fingerprint recognition with the ability to be close to a single molecule level [26–29]. The main contribution of amplification of Raman signal intensity arises from the local electromagnetic field enhancement due to the surface plasmon resonance (SPR) of the metal nanoparticles, which has been extensively reported [15–21]. In a conventional SERS detection technique, the target molecules need to stay on the "hot spots" or within the strong electromagnetic field enhancement areas above the SERS substrates [30, 31]. Traditionally, the SERS-based detection is limited to the target molecules, which have high affinity with the metal (gold or silver, etc.) surfaces. It is thus a prerequisite that the substrates can capture or adsorb the target molecules within the strong field-enhancement areas or on the hot spots for the SERS-based detection. However, the organophosphorus nerve agents (including sarin, cyclosarin and soman, etc.) have only very weak interaction (or even no affinity) with highly SERS-activated noble metal substrates and are hardly adsorbed on the substrates or have only very short residence time on them. Obviously, in this case, the normal SERS-based technique is difficult to realize the effective detection of such molecules [13, 32]. For instance, the organophosphorus molecule dimethyl methylphosphonate (DMMP, a typical sarin simulant agent), which is hardly adsorbed on the noble metal surfaces [33], is very difficult to be detected by the SERS-based technique with reproducible measurements and trace level [33, 34]. This is the reason why there have only been very limited reports on the SERS-based detection of the organophosphorus molecules that could only weakly interact with the noble metals [35, 36].

The SERS substrates after surface modification could selectively adsorb and enrich such molecules on their surfaces. However, it is generally difficult to obtain proper modifying agents for the given target molecules and to realize the reproducible measurements and quantitative detection of them without interfering effect. Besides, for the organic modifying agents, they may induce the complicacy of Raman spectral pattern and hence misidentification. Recently, there have been some new approaches developed for the SERS-based detection of the organophosphorus molecules. In this chapter, we introduce the progresses in this field, mainly including (1) the thin water film confinement and evaporation concentrating strategy and (2) the surface modification of the SERS substrates and amidation reaction. These works have provided new ways for highly efficient SERS-based detection of the organophosphorus nerve agents and some other target molecules that weakly interact with the coin metal substrates.

2. Thin water film confinement and evaporation concentrating strategy

For the SERS-based detection of the organophosphorus nerve agents that can only weakly interact with the coin metal substrates, a new and effective route has been developed to capture the hydrosoluble organophosphorus molecules based on the thin water film confinement and evaporation concentrating strategy [33]. DMMP, which is very difficult to be captured by the noble metal substrates, was used as the target molecule, and the gold micro-/nanostructured array was employed as the SERS substrate to demonstrate the validity of this strategy. It has been confirmed that by such strategy, the noble metal SERS substrate can effectively capture the DMMP molecules, realizing SERS-based trace detection of them.

2.1. Strategy and model

2.1.1. Strategy

Normally, in the conventional detections based on the SERS effect, the SERS substrates are firstly soaked in the solution containing target molecules for a certain duration to make the molecules adsorbed on the substrate's surfaces, and then taken out and dried before spectral measurement. If the target molecules can only weakly interact with the substrates, however, such procedures would be of no avail because of no or too less molecules on the substrate's surface after drying. Here, a water film confinement and evaporation concentrating strategy could overcome the abovementioned problem, as schematically illustrated in **Figure 1**.

First, the aqueous solution containing target molecules is dropped onto a SERS-active substrate with hydrophilic surface (**Figure 1a**), which will then spread out and form a water film on the surface. The target molecules (or the solute molecules) are accordingly confined within the thin water film (see **Figure 1b**). The subsequent solvent (water) evaporation induces the thinner and thinner water film, and continuous concentrating of the solutes within the film, assuming that solute volatilization is insignificant or negligible compared with the water evaporation (see **Figure 1c**). When the water film decreases in thickness down to the nanometer level, all the target molecules confined in the water film are localized within the region above the substrate, within which the electromagnetic field can be enormously enhanced under external excitation (see **Figure 1d**). If Raman spectra are measured for this film at this moment, the Raman signal of the target molecules should be enhanced by both the concentrated solutes and the substrate surface. After complete drying, no target molecules will be left on the substrate due to their weak interaction with the substrate and hence we cannot obtain their Raman signals.

2.1.2. Concentrating factor

For quantitative analysis, the evaporation-induced concentrating factor (*CF*) of the target molecules in the water film is defined as the ratio of solute concentration (C_m) in the water film at the moment of Raman spectral measurement to that at the initial stage (C_0), or

$$CF = \frac{C_m}{C_0} \tag{1}$$

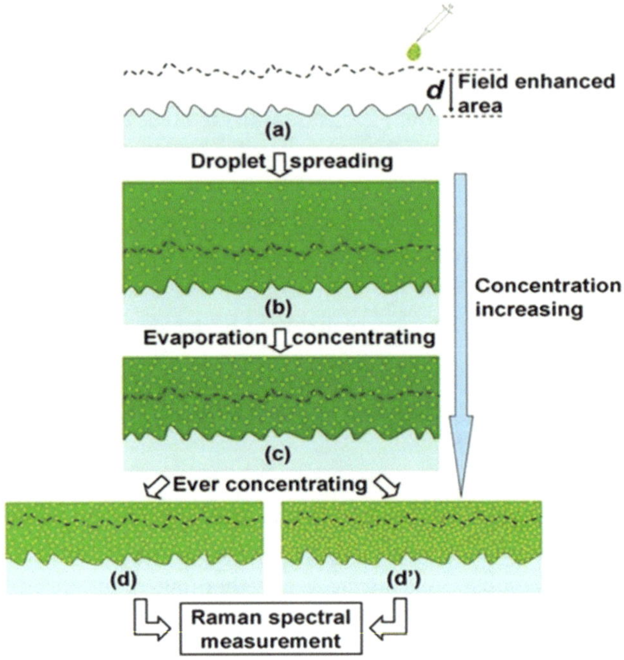

Figure 1. The schematic illustration for the water film confinement and evaporation-concentrating process. (a) A clean SERS substrate. (b) A liquid film formed on the substrate after dropping a certain amount of the solution on it. (c) The liquid film is decreased in thickness and solute is concentrated due to solvent's evaporation. (d) and (d') The liquid film is comparable in thickness to the field-enhanced space for the substrates with hydrophilic and hydrophobic surfaces, respectively [33].

Based on the above described strategy (or in **Figure 1**), if the substrate surface is hydrophilic, the concentrating factor can be approximately written as

$$CF = (1 - S) \cdot \frac{H_0}{H_m} \tag{2}$$

in which

$$S = \frac{M_0 - M_m}{M_0} \quad \text{and} \quad H_m \geq d \tag{3}$$

where $\boldsymbol{H_0}$ is the initial thickness of water film after the water droplet is spreading out on the substrate, $\boldsymbol{H_m}$ is the corresponding value when the Raman spectral measurement is performed, $\boldsymbol{d}$ is the distance (in nanometer scale) from the substrate surface and within which the electromagnetic field can be significantly enhanced during laser excitation, S is the loss-ratio of the target molecules in the water film due to the possible volatilization during solvent evaporation and $\boldsymbol{M_0}$ and $\boldsymbol{M_m}$ are the mole numbers of the target molecules in the water film at the initial and the spectral measurement stages, respectively.

Generally, compared with the evaporation of the solvent, volatilization of the molecules like DMMP in solutions is significantly slower due to the much heavier molecular weight than water and could be ignorable. For better understanding, here, let us semiquantitatively evaluate the ***CF*** value. If letting $S \approx 0$ and $H_0 = 1$ mm, we can obtain $CF = 10^5$ when $H_m = 10$ nm, from Eq. (2). Such values should be credible in the order of magnitude. So, the solvent evaporation-induced concentrating effect is quite significant.

In addition, if the water film is on the SERS substrate with hydrophobic surface and reduced to an enough thin thickness due to the evaporation, it could horizontally shrink and decrease the coverage area on the substrate, which would induce further concentrating. Therefore, the *CF* value would be higher in the local water films than that for the substrate with hydrophilic surface, as shown in **Figure 1(d)′**. In other words, the substrate with hydrophobic surface should be of better concentrating effect.

Furthermore, according to Eq. (2), the strategy shown in **Figure 1** should be more effective for the target molecules with lower volatility and/or when the target molecule concentration in the initial aqueous solution is very low due to the evaporation-induced concentrating effect.

2.1.3. *Effects of the thin water film*

In this strategy, the thin water film would function as follows: (1) Limitation of the target molecules to the small region above the SERS substrates. When the water film is becoming very thin in thickness, the molecules are localized in the field-enhanced space although they are not adsorbed on the substrate's surface; (2) Enrichment of the target molecules. Solvent's evaporation will induce the concentrating or enrichment of the target molecules in the water film and the increase of the target molecules' number in the field-enhanced space above the substrate; (3) Decrease of laser-induced thermal effect. The water film can protect the target molecules from laser-induced damage. So, the Raman intensities can also be enhanced by increasing the laser excitation power and (4) The evaporation-induced reorientation of the target molecules. During solvent evaporation, the target molecules could re-orientate [37]. This would generate a significant Raman signal enhancement due to the charge-coupling between the molecules and the metallic surface.

2.2. Application in SERS-based detection of DMMP

Here, the gold hierarchically micro-/nanostructured bowl-like array was chosen as the SERS substrate and the DMMP as the target molecules to demonstrate the validity of the above thin water film confinement and evaporation concentrating strategy.

2.2.1. *Surface morphology and wettability of the SERS active substrate*

The fabrication process of the SERS substrate was prepared by the electrodeposition on a preformed monolayer polystyrene (PS) colloidal crystal (2 μm in PS sphere-diameter) in the $HAuCl_4$ aqueous solution at room temperature, as previously described. [38]. **Figure 2(a)** shows the typical morphology. The SERS substrate is the gold array consisting of the hexagonally arranged bowl-like pores with 2 μm in period. The skeleton among the pores in the array is built of nearly vertical quasi rod-shaped nanoparticles. The static contact angle of such substrate surface is about 105°, exhibiting the slightly hydrophobic surface (see **Figure 2b**).

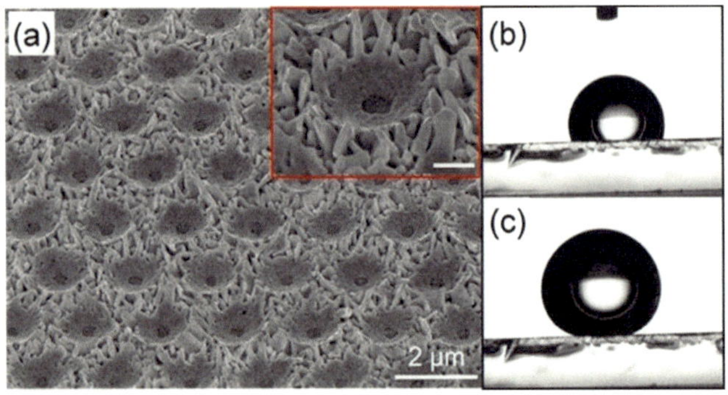

Figure 2. The morphology and wettability of a gold micro-/nanostructured array. (a) The scanning electron microscopic (SEM) image. The inset: A local magnified image and the scale bar = 500 nm. (b) and (c) The photos of the water droplets on the Au array before and after surface modification with thiol, respectively. The water droplets are 2 and 5 μL in volume for (b) and (c), respectively [33].

2.2.2. Raman spectral measurements

The DMMP aqueous solutions with different concentrations were firstly prepared and stored in a refrigerator before use. A droplet of the DMMP aqueous solution was then dropped on the SERS substrate or the gold array and spread out to form the thin water film. After evaporation at room temperature for different time intervals, the Raman spectra for the thin water film were measured on a confocal microprobe Raman spectrometer (Renishaw inVia Reflex) with a laser beam of 632.8 nm in wavelength. The Raman spectral integral time is 10 s.

Raman spectral peaks and identification. When a droplet of aqueous solution, with 20 μL in volume and 10^{-2} M DMMP in initial content, was dropped on the Au bowl-like array without surface modification, it spread out and formed a water film with about 3 mm in thickness. During the initial evaporation, no distinct Raman signal was detected for such water film on the substrate, except the background [see curve (I) in **Figure 3(a)**]. After evaporation for sufficiently long time (but without completely drying), however, a significant Raman spectrum could be observed. Typically, curve (II) in **Figure 3** shows the result corresponding to the evaporation for ~300 s after starting timing at which the distinct signals are very weak but detectable, in this experimental condition (see next subsection). At this moment, the water film was enough thin in thickness. There are one dominant Raman peak at 2936 cm^{-1} and the other three smaller peaks at 2861, 2966 and 3006 cm^{-1}, respectively. For reference, the Raman spectrum of the pure DMMP (liquid state in a quartz cell) was shown in curve (IV) of **Figure 3**, which is in good agreement with the previous report [34]. By comparison between curves (II) and (IV), there are about five wavenumbers lower for the water film than the pure DMMP. These four peaks are assigned to the C—H stretching modes of DMMP molecules [39]. Taking into consideration of the possible influence of water on the DMMP, which leads to a slight peak shift, the peaks in curve (II) of **Figure 3** can thus be attributed to the DMMP in the water film. It is well known that DMMP molecule is weak Raman scatterer with a very small cross-section on the order of 1 × 10–30 cm^2 under 514.5 nm excitation for the strongest line [40]. Since it is difficult to adsorb on the SERS active metal substrates, the characteristic vibration

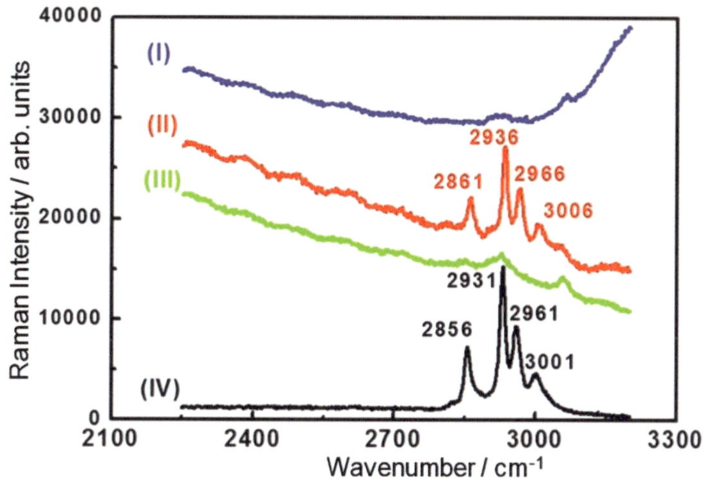

Figure 3. The Raman spectra of DMMP molecules. Curves (I), (II) and (III) are for a droplet of 10^{-2} M DMMP aqueous solution on the SERS substrate after evaporation for a very short time, a certain time and sufficiently long time (completely dried), respectively. Curve (IV): The Raman spectrum of the pure DMMP liquid. [33].

peaks of DMMP molecules in the aqueous solution are usually difficult to be detected by the SERS effect. No characteristic Raman peak from DMMP was observed after the water film was completely evaporated (see curve (III) in **Figure 3**). Obviously, such method is quite valid to detect them.

Spectral evolution with the evaporation. Further, the Raman spectral evolution of the DMMP-contained water film on the substrate with the evaporation time was measured. **Figure 4(a)** shows the spectra after evaporation for different durations for the water film with 10^{-2} M DMMP in initial concentration. During initial stage of the evaporation, no characteristic Raman signals were detected and the film is opaque in the field of optical microscope, as shown in **Figure 4(b)**. With the evaporation, the water film gets thinner and thinner. When evaporation is for t_0, which varies from several minutes to few 10 min, depending on the ambient conditions (temperature and relative humidity), the water film on the substrate was slightly transparent, and the characteristic peaks at 2936 and 2966 cm^{-1} are very weak but detectable (see curve 1) in **Figure 4(a)**. Timing begins at this moment. The further evaporation leads to appearance of the other two characteristic peaks (at 2861 and 3006 cm^{-1}) and continual enhancement of the four Raman peaks [see curves 2–6 in **Figure 4(a)**]. When the evaporation duration was up to 300 s after timing, the intensity of the characteristic Raman peaks were enhanced to the maximum, as illustrated in curve 7 in **Figure 4(a)**. The corresponding photo of the water film is given in **Figure 4 (c)**, showing translucency or semitransparence. The gold pattern on the substrate looms up. However, the longer the evaporation induced, the rapid is the decrease of the Raman signals and vanishing within ~20 s, corresponding to complete evaporation or drying. At this moment, the gold array on the substrate is clearly seen [**Figure 4(d)**]. Representatively, **Figure 5(a)** shows intensity of the dominant peak at 2936 cm^{-1} as a function of evaporation time [the data from **Figure 4(a)**] and clearly shows such evolution of the peak intensity.

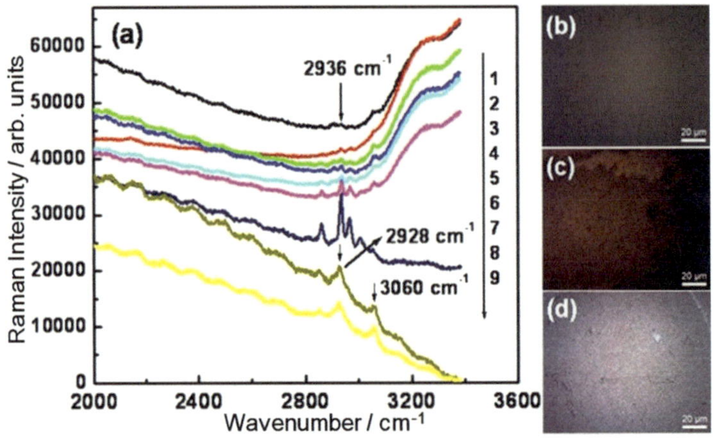

Figure 4. Raman spectra and photos of the gold array with a droplet of 10^{-2} M DMMP aqueous solution during evaporation. (a) The spectra after evaporation for different intervals. Curves 1–9 correspond to the interval Δt = 0, 112, 155, 199, 243, 279, 301, 323 and 348 s, respectively. (b)–(d) The photos of the gold array covered with water film in the initial stage, microsized-thickness and completely dried stage, respectively [33].

2.3. Concentrating-enhanced Raman scattering (CERS) effect

2.3.1. *A quantitative description*

The above successful observations of Raman characteristic peaks for the DMMP molecules in an aqueous solution are easily understood. This is mainly attributed to the thin water film confinement and subsequent evaporation-induced DMMP concentrating or enrichment, in addition to the electromagnetic enhancement mechanism from the Au bowl-like array. Because of the space confinement of the thin film, water evaporation induces the concentrating of DMMP within the thinner and thinner film. That is to say, more and more DMMP molecules in the water film are confined within the region with significant local electromagnetic field enhancement above the substrate, exhibiting ever-increasing Raman signal with the increasing evaporation duration, which we could call the concentrating-enhanced Raman scattering (CERS) effect, as demonstrated in **Figure 5(a)**. Obviously, after the water film is completely evaporated, the confinement effect thus vanishes. At this moment, the DMMP molecules cannot stay on the substrate due to the weak interaction, and the corresponding Raman peaks disappear. Besides, the evaporation-induced reorientation of the DMMP molecules could also induce an additional enhancement of the Raman signal owing to the charge-coupling between the molecules and the metallic surface [37].

According to the evolution of the Raman intensity with the evaporation duration shown in **Figure 5(a)**, we could semiquantitatively describe the concentrating kinetics of DMMP molecules in the water film during evaporation. First, under a given ambient condition (temperature and humidity) and a certain volume of solution droplet on the substrate with hydrophilic surface, we have the water film thickness (H) as a function of the water evaporation duration $\boldsymbol{t}$:

$$H = H_0 - V \cdot t \quad (t < t_S) \tag{4}$$

in which

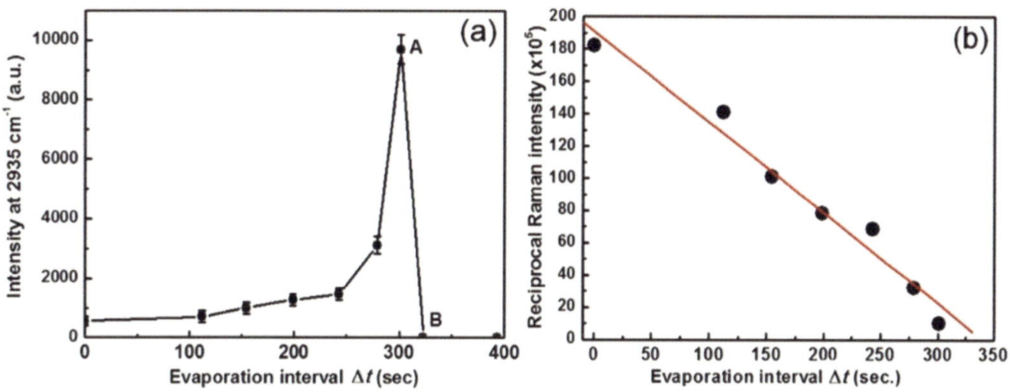

Figure 5. (a) Intensity of the Raman peak at 2936 cm^{-1} versus evaporation duration [data from **Figure 4(a)**]. Point A: The maximal measured value. Point B: The peak vanishes. (b) Plot of the reciprocal Raman intensity versus evaporation duration [data from (a)]. The solid line is the linear fitting result [33].

$$t_S = \frac{H_0}{V} \tag{5}$$

where $\boldsymbol{V}$ is the water evaporation speed ($m{\cdot}s^{-1}$ in dimension) and should be a constant independent of t, and $\boldsymbol{t_s}$ is the duration from the initial to complete evaporation.

During evaporation of solvent (water), volatilization of the solute or DMMP in water film inevitably takes place. Here, it can be assumed that its volatile rate is directly proportional to its concentration $\boldsymbol{C}$ in the water film, or

$$F = p \cdot C \tag{6}$$

where the volatile rate $\boldsymbol{F}$ is defined as the mole number volatilized in unit time and area ($mol\ s^{-1}\ m^{-2}$) and the parameter $\boldsymbol{p}$ is the proportional constant. So, after evaporation for $\boldsymbol{t}$, from Eq. (6), the volatile-loss ΔM of the solute (DMMP) in the film can be described by

$$\Delta M = A \cdot p \int_0^t C \cdot dt \tag{7}$$

where $\boldsymbol{A}$ is the surface area of the water film. Finally, from Eq. (7), we can quantitatively establish the relation between the solute concentration and evaporation duration of the water film, or

$$C = \frac{C_0 \cdot A \cdot H_0 - \Delta M}{A \cdot H} = \frac{C_0 \cdot H_0 - p \int_0^t C \cdot dt}{H_0 - V \cdot t} \tag{8}$$

Obviously, if $\boldsymbol{p} << 1$ or the solute is nearly involatile in the water, and/or the evaporation speed V value is relatively large enough, the volatile loss $\Delta M \approx 0$. In this case, Eq. (8) can be approximately written as

$$C \approx \frac{C_0 \cdot H_0}{H_0 - V \cdot t} \tag{9}$$

or

$$\frac{1}{C} \approx \frac{1}{C_0} - \frac{V}{C_0 H_0} \cdot (t_0 + \Delta t) \quad t = t_0 + \Delta t \tag{10}$$

where Δt is the evaporation interval after starting timing at $\boldsymbol{t_0}$ (as mentioned earlier). It means that the reciprocal solute concentration in the water film is nearly directly proportional to the evaporation duration of water. Eqs. (8) and (9) clearly indicate concentrating of the solute during evaporation.

When the solute concentration (**C**) in the water film is low enough, intensity (***I***) of its Raman-shift peaks should be directly proportional to **C**. Here, DMMP is of good water solubility and much heavier molecular weight (124.08) than water [41]. Therefore, Eq. (9) could be a good description of the solute concentration evolution during water evaporation. In other words, from Eq. (10), the reciprocal intensity of the characteristic Raman peaks should be of nearly linear relation with the evaporation duration (***t*** or Δt) of water, which has been confirmed by further spectral analysis. **Figure 5(b)** shows the results corresponding to the reciprocal Raman intensity of the peak at 2936 cm^{-1} versus the evaporation interval Δt [data from **Figure 5(a)**], showing a good linear relation between them and significant CERS effect. This also implies that the volatile loss of DMMP in water was very small or negligible during water film evaporation at ambient environment. By linear fitting, the plot in **Figure 5(b)** can be described as

$$\frac{1}{I} = 190.88 - 0.5608 \cdot \Delta t \tag{11}$$

in which the intensity ***I*** was multiplied by 10^{-5} and Δt is in second.

2.3.2. Factors influencing CERS effect

The CERS effect mentioned earlier would be influenced by some factors such as SERS active substrates, evaporation conditions and volatility of the solute, and so on. Obviously, the highly SERS active substrates, low solute's volatility and appropriate solvent's evaporation rate would be beneficial to exhibiting significant CERS effect. In addition, further experiments have revealed that the surface wettability of SERS substrate and the laser excitation power are also important to induce the strong CERS effect.

Wettability of SERS substrates surface. As mentioned in Section 2.1.2, the substrate with hydrophobic surface should be of stronger concentrating effect than that with hydrophilic surface. For confirmation, the Au micro-/nanostructured bowl-like array shown in **Figure 2(a)** was surface-modified with thiol. Correspondingly, the static contact angle was increased up to about 160° [**Figure 2(c)**], exhibiting superhydrophobic surface. Using such modified array as SERS substrate, the CERS effect was really significantly enhanced. Typically, **Figure 6** shows the Raman spectrum of DMMP on the gold array with and without surface modification,

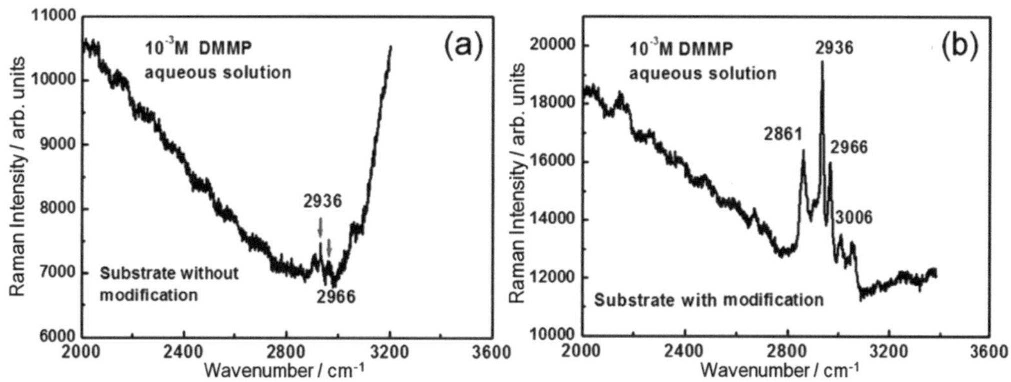

Figure 6. Raman spectra of 10^{-3} M DMMP aqueous solution on the substrates without (a) and with (b) surface modification after an optimal evaporation (the details are given in the text) [33].

measured after evaporation for an optimal duration, for the water film with 10^{-3} M DMMP in initial concentration. For the substrate without modification, the characteristic peaks of DMMP are very weak [**Figure 6(a)**]. However, the surface modification induced much stronger Raman-shift peaks of DMMP [**Figure 6(b)**], and the intensity of the dominant peak at 2936 cm^{-1} was one order of magnitude (~12 times) higher than that without modification. Such Raman enhancement should be mainly attributed to the superhydrophobic surface-induced much higher *CF* value, which led to stronger CERS effect.

Evaporation conditions. As mentioned in Section 2.1.1, only when the thickness of the water film is reduced to the nanoscale by evaporation, the concentrating effect, that is, the CERS effect can reach the maximum. However, for the measurements in the normal ambient conditions, as shown in **Figure 4**, the thickness of the water film, corresponding to the maximal measured Raman signals could be estimated to be in micron scale (~ 8–30 μm) according to its initial thickness and the whole time for evaporating the water droplet on the substrate. In the normal ambient conditions, it is too late to measure the Raman spectra when the water film was reduced to the nanoscale in thickness, because such thin water film would be completely evaporated or dried within one millisecond. This is the reason why the Raman peaks disappear immediately after the maximal measured value, as demonstrated in **Figure 5(a)** (points A and B). It means that the maximal measured value could be much lower than the real maximal one. Although we can obtain the *CF* value about 10^3 in order of magnitude when the water film was reduced to micron scale in thickness according to Eq. (2) and enough strong CERS effect, the real maximal or optimal effect is far from reached in the normal ambience.

Obviously, to further increase the CERS effect, we should decrease evaporation speed of the water film, especially, since $t = t_0$ at which the Raman shift peaks are weak but detectable (if the solute volatilization is neglectable). In fact, control of the evaporation speed is easily achieved. For example, we can control the evaporation rate by putting the substrate with water film into a quartz cell with a controllable opening.

Laser excitation power. Generally, the Raman scattering intensity is directly proportional to the excited laser power. But too high laser power would break down the molecules due to the thermal effect, leading to the low Raman signal instead. However, the case here is an exception. The thin water film could protect the target molecules from the laser-induced damage because the water film can remarkably reduce the laser-induced thermal effect, as demonstrated in **Figure 7(a)**, corresponding to the Raman spectra of DMMP in the water film on the substrate, excited with different laser powers under the same evaporation duration. We could use the maximal power (P_{max} = 17 mW) of the equipment in this case, while in the conventional measurement only 5 mW or less is usually used. It has been shown that the intensity of the peak at 2936 cm^{-1} has a good linear relation with the power in whole power range, as indicated in **Figure 7(b)**. The straight line passes through the origin. So, for this thin water film confinement and evaporation concentrating strategy, one can use enough high laser excitation power (>17 mW) to further increase Raman scattering intensity, exhibiting the stronger CERS effect.

2.4. Suitability of the strategy

Based on the abovementioned text, using the thin water film confinement and evaporation concentrating strategy, one can effectively capture the hydrosoluble and weak affinity molecules within the strong electromagnetic field enhanced space above the SERS substrate and realize the SERS-based detection of them. The thin water film not only confines the target molecules within a limited space but also protects the target molecules from laser-induced damage.

It should be mentioned that the hydrophobic substrate surface, slower evaporation and stronger excitation power can further increase CERS effect. Especially, the slow and controlled evaporation in the anaphase would lead to several orders of magnitude in higher CERS effect. The strategy given here is an effective route to the SERS-based detection of the soluble molecules, which are of small Raman scattering cross-section and hardly adsorbed on the SERS substrates, by choosing proper solvents, but not suitable for the volatile soluble molecules as the liquid film cannot confine these molecules.

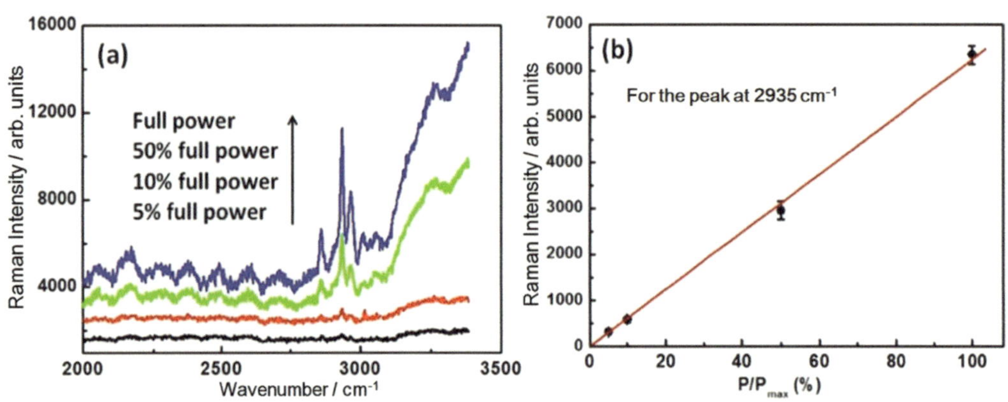

Figure 7. (a) The Raman spectra of DMMP aqueous solution droplet on the substrate, under the excitation with different laser powers (***P***), after the evaporation for the same duration. The maximal excitation power of the equipment P_{max} = 17 mW. (b) The plot of the intensity of the peak at 2936 cm^{-1} versus the laser excitation power [the data are from (a)]. The solid line is the linear fitting results [33].

http://dx.doi.org/10.5772/intechopen.72630

3. The surface modification and amidation reaction

In addition to the abovementioned thin water film confinement strategy, here, we introduce another approach to the SERS-based ultrasensitive detection of metal weakly interacted organophosphorus nerve agent sarin based on surface modification of the SERS substrates and amidation reaction [42]. The methanephosphonic acid (MPA) was chosen as the sarin simulation agent (or the target molecule). The Au-coated Si nanocone array was surface-modified with 2-aminoethanethiol molecules and used as the SERS-substrate for detection of MPA. It has been demonstrated that the modified substrate can selectively capture MPA in the solution under the existence of the coupling agent, and hence realize the SERS-based detection of the MPA in the solution with good selectivity and high sensitivity.

3.1. Surface modification-based SERS detection strategy

3.1.1. Choice of sarin-simulated agent

For convenient study of SERS-based detection of sarin, its simulation agent should be chosen. Such simulation agent should be of less or moderate toxicity but the chemical properties and especially the Raman spectrum should be similar to sarin. It has been found that methanephosphonic acid (MPA) is also a suitable simulation agent for sarin, in addition to the commonly used DMMP. The molecular formula of sarin and MPA are $(CH_3)_2CHOOPF(CH_3)$ and CH_5O_3P, respectively. Both have the C-P bonds and the similar bond length, chemically belonging to the organophosphorus group. Both MPA and sarin can produce amidation reaction with amino compounds [43]. It is expected that these similarities in chemical structures could have similar Raman spectral pattern to each other.

The Raman spectra of sarin and MPA were simulated based on density functional theory (DFT) by means of the Gaussian 09 software [44]. **Figure 8(a)** is the measured Raman spectrum for the pure MPA. The simulated Raman spectrum is very similar in the primary and minor peaks except the small difference in the peak positions, demonstrating the validity of the spectral simulations. **Figure 8(b)** shows the simulated Raman spectrum of sarin. Correspondingly, the

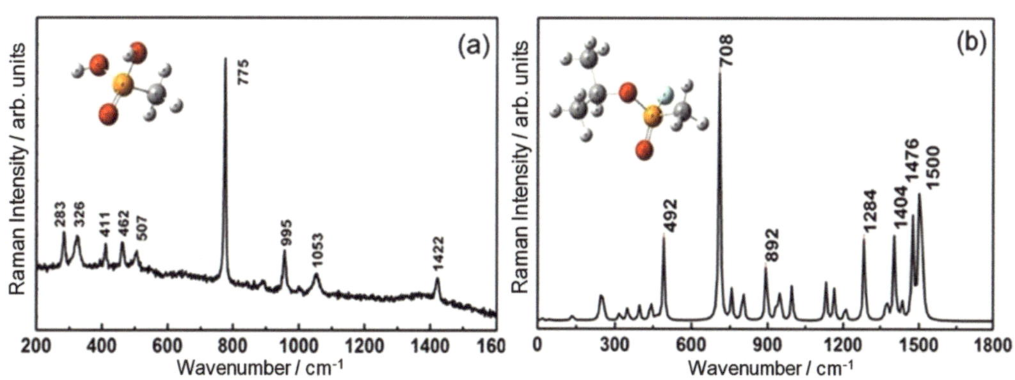

Figure 8. The Raman spectra for MPA (a) and sarin (b). (a) The measured Raman spectrum of pure MPA (excited by 785 nm laser). (b) The simulated Raman spectrum of sarin based on DFT calculations [42].

vibrational peaks can be assigned according to the DFT calculations [42]. The Raman spectral bands are mostly similar in wavenumbers for sarin and MPA. The MPA can thus be used as a sarin-simulated agent.

3.1.2. Surface modification of SERS substrate

Generally, the Raman signals could be detected only when the MPA molecules are adsorbed on the SERS substrates. However, the MPA molecules can hardly be adsorbed on the noble metals, due to the weak interaction between them. The surface modification strategy was used to overcome such problem. A surface modifier should be chosen in such a way that it can strongly interact with both the SERS substrate and MPA molecules.

The 2-aminoethanethiol molecule contains two-head groups such as amino and thiol groups. It is well known that there is a strong covalent bond interaction between thiol and gold according to the theory of hard and soft (Lewis) acids and bases [45, 46]. As for the amino group, it can react with phosphonic group to generate phsophonamidate in the presence of coupling agents [such as dicyclohexylcarbodiimide, N, N-diisopropylcarbodiimide, 1-ethyl-3-(3-(dimethylamino) propyl) carbodimide] [47]. Therefore, the thiol groups in 2-aminoethanethiol molecules would tend to be bound with gold substrate to form Au—S covalent bonds, and the amino groups would selectively capture phosphonic groups in MPA molecules in the solution, as schematically shown in **Figure 9**. 2-Aminoethanethiol could thus be a suitable modifying agent of the SERS substrate. It is expected that the surface-modified SERS substrate would selectively capture the organophosphorus molecules (such as sarin, MPA), as demonstrated in **Figure 9**. In this case, we could realize detection of MPA or sarin based on the SERS effect.

3.2. SERS measurements

3.2.1. Surface-modified SERS substrate

Au-coated Si nanocone array was used as the SERS substrate, which was prepared by sputtering deposition of gold on the Si nanocone array induced by PS colloidal monolayer and plasma etching strategy, as previously described in detail [42]. **Figure 10** shows the Si nanocone array before and after sputtering deposition with a gold layer about 10 nm in thickness. Such array is of good uniformity in structure.

Such Au-coated Si nanocone array was immersed into the ethanol solution of 2-aminoethanethiol (1 mM) for surface modification. FTIR spectral measurement has confirmed that the modified

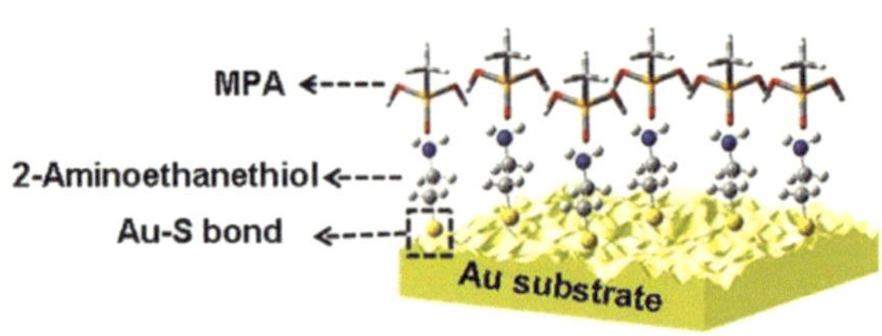

Figure 9. Schematic illustration for the interaction of the modifying agent (2-aminoethanethiol) with MPA molecules and SERS substrate (gold) [42].

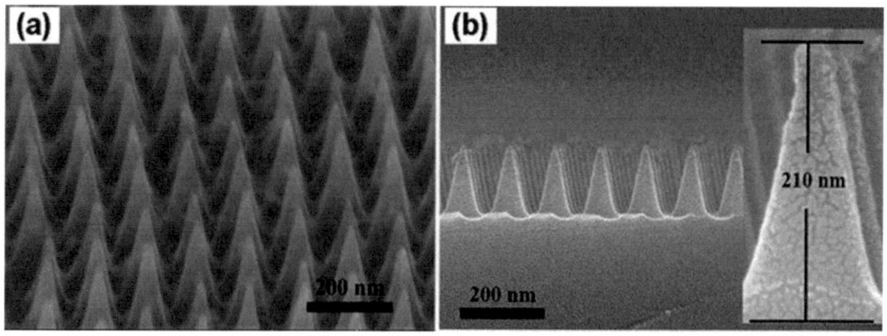

Figure 10. The morphology of the as-prepared nanocone array. (a) The FESEM image of the Si nanocone array induced by plasma etching the PS colloidal monolayer on Si wafer. (b) The cross sectional image after sputtering deposition of gold on the Si nanocone array. The inset: A magnified image of a single Au-coated cone [42].

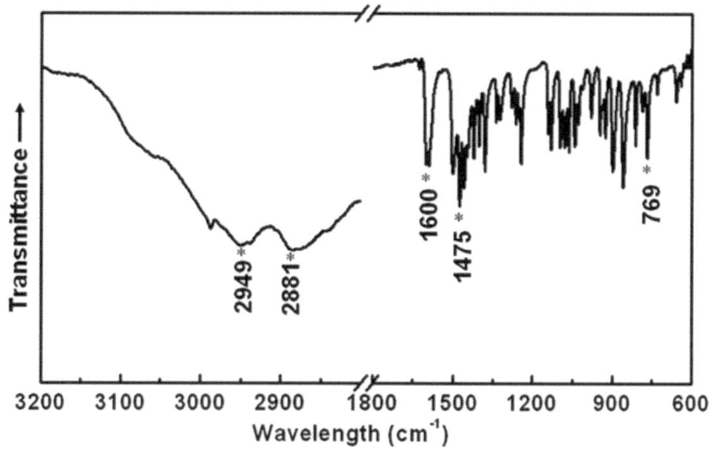

Figure 11. FTIR spectrum of the surface-modified au-coated Si cone array [42].

2-aminoethanethiol molecules were bound with the Au film on the array's surface by the thiol group, as shown in **Figure 11**. The peaks at 2881 and 2949 cm^{-1} are attributed to the symmetry stretching and asymmetry vibration of the CH_2 in 2-aminoethanethiol [48], while the peak at 1600 cm^{-1} is ascribed to the in-plane bending vibration of the NH_2 groups in 2-aminoethanethiol [48]. In addition, the peak at 1475 cm^{-1} originates from the shear vibration of the CH_2 in 2-aminoethanethiol [48]. Furthermore, the peak at 769 cm^{-1} is assigned to swing plane vibration of CH_2 chains, corresponding to the two methylene groups in 2-aminoethanethiol molecules [48]. It could thus be concluded that the modified array's surface was rich of 2-aminoethanethiol molecules.

3.2.2. *Raman spectral measurements*

The modified substrate was then immersed into the ethanol solution of MPA in the presence of the coupling agent 1-ethyl-3-(3-(dimethylamino) propyl) carbodimide (or EDC for short) (2 mM) for 3 h which was long enough to reach the equilibrium adsorption of the MPA on the SERS substrate. Here, the coupling agent EDC was employed to activate phosphonic groups in MPA molecules for coupling with the primary amines in the modifying agent. Finally, the soaked substrate was taken out and cleaned with deionized water and ethanol to remove any unbound molecules, and dried in the flow of N_2 prior to the SERS spectral measurement under excitation at 785 nm and exposure time 10 s.

The Raman spectral pattern. Figure 12 demonstrates the Raman spectrum of the surface-modified Au-coated Si nanocone array after immersion in the MPA solution (10^{-3} M) with EDC. The Raman peaks at 643, 725, 976, 1025, 1212, 1412, and 1445 cm^{-1} are clearly observed, as shown in curve (I) of **Figure 12**. On the contrary, for the surface-modified array without immersion or after immersion in the EDC solution without MPA or in the MPA solution without EDC, no Raman peak was detected, as illustrated in curves (II, III, IV) of **Figure 12**. So the Raman peaks in curve (I) of **Figure 12** should be associated with the coexistence of MPA and EDC. By comparing with **Figure 8(a)**, however, we can know that the Raman spectrum in curve (I) is completely different from that of the pure MPA. It means that the Raman peaks in **Figure 12** are not attributed to MPA directly.

Concentration dependence. Further, the concentration-dependent Raman spectra were measured for the surface-modified SERS substrate (or Au-coated Si nanocone array) after soaking in the MPA solutions with different concentrations in the presence of EDC, as shown in **Figure 13(a)**. The intensities of all Raman peaks increase with the rising MPA concentration in the solutions. The peak intensity I is approximately subject to the linear double logarithmic relation with the MPA concentration $\mathbf{C}$ from 10^{-8} to 10^{-2} M (or ~1 to ~1000 ppm), or

$$LogI = A_0 + B_0 \cdot LogC \tag{12}$$

where A_o and B_0 are the constants independent of the concentration. **Figure 13(b)** shows the typical result corresponding to the main peak at 976 cm^{-1}, exhibiting a good linear relation.

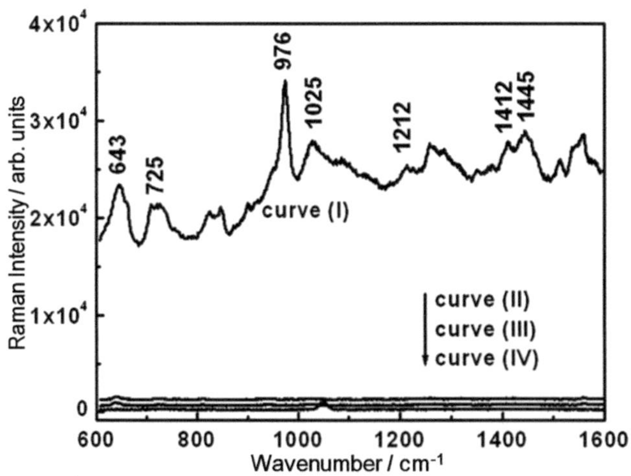

Figure 12. The Raman spectra for the Au-coated Si nanocone array after soaking in different solutions for 3 h. Curve (I): After soaking in the MPA solution (10^{-3} M) with the EDC. Curve (II): The array without immersion; curve (III): After soaking in the EDC solution without MPA; curve (IV): After soaking in the MPA solution without EDC [42].

http://dx.doi.org/10.5772/intechopen.72630

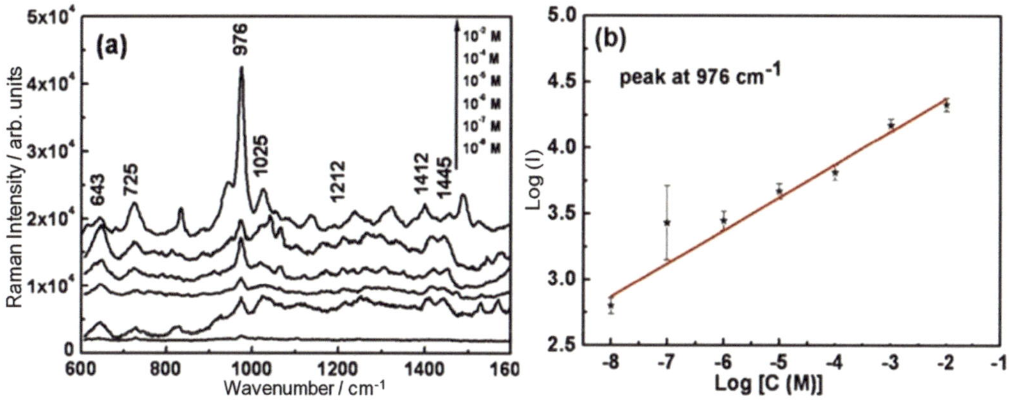

Figure 13. (a) The Raman spectra for the surface-modified Au-coated Si nanocone array after soaking in the MPA solutions with different concentrations in the presence of EDC. (b) Plot of the logarithmic peak intensity (*I*) at 976 cm^{-1} versus the logarithmic MPA concentration (*C*) [the data are from (a)]. The straight line is the linear fitting results [42].

The parameter B_0 value is about 0.25 by fitting. We can thus rewrite Eq. (12) as a power function, or

$$I = M \cdot C^{0.25} \tag{13}$$

where **M** is the constant independent of the concentration **C**.

3.3. Amidation reaction

Regarding the Raman spectral origin and their evolution, it can be attributed to adsorption of MPA molecules and subsequent amidation reaction on the SERS substrate's surface, as schematically illustrated in **Figure 14**.

3.3.1. Amidation reaction-induced Raman spectra

After the coupling agent EDC was added to the ethanol solution of MPA, the coupling between them would activate the phosphonic groups in MPA [see **Figure 14(a)**] [47]. When the surface-modified SERS substrate was subsequently immersed into the MPA solution, because of the strong interaction between the amino groups in 2-aminoethanethiol and the phosphonic groups in the activated MPA molecules [47], the activated MPA molecules would diffuse onto and be adsorbed on the substrate's surface, as illustrated in **Figure 14(b)**.

At this time, the adsorbed MPA molecules could react with the 2-aminoethanethiol molecules on the Au-coated Si cone array due to the amino groups in the 2-aminoethanethiol [47], or the amidation reaction between them

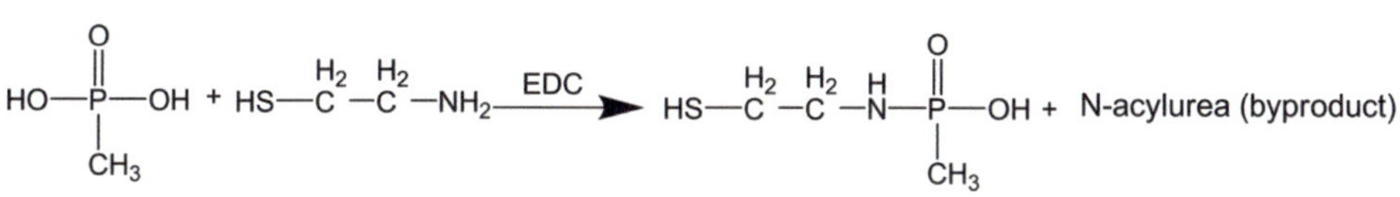

(14)

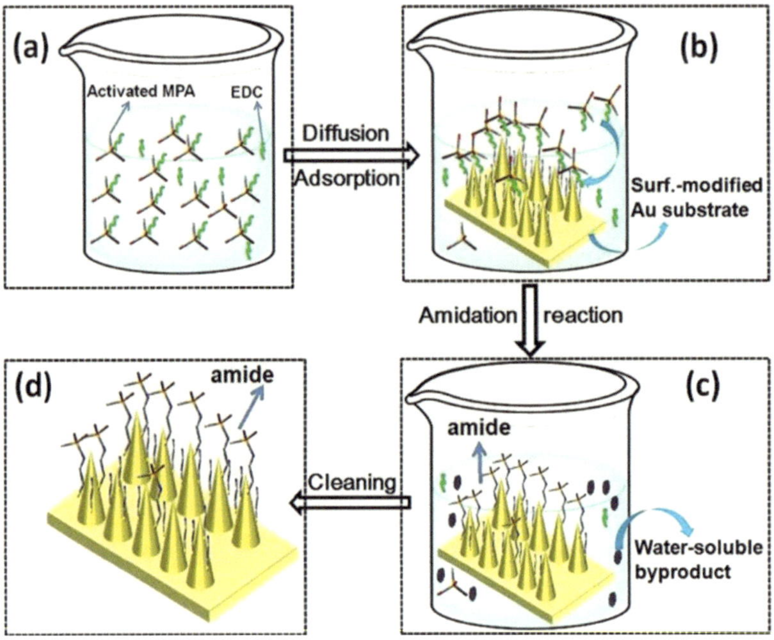

Figure 14. The schematic illustration for the MPA molecules' adsorption and amidation reaction on the Au-coated Si nanocone array. (a) Ethanol solution of the EDC-activated MPA. (b) The activated MPA molecules diffuse into and are adsorbed on the surface-modified array. (c) The amidation reaction on the substrate is finished. (d) The reaction products (organic phosphorus amide molecules) are bound to the substrate [42].

would occur on the surface of the array. The reaction products N-(2-mercaptoethyl)-P-methylphosphonamidic acid $C_3H_{10}O_2NPS$ should be formed and bound or anchor on the surface of the array [see **Figure 14(c)**]. Besides, the byproduct N-acylurea was also produced in the solution. It is water-soluble and removable before Raman spectral measurement [see **Figure 14(d)**]. So, the Raman spectrum shown in curve (I) of **Figure 12** could be ascribed to the product $C_3H_{10}O_2NPS$ on the substrate.

Evidently, the higher MPA content in the solution would induce the more activated MPA molecules adsorbed on the modified Au-coated Si nanocone array, and the more reaction products $C_3H_{10}O_2NPS$ bound on the array. This would result in higher Raman peak intensity, showing increase of the Raman peak intensity with the rising MPA content in the solutions, as illustrated in **Figure 13(a)**.

Quantitatively, as mentioned earlier, the concentration-dependent Raman intensity can be described by a power function [see Eq. (13)]. It should be associated with the adsorption behavior of the MPA molecules on the modified substrate. According to the Freundlich theory [49], the adsorption of molecules on a heterogeneous surface could be described by:

$$q_e = K_F \cdot C^{\frac{1}{n}} \tag{15}$$

where $\boldsymbol{q_e}$ is the equilibrium adsorption amount, the parameters $\boldsymbol{K_F}$ and $\boldsymbol{n}$ are the parameters reflecting the adsorption capacity and adsorption intensity, respectively. Obviously, the intensity $\boldsymbol{I}$ of a Raman peak should be proportional to the number density of the molecules adsorbed on the substrate within the area of a laser spot or show a linear relation with the adsorption amount $\boldsymbol{q}$:

$$I = K_0 \cdot q \tag{16}$$

where $\boldsymbol{K_0}$ is the constant. By combining Eqs. (15) and (16), we have the relationship between the intensity of Raman signal and the MPA concentration in the soaking solution:

$$I = K \cdot C^{\frac{1}{n}} \tag{17}$$

where $K = K_0 \cdot K_F$. Eq. (17) is in complete agreement with Eq. (13), which has also confirmed the Freundlich-typed adsorption of the MPA molecules on the substrate. By combining Eqs. (13) and (17), the value of MPA adsorption parameter ($\boldsymbol{n}$) can thus be estimated to be $\boldsymbol{n} = 4$. This also presents a simple way to measure the adsorption parameters, which are normally acquired by the time-consuming measurement of the adsorption isotherms.

3.3.2. Confirmation of the amidation reaction

For confirmation of reaction (9) occurring on surface of the substrate, the amidation reaction experiment was carried out, according to Vijay et al.'s method [50], by preparing the ethanol solution with EDC, MPA and 2-aminoethanethiol and continuously stirring it at room temperature for 15 h, as previously described in detail [42]. The pure amidation compound was thus acquired. The FTIR measurement was conducted for this compound, as shown in **Figure 15**. All peaks can be ascribed to the vibrations of N-(2-mercaptoethyl)-P-methylphosphonamidic acid ($C_3H_{10}O_2NPS$) [51]. For instance, the peaks at 731, 768 and 812 cm^{-1}correspond to twisting vibrations of the carbon chains $(CH_2)_2$ in $C_3H_{10}O_2NPS$; the peak at 894 cm^{-1} is assigned to the stretching vibration of (P—CH_3) + (P—O); and the peaks at 1041 and 1064 cm^{-1} are from the stretching vibrations of (C—N) [48]. These indicated that pure amide $C_3H_{10}O_2NPS$ was obtained.

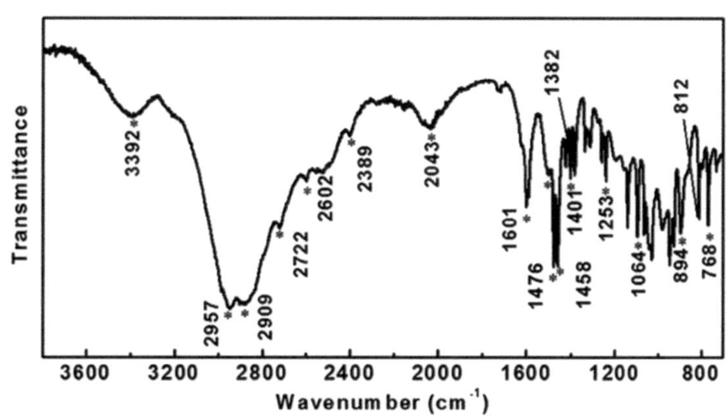

Figure 15. FTIR spectrum of the products after amidation reaction [42].

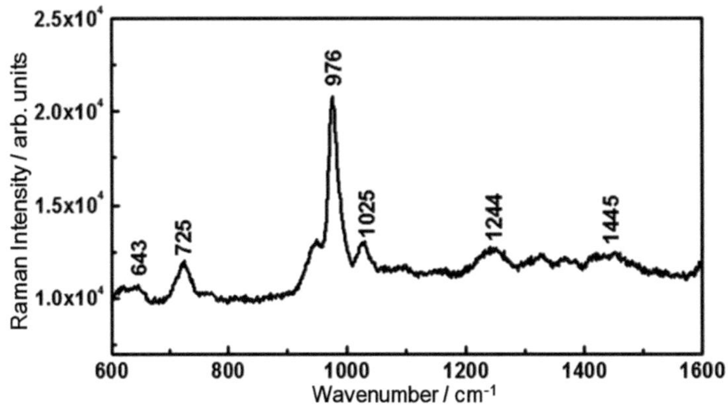

Figure 16. The Raman spectrum for the Au-coated Si nanocone array after immersion in the ethanol solution with 1.0×10^{-3} M $C_3H_{10}O_2NPS$ [42].

For the Raman spectral measurements, the pure amide ($C_3H_{10}O_2NPS$) was diluted, with ethanol, to a given concentration. The Au-coated Si nanocone array was then immersed into the $C_3H_{10}O_2$-NPS-contained ethanol solution before the Raman spectral measurements. **Figure 16** shows the results corresponding to the solution with 1.0×10^{-3} M in $C_3H_{10}O_2NPS$ concentration. The spectral pattern is in good agreement with that shown in curve (I) of **Figure 12**. So, the Raman spectrum in curve (I) of **Figure 12** should be attributed to the amidation compound $C_3H_{10}O_2NPS$. These results have confirmed that the amidation reaction occurred on the surface of the modified Au-coated Si nanocone array during its immersion in the MPA solutions with EDC, and that the reaction products $C_3H_{10}O_2NPS$ molecules were formed on and bound with the array's surface.

3.4. Quantitative SERS-based detection of MPA

As mentioned earlier, the Au-coated Si nanocone array modified with 2-aminoethanethiol can capture selectively MPA in the solution in the presence of EDC via diffusion and adsorption, leading to the amidation reaction and the formation of $C_3H_{10}O_2NPS$ molecules which were still bound on the array's surface. The bound $C_3H_{10}O_2NPS$ molecules were corresponding to the MPA molecules adsorbed on the SERS substrate. Therefore, by using the 2-aminoethanethiol-modified Au-coated Si nanocone array, we can realize the SERS-based ultrasensitive and quantitative detection of MPA in the solution. The obtained Raman spectra are from the $C_3H_{10}O_2NPS$ molecules but corresponding to the MPA, which exhibits a linear double logarithmic relation between the Raman peak intensity and the MPA concentration, as described in Eq. (4). Since there exist similarities between MPA and sarin in chemical properties and Raman spectral pattern, as mentioned in Section 3.1.1, it is thus expected that the abovementioned method is also suitable for sarin detection.

Finally, it should be mentioned that the method introduced here relies on the activation of phosphonic groups by the coupling agent EDC which creates reactive phosphonamide. Both EDC and the by-product N-acylurea can be removed by subsequent substrate cleaning before Raman spectral measurement.

4. Conclusions and outlook

We have introduced some recent progresses in the SERS-based detection of the organophosphorus nerve agents, including the thin water film confinement evaporation concentrating strategy and the SERS substrates' surface modification/amidation reaction. For the former, when the solution containing target molecules is dropped on the SERS substrate and forms a thin water film on it, the target molecules are limited within the film. Subsequent water evaporation leads to the enrichment or concentrating of the target molecules within the region of strongly enhanced electromagnetic field above the substrate, and hence significantly enhances the Raman signal or induces the CERS effect. The validity of this strategy has been demonstrated by taking the sarin simulant DMMP as the target molecule, which are hardly adsorbed on the gold substrates, exhibiting significant CERS effect during water film evaporation and showing a good linear relation between the reciprocal intensity for the Raman characteristic peak and the evaporation interval, which is in agreement with quantitative description of evaporation-induced solute concentrating. The thin water film not only confines the target molecules within a limited space but also protects the target molecules from laser-induced damage. This approach should also be suitable for the other soluble molecules with low volatility. For the latter, because the 2-aminoethanethiol molecules possess two-head groups: amino and thiol groups: one can be bound with gold film and the other can capture the phosphonic groups in sarin simulation agent MPA in presence of the coupling agent EDC, the 2-aminoethanethiol-modified SERS substrate could selectively capture MPA molecules in the solution, which thus induces the amidation reaction on the substrate's surface. The reaction products or $C_3H_{10}O_2NPS$ molecules are still bound on the substrate's surface. Correspondingly, we could obtain the Raman spectra of amide $C_3H_{10}O_2NPS$, which correspond to the MPA molecules adsorbed on the substrate. The Raman peak intensity shows a good linear double logarithmic relation with the MPA concentration in a large range, which could be attributed to Freundlich adsorption behavior of MPA on the surface-modified SERS substrate. The minimum detection level of MPA is down to $\sim$1 ppb. We can thus quantitatively detect MPA or sarin in solutions based on the SERS effect. This route could also be suitable for the other organophosphorus nerve agents and some other molecules weakly interacted with the coin metal substrates by choosing appropriate modifiers. In a word, the abovementioned progresses provide new ways for highly efficient SERS-based detection of the organophosphorus nerve agents and some other target molecules that weakly interact with the coin metal substrates.

Acknowledgements

This work is financially supported by the National Key Research and Development Program of China《Fundamental Research on nano sensing materials and high performance sensors focused on pollutants detection》(Grant No. 2017YFA0207101), Natural Science Foundation of China (Grant No. 51531006, 11574313, 11374300 and 51571188) and the CAS/SAF International Partnership Program for Creative Research Teams.

Author details

Qian Zhao, Guangqiang Liu* and Weiping Cai*

*Address all correspondence to: liugq@issp.ac.cn and wpcai@issp.ac.cn

Key Lab of Materials Physics, Anhui Key Lab of Nanomaterials and Nanotechnology, Institute of Solid State Physics, Chinese Academy of Sciences, Hefei, PR China

References

[1] Federation of American Scientists. Types of Chemical Weapons [Internet]. 2015. Available from: https://fas.org/cw/documents/cwagents.pdf

[2] Yang YM, Ji HF, Thundat T. Nerve agents detection using a Cu^{2+}/L-cysteine bilayer-coated microcantilever. Journal of the American Chemical Society. 2003;**125**:1124-1125. DOI: 10.1021/ja028181n

[3] Guilbault GG, Kristoff J. Detection of organophosphorus compounds with a coated piezoelectric crystal. Analytical Chemistry. 1985;**57**:1754-1756. DOI: 10.1021/ac00285a057

[4] Riter LS, Peng YN, Noll RJX. Analytical performance of a miniature cylindrical ion trap mass spectrometer. Analytical Chemistry. 2002;**74**:6154-6162. DOI: 10.1021/ac0204956

[5] Steiner WE, Klopsch SJ, English WA. Detection of a chemical warfare agent simulant in various aerosol matrixes by ion mobility time-of-flight mass spectrometry. Analytical Chemistry. 2005;**77**:4792-4799. DOI: 10.1021/ac050278f

[6] Fleischmann M, Hendra PJ, McQuilla AJ. Raman spectra of pyridzne adsorbed at a silver electrode. Chemical Physics Letters. 1974;**26**:163-166. DOI: 10.1016/0009-2614(74)85388-1

[7] Albrecht MG, Creighton JA. Anomalously intense Raman spectra of pyridine at a silver electrode. Journal of the American Chemical Society. 1977;**99**:5215-5217. DOI: 10.1021/ja00457a071

[8] Jeanmaire DL, Vanduyne RP. Surface Raman spectroelectrochemistry: Part I. Heterocyclic, aromatic, and aliphatic amines adsorbed on the anodized silver electrode. Journal of Electroanalytical Chemistry. 1977;**84**:1-20. DOI: 10.1016/S0022-0728(77)80224-6

[9] Lin EC, Fang J, Park SCX. Effective localized collection and identification of airborne species through electrodynamic precipitation and SERS-based detection. Nature Communications. 2013;**4**:1-8. DOI: 10.1038/ncomms2590

[10] Cecchini MP, Turek VA, Paget JX. Self-assembled nanoparticle arrays for multiphase trace analyte detection. Nature Materials. 2013;**12**:165-171. DOI: 10.1038/nmat3488

[11] Lim DK, Jeon KS, Hwang JH. Highly uniform and reproducible surface-enhanced Raman scattering from DNA-tailorable nanoparticles with 1-nm interior gap. Nature Nanotechnology. 2011;**6**:452-460. DOI: 10.1038/nnano.2011.79

[12] Wang YQ, Yan B, Chen LX. SERS tags: Novel optical nanoprobes for bioanalysis. Chemical Reviews. 2013;**113**:1391-1428. DOI: 10.1021/cr300120g

[13] Alvarez-Puebla RA, Liz-Marzan LM. Traps and cages for universal SERS detection. Chemical Society Reviews. 2012;**41**:43-51. DOI: 10.1039/C1CS15155J

[14] Wu DY, Li JF, Ren BX. Electrochemical surface-enhanced Raman spectroscopy of nanostructures. Chemical Society Reviews. 2008;**37**:1025-1041. DOI: 10.1039/B707872M

[15] Tao A, Kim F, Hess CX. Langmuir–Blodgett silver nanowire monolayers for molecular sensing using surface-enhanced Raman spectroscopy. Nano Letters. 2003;**3**:1229-1233. DOI: 10.1021/nl0344209

[16] Xiong YJ, McLellan JM, Chen JY. Kinetically controlled synthesis of triangular and hexagonal nanoplates of palladium and their SPR/SERS properties. Journal of the American Chemical Society. 2005;**127**:17118-17127. DOI: 10.1021/ja056498s

[17] Li JF, Huang YF, Ding YX. Shell-isolated nanoparticle-enhanced Raman spectroscopy. Nature. 2010;**464**:392-395. DOI: 10.1038/nature08907

[18] Stuart DA, Biggs KB, Van-Duyne RP. Surface-enhanced Raman spectroscopy of half-mustard agent. The Analyst. 2006;**131**:568-572. DOI: 10.1039/B513326B

[19] Habouti S, Mátéfi-Tempfli M, Solterbeck CH. On-substrate, self-standing au-nanorod arrays showing morphology controlled properties. Nano Today. 2011;**6**:12-19. DOI: 10.1016/j.nantod.2010.11.001

[20] Moskovits M. Surface-enhanced Raman spectroscopy: A brief retrospective. Journal of Raman Specroscopy. 2005;**36**:485-496. DOI: 10.1002/jrs.1362

[21] SSR D, Singh AK, Senapati DX. Gold nanoparticle based label-free SERS probe for ultrasensitive and selective detection of trinitrotoluene. Journal of the American Chemical Society. 2009;**131**:13806-13812. DOI: 10.1021/ja905134d

[22] Camden JP, Dieringer JA, Wang YM. Probing the structure of single-molecule surface-enhanced Raman scattering hot spots. Journal of the American Chemical Society. 2008; **130**:12616-12617. DOI: 10.1021/ja8051427

[23] Huang Y, Dai L, Song L. Engineering gold nanoparticles in compass shape with broadly tunable plasmon resonances and high-performance SERS. ACS Applied Materials & Interfaces. 2016;**8**:27949-27955. DOI: 10.1021/acsami.6b05258

[24] Sajanlal PR, Pradeep T. Functional hybrid nickel nanostructures as recyclable SERS substrates: Detection of explosives and biowarfare agents. Nanoscale. 2012;**4**:3427-3437. DOI: 10.1039/C2NR30557G

[25] Tang HB, Meng GW, Huang QX. Arrays of cone-shaped ZnO nanorods decorated with Ag nanoparticles as 3D surface-enhanced Raman scattering substrates for rapid detection of trace polychlorinated biphenyls. Advanced Functional Materials. 2012;**22**:218-224. DOI: 10.1002/adfm.201102274

[26] Dieringer JA, RB L-II, Scheidt KA. A frequency domain existence proof of single-molecule surface-enhanced Raman spectroscopy. Journal of the American Chemical Society. 2007;**129**:16249-16256. DOI: 10.1021/ja077243c

[27] Bell SEJ, Sirimuthu NMS. Surface-enhanced Raman spectroscopy (SERS) for sub-micromolar detection of DNA/RNA mononucleotides. J. Am. Chem. Soc. 2006;**128**: 15580-15581. DOI: 10.1021/ja066263w

[28] Nie S, Emory SR. Probing single molecules and single nanoparticles by surface-enhanced Raman scattering. Science. 1997;**275**:1102-1106. DOI: 10.1126/science.275.5303.1102

[29] Kneipp K, Wang Y, Kneipp HX. Single molecule detection using surface-enhanced Raman scattering (SERS). Physical Review Letters. 1997;**78**:1667-1670. DOI: 10.1103/PhysRevLett.78.1667

[30] Moskovits M. Surface-enhanced spectroscopy. Rev. Mod. Phys. 1985;**57**:783-826. DOI: 10.1103/RevModPhys.57.783

[31] Lee SJ, Morrill AR, Moskovits M. Hot spots in silver nanowire bundles for surface-enhanced Raman spectroscopy. Journal of the American Chemical Society. 2006;**128**: 2200-2201. DOI: 10.1021/ja0578350

[32] Sharma B, Frontiera RR, Henry AI. SERS: Materials, applications, and the future. Materials Today. 2012;**15**:16-25. DOI: 10.1016/S1369-7021(12)70017-2

[33] Wang JJ, Duan GT, Liu G. Detection of dimethyl methylphosphonate by thin water film confined surface-enhanced Raman scattering method. Journal of Hazardous Materials. 2016;**303**:94-100. DOI: 10.1016/j.jhazmat.2015.10.022

[34] Taranenko N, Alarie JP, Stokes DL. Surface-enhanced Raman detection of nerve agent simulant (DMMP and DIMP) vapor on electrochemically prepared silver oxide substrates. Journal of Raman Specroscopy. 1996;**27**:379-384. DOI: 10.1002/(SICI)1097-4555(199605)27:5<379::AID-JRS925>3.0.CO;2-G

[35] Pearman WF, Fountain AW. Classification of chemical and biological warfare agent simulants by surface-enhanced Raman spectroscopy and multivariate statistical techniques. Applied Spectroscopy. 2006;**60**:356-365. DOI: 10.1366/000370206776593744

[36] Yan F, Tuan VD. Surface-enhanced Raman scattering detection of chemical and biological agents using a portable Raman integrated tunable sensor. Sensors and Actuators B: Chemical. 2007;**121**:61-66. DOI: 10.1016/j.snb.2006.09.032

[37] Kim NJ. Physical origins of chemical enhancement of surface-enhanced Raman spectroscopy on a gold nanoparticle-coated polymer. Journal of Physical Chemistry C. 2010;**114**:13979-13984. DOI: 10.1021/jp103360m

[38] Wang JJ, Duan GT, Liu G. Gold quasi rod-shaped nanoparticle-built hierarchically micro/nanostructured pore array via clean electrodeposition on a colloidal monolayer and its structurally enhanced SERS performance. Journal of Materials Chemistry. 2011;**21**:8816-8821. DOI: 10.1039/C1JM10773A

[39] Van-der-Veken B, Herman M. Vibrational spectra of $CH_3PO(OCH3)_2$ and isotopically substituted derivatives. Phosphorus. Sulfur. 1981;**10**:357-367. DOI: 10.1080/03086648108077388

[40] Christesen SD. Raman cross sections of chemical agents and simulants. Applied Spectroscopy. 1988;**42**:318-321. DOI: 10.1366/0003702884428220

[41] Tomchenko AA, Harmer GP, Marquis BT. Detection of chemical warfare agents using nanostructured metal oxide sensors. Sensors and Actuators B: Chemical. 2005;**108**:41-45. DOI: 10.1016/j.snb.2004.11.059

[42] Zhao Q, Liu GQ, Zhang HW. SERS-based ultrasensitive detection of organophosphorus nerve via substrate's surface modification agents. Journal of Hazardous Materials. 2017;**324**:194-202. DOI: 10.1016/j.jhazmat.2016.10.049

[43] Montalbetti CAGN, Falque V. Amide bond formation and peptide coupling. Tetrahedron. 2005;**61**:10827-10852. DOI: 10.1016/j.tet.2005.08.031

[44] Costa JCS, Ando RA, Sant'Ana AC. Surface-enhanced Raman spectroscopy studies of organophosphorous model molecules and pesticides. Physical Chemistry Chemical Physics. 2012;**14**:15645-15651. DOI: 10.1039/C2CP42496G

[45] Pearson RG. Hard and soft acids and bases, HSAB, part 1: Fundamental principles. Journal of Chemical Education. 1968;**45**:581-587

[46] Pearson RG. Hard and soft acids and bases, HSAB, part II: Underlying theories. Journal of Chemical Education. 1968;**45**:643-648

[47] Raghav R, Srivastava S. Immobilization strategy for enhancing sensitivity of immunosensors: L-asparagine-AuNPs as a promising alternative of EDC-NHS activated citrate-AuNPs for antibody immobilization. Biosensors & Bioelectronics. 2016;**78**:396-403. DOI: 10.1016/j.bios.2015.11.066

[48] Speight JG. Section 3(3.1) Infrared absorption spectroscopy. In: Speight JG, editor. Lange's Handbook of Chemistry. 16th ed. Wyoming: The McGraw-Hill Companies Inc.; 2005. pp. 3.3-3.36

[49] Freundlich H, Heller W. The adsorption of cis- and trans-azobenzene. Journal of the American Chemical Society. 1939;**61**:2228-2230

[50] Sehgal D, Vijay IK. A method for the high-efficiency of water-soluble carbodiimide-mediated amidation. Analytical Biochemistry. 1994;**218**:87-91. DOI: 10.1006/abio.1994.1144

[51] Speight JG. Section 3(3.1) Infrared absorption spectroscopy. In: Speight JG, editor. Lange's Handbook of Chemistry. 16th ed. Wyoming: The McGraw-Hill Companies Inc; 2005. pp. 3.3-3.23

Section 3

SERS and Raman Spectroscopy: Carbon Nanomaterials

Chapter 8

Raman Spectroscopy of Graphitic Nanomaterials

Daniel Casimir, Iman Ahmed, Raul Garcia-Sanchez,
Prabhakar Misra and Fabiola Diaz

Additional information is available at the end of the chapter

http://dx.doi.org/10.5772/intechopen.72769

Abstract

This chapter showcases some of the versatility of Raman spectroscopic data as applied to the characterization of single (SWNT) and multi-walled (MWNT) carbon nanotubes, few layer graphene and functionalized graphene nanoplatelets, with an emphasis on gas-sensing applications. Specifically, water vapor and a variety of toxic gases (NO, NO_2, and SO_2 at 500 ppm in gaseous nitrogen) have been targeted for detection over the temperature range 24–200°C. The structure of sp^2-hybridized carbon allotropes is reviewed and scanning electron microscopy (SEM) imagery utilized in conjunction with Raman spectroscopy to physically and spectrally characterize the various graphitic nanomaterials studied. A Kataura plot analysis associated with the Radial Breathing Mode (RBM) vibrations of SWNT has been used to identify possible chiralities in the graphitic samples employing 455, 532 and 780 nm laser excitation wavelengths to record the Raman spectra. The effect of temperature on the various Raman vibrational modes (RBM, G^+ and G^-) has been investigated, along with a determination of the thermal conductivity of SWNT samples and correlation between the purity of the sample and the variation of the slope of the G^+ band with increasing laser power.

Keywords: Raman spectroscopy, carbon nanotubes, graphene, graphene nanoplatelets, nanomaterials

1. Introduction

The first class of materials whose properties and characterization via Raman spectroscopy we discuss here are the graphitic allotropes—single and multi-walled carbon nanotubes (SWNT, MWNT), followed by graphene and graphene nano flakes, specifically plasma functionalized graphene nanoplatelets. The chapter will begin with some discussion of the rich Raman spectral features of sp^2 carbon allotropes, which will be necessary since there will be an emphasis on those particular Raman bands and features that provide useful structural/thermal data about carbon nanotube samples. It will be followed by an overview of graphene and graphene nanoplatelets and their usefulness for gas-sensing applications utilizing Raman spectroscopy.

The one-dimensional graphite allotrope, carbon nanotube, is conceptually described as being a rolled-up graphene sheet, yielding the cylindrical nanomaterials that have diameters of a few nanometers. The multi-walled varieties contain several concentric cylindrical shells. The Raman bands and the variations under various external perturbations of the sp^2 graphitic materials cited above include the graphite G-band common to all sp^2 carbons at around 1580 cm^{-1} due to the intraplanar bond stretching of the two carbons in the hexagonal lattice unit cell, and the carbon nanotube specific radial breathing mode (RBM)—which arises as a consequence of their cylindrical geometry. Lastly, there is the defect D-Raman band, which arises due to defects, finite size effects, or any other cause of departure from perfect crystalline regularity, and the 2-D band. The characterization topics discussed in connection with the graphitic materials will be the identification of the chirality types present in carbon nanotube samples using the resonant RBM mode of carbon nanotubes and its connection to the interesting quasi 1-dimensional character of their electronic structure. The other properties obtained via Raman spectroscopy discussed will also be the anomalous thermal expansion and thermal conductivity of the sp^2 graphitic materials investigated.

We have also utilized Raman spectroscopy to understand the behavior of vibrational modes associated with graphene following gas exposure. Specifically, we have studied the effects of water vapor and toxic gases (SO_2, NO_2, NO), via variable humidity levels, gas concentrations, exposure times, and thermal loading, on the Raman spectra of graphene.

Functionalized graphene nanoplatelets are comprised of an amorphous mixture of graphene sheets. Their thicknesses range from 6 to 8 nm, and the overall density usually lies between 0.03 and 0.1 g/cc. The oxygen content of the majority of samples normally are <1%, with the remaining carbon content exceeding 99.5 wt % (STREM). The morphology of this amorphous material plays a large and significant role in its enhanced mechanical properties, such as stiffness, strength, and surface hardness. By incorporating a small number of certain atoms that differ in the number of valence electrons into the pure crystal, the doping of graphene nanoparticles can lead to an enhancement in conductivity.

For the thermal conductivity measurements of the carbon allotropes, we have used the G-Raman band and its variation with increased sample temperature through laser heating. The method and the useful information it provides is due to Terekhov et al. [1]. Also, edge defect characterization of graphene nanoplatelets based on Eq. (1) due to Cancado et al. [2] has also been included, where L is the characteristic in-plane crystallite size of the graphene nano flake, λ is the laser wavelength, and I_D and I_G are the intensities of the Raman D-band and G-band, respectively.

$$L = \left(2.4 \times 10^{-10}\right) \times \lambda^4 (I_D/I_G)^{-1} \tag{1}$$

Our research presented here is aimed at extending the knowledge regarding the nature of graphitic nanomaterial-gas sensing interactions and help develop better models for their enhanced understanding, which in turn would make the development and production of more effective *in situ* gas sensors feasible.

http://dx.doi.org/10.5772/intechopen.72769

2. Structure of sp^2 nanocarbons

This introductory section presents a cursory discussion of SWNT and MWNT structures. This will be done by first looking at the unit cell of planar graphene and also graphite since the former material is considered to be the conceptual parent material of all sp^2 graphitic materials, including SWNTs through the application of a simple rolling up operation. Since the molecular/electronic and geometric structures are highly dependent on graphene, the majority of a SWNT's structural features are expressed via the lattice vectors $\vec{a_1}$, and $\vec{a_2}$ of the graphene unit cell shown in **Figure 1**.

The two unique Carbon atoms A and B in each unit cell are located respectively at (0, 0) and at 1/3* $\left(\vec{a_1} + \vec{a_2}\right)$ (**Figure 1** adapted from Wong and Akinwande [3]). The progression to the first related graphitic material, three dimensional graphite, is accomplished through the stacking of several layers of 2-dimensional graphene layers, where in the A-B Bernal stacking structure, there are 2*N atoms per unit cell, N being the number of layers [4]. A major structural factor of graphite that results in the electronic structure of 2-dimensional graphene being a reasonable first order approximation of the former is the average inter-layer spacing of 3.35 Angstroms. This distance is much larger than the nearest neighbor Carbon–Carbon distance of 1.42 Å, hence resulting in much weaker overall attractive interaction between layers compared to intra-planar interactions [5].

Moving now to one of major foci of the chapter single-walled carbon nantoubes (SWNTs), the conceptual operation performed on the single 2-dimensional graphene sheet is "rolling" it up into a cylinder. The diameter distribution of most SWNTs produced by various techniques is dominated by tubes with diameters less than 2 nm, although diameters in the range of 0.7–10.0 nm are possible [5]. Ignoring the two ends and exploiting the very large length to diameter ratio (~10^4–10^5) of SWNTs allows one to safely view these sp^2 nanocarbons as quasi 1-dimensional objects [5].

The concept of chirality is essential in the description of SWNT structure. It is defined by the chiral vector, denoted by C_h in **Figure 1**, and several equivalent interpretations of this structural quantity are usually given. For example one may consider the fact that the chiral vector determines the arrangement of the six-sided carbon hexagons in the curved planar wall of the SWNT [5]. Alternatively, one may also view the chirality of a SWNT in terms of the overall symmetry of the constructed SWNT, specifically whether or not the SWNT has vertical mirror plane reflection

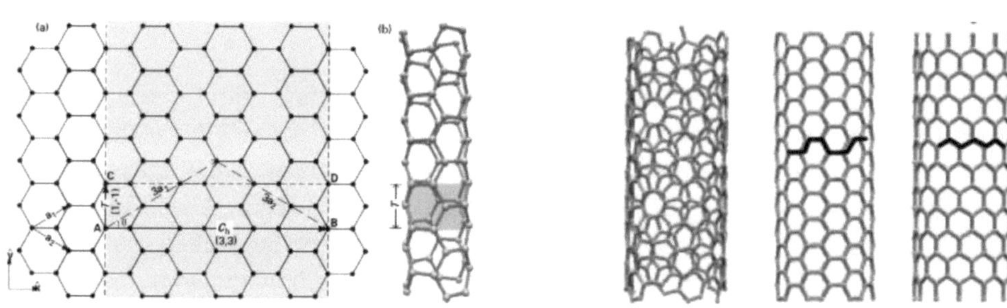

Figure 1. Graphene unit cell.

symmetry across planes containing the tube axis [3]. An additional benefit of the latter viewpoint is its more direct path to the discovery of there being only three overall structural categories of SWNT as shown in **Figure 1**.

Chiral SWNTs are formed such that the orientation of the carbon hexagons on the tube surface do not allow the sides of the tube across a vertical mirror plane to be superimposed on one another. The remaining two subcategories achiral SWNTs, armchair and zig-zag SWNTs however, do allow for such reflection symmetry based on the arrangement of the Carbon hexagons along the cylinder walls. The names armchair and zig-zag refer to the circular cross-sections of each of these achiral SWNT types shown by the bold lines in **Figure 1**. We conclude this section dealing with the structural properties of primarily SWNTs with the actual construction of a SWNT, starting just the two SWNT graphene lattice vectors, $\vec{a_1}$ and $\vec{a_2}$, which have the following Cartesian components $\left[\frac{a\sqrt{3}}{2}, \frac{a}{2}\right]$, and $\left[\frac{a\sqrt{3}}{2}, -\frac{a}{2}\right]$ respectively.

Relying on the example of **Figure 1** which demonstrates the formation of a (3, 3) armchair nanotube, the planar unit cell is formed by rolling the gray shaded region along the chiral vector such that points C and D coincide respectively with points D and B. After performing the previous conceptual rolling operation the pertinent quantity that defines the SWNT unit cell in the resultant nanotube is the translational vector $\vec{T}$ as shown in the right hand portion of **Figure 1**. As its name suggests this vector is the shortest vector that is perpendicular to the chiral vector, and represents the axial component of the SWNT unit cell that is repeated in this same direction. This formalism of SWNT construction that begins from the planar graphene lattice provides the readily obvious interpretation for the magnitude of the chiral vector, namely its magnitude equaling the nanotube circumference given by the expression, $\vec{C_h} = a\sqrt{m^2 + n^2 + nm}$ where a ≈ 2.46 Å is the graphene hexagonal lattice constant equal to $\sqrt{3}$ times the nearest neighbor C-C distance of 1.42 Å. The remaining structural parameter in **Figure 1**, the angle q, is the chiral angle, conventionally chosen to be the angle between the chiral vector and the $\vec{a_1}$ lattice vector. This angle ranges from 0° ≤ q ≤ 30° with the lower bound corresponding to zig-zag SWNTs and the upper bound corresponding to armchair SWNTs. The integers, n and m simply refer to the number of $\vec{a_1}$ and $\vec{a_2}$ lattice vectors used in the construction of the chiral vector usually with the convention of n ≥ m. For zig-zag SWNTs m = 0, and in the case of armchair SWNTs both chiral indices are identical [3].

3. Experimental

In Raman spectroscopy, a laser diode emits photons, which interacts with the sample, most of the light bouncing off unchanged with the same frequency as the source (Rayleigh scattering). However, a small amount of light experiences an energy shift (Raman scattering) and is filtered to allow only the Raman scattered light to be collected by the detector. The sample vibrates uniquely to its structure and each vibration mode uniquely alters the emitted photons wavelength and that change is graphed as intensity per wavelength. An unknown sample's Raman spectrum can be compared to the known Raman spectral graph.

3.1. Thermo fisher scientific DXR smart Raman spectrometer

The primary instrument used to record the majority of the Raman spectra was a DXR SmartRaman spectrometer (that uses 780, 532, and 455 nm laser sources). The first wavelength (780 nm) was used for the bulk of the recorded spectra and utilized a high brightness laser of the single mode diode (as does the 532 nm light source), while the 455 nm source is a diode-pumped solid state laser. This instrument employs the 180-degree backscattering geometry, full range grating and triplet spectrograph, coupled with automated entrance slit selections in order to provide the Stokes-shifted Raman bands.

3.2. Renishaw inVia Raman spectrometer

The Renishaw inVia Raman spectrometer uses a 532-nm laser source and was used to obtain the Stokes spectra of the graphene and functionalized Nanoplatelets samples. It consists of a microscope to shine light on the sample and collecting the scattered light, filtering all the light except for the tiny fraction that has been Raman scattered, together with a diffraction grating for splitting the Raman scattered light into component wavelengths, and a CCD camera for final detection of the Raman spectrum.

3.3. Ventacon heated cell

The Raman spectral data in this study of the two different SWNT samples were obtained under thermal loading from room temperature to 200°C in steps of 10°C. Both powdered samples were heated externally via a Ventacon™ model H4–200 heat cell that is diagrammed in **Figure 2**. The first SWNT sample was produced by Unidym™ Carbon Nanotubes. The Hipco technique was used in the production of this sample, which involves the nucleation of SWNTs on $Fe(CO)_5$ catalyst material using high pressure CO, followed by various quality control methods (Misra et al., 2013) [6]. According to manufacturer specifications, the diameters and lengths of the nanotubes in this sample ranged between 0.8 to 1.2 nm, and 100 to 1000 nm, respectively. The sample data also claimed a purity level of only 8% residual Fe catalyst by weight present.

Information about the method of production or purity levels of the second SWNT sample was not available. In the data sets for both samples each point in the ω_{RBM} vs. temperature plots is the mean value from two separate Raman collections. The standard error of each data point obtained from both Raman spectra collected at each temperature is also displayed for both samples. The spectra recorded at each pre-set temperature were obtained with a temperature variation of ±0.1°C.

3.4. Aluminum disk cell for graphene gas exposure Raman spectroscopy

Figure 3 is a diagram of the components of the sample cell and how it is placed underneath the Renishaw Raman spectrometer. The cell contains apertures that connect to the center of the cell, where the sample is placed and sealed through means of a glass disk and an O-ring. In addition, the cell has an aperture to place a thermocouple to read the temperature of the cell and another one where we place a voltage-induced heating cylinder. The gas flow comes from the gas cylinder into the rotameter and then through a series of tubing to the cell. These tubes

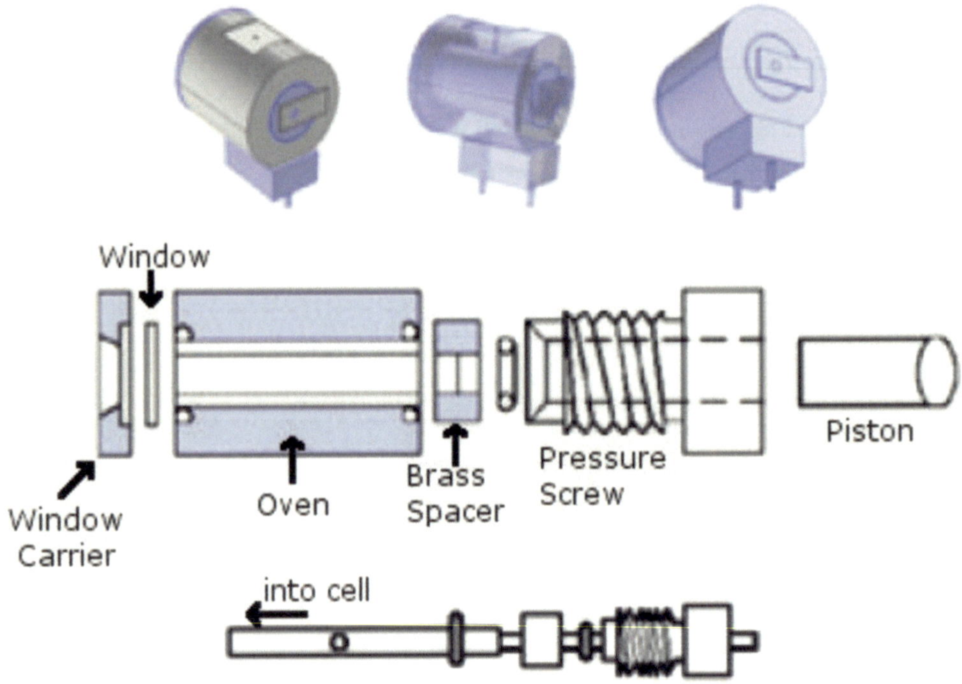

Figure 2. Top: Ventacon H4-200 heat cell. Middle: Cross sectional layout of oven. Bottom: Cylindrical powdered/solid sample holder.

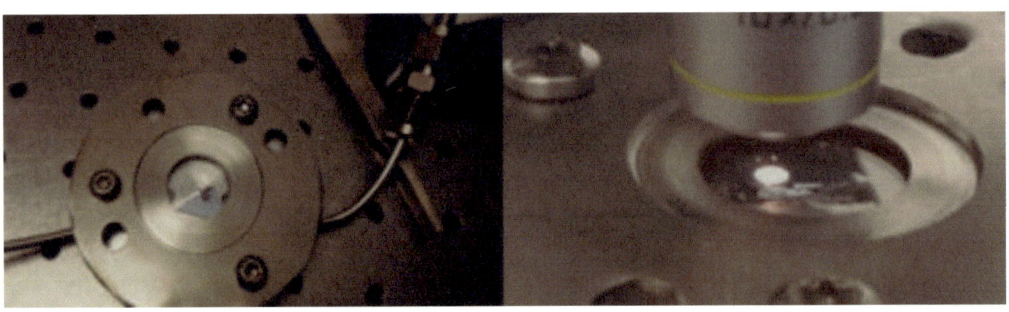

Figure 3. Aluminum disk cell used for gathering Raman spectral data during gas exposure of graphene samples; top view (left) | under Raman microscope (right) [7].

connect to holes within the sealed sample chamber and flow out through the other output tube and into the bubbler. The bubbler is submerged in liquid nitrogen in order to separate the more dangerous components of the gases used and flow the rest through an inlet fan connected do an exhaust. At the end of the experiment, the bubbler is placed in a fume hood and left until the following day. **Figure 4** shows multiple components of this setup that were used for the gas exposure experiments.

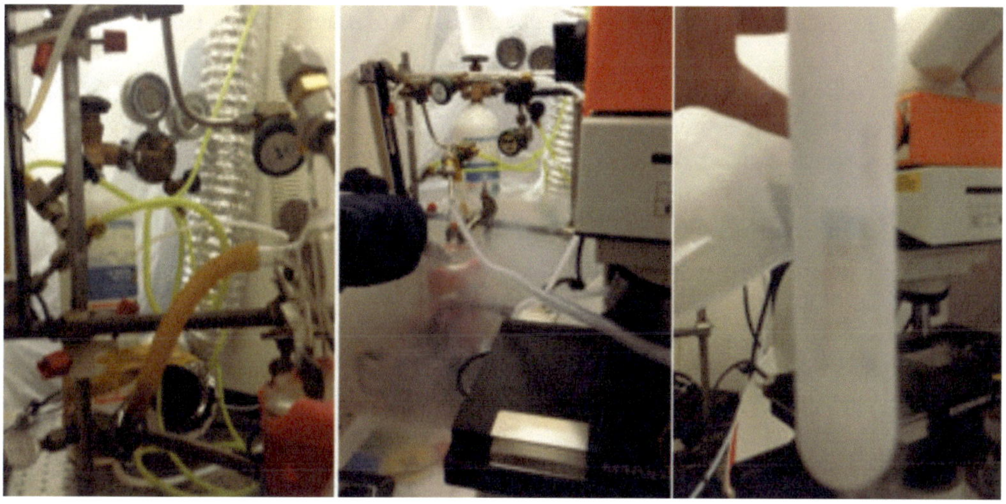

Figure 4. Left: Setup showing the tubing layout for the gas exposure Raman spectroscopy experiments; center: Filling the bubbler with liquid nitrogen to break up dangerous gas components; right: Bubbler after experiments had concluded [7].

The measurement sequence for the gas exposure of graphene samples is as follows:

1. The sample is placed in the cell, nitrogen gas flows into the sample chamber and heated up to the 130–150°C temperature range. Raman spectroscopy data are taken as the cell heats up. Typically, 60 scans are taken, with a 20-second exposure for graphene. The voltage is adjusted in order to control the temperature increase.

2. Voltage is reduced and the Raman spectral data recorded, as the temperature drops back to near ambient value.

3. Nitrogen flow stops and the toxic gas for that experiment flows into the sample chamber. We take sets of 120 exposures with laser exposure and delays that result in a 30-minute exposure. This is repeated for different gas flows, with the rotameter reading 50, 100, 150 and maximum values for a total of 2-hour exposure.

4. We repeat Steps 1 and 2 to see if the Raman features return back to normal; this allows us to verify the effects of the gas exposure on the graphene sample.

3.5. JEOL JSM-7600F scanning electron microscope

Images of functionalized graphene nanoplatelets were taken with the JEOL JSM-7600F scanning electron microscope [8]. The secondary electron detector on the SEM uses an EMI current of 138.20 nA. Beam current has a range of 1 pA to 200 nA. The JEOL JSM-7600F SEM contains a large variety of detectors that can be used on specimen samples up to 200 mm in diameter. Various magnifications were selected when appropriate to accurately display the sample structure; SEM magnifications range between 25 and 1000000×. The modular software program Gwyddion was used to generate 3-dimensional visualization of the nanoplatelet aggregate structures.

4. Raman spectra of single-walled carbon nanotubes (SWNTs)

The current section is an overview of the Raman active modes of SWNTs. The current treatment is meant only to preview the Raman bands and features relied on most heavily during the authors' characterization of SWNT samples. The first of the Raman bands featured in this section is the G-band, so called since it is common to all sp^2 nanocarbons, centered at ~1580 cm^{-1}. This high energy band is a consequence of the in-plane C-C bond stretching. Interestingly due to the induced strain from the curvature of SWNT walls the G-band for these structures is split into several peaks, with the two most prominent ones being the symmetric A_1 denoted as G^+ and G^- at ~1590 and ~1560 cm^{-1}, respectively. The variation of the smaller of these two peaks with diameter is given by [4]:

$$\omega_G = 1591 + \frac{C}{d_t^2} \tag{2}$$

where d_t is the nanotube diameter, and the value C corresponding to the G^- band for semiconducting and metallic nanotubes respectively are 47.7 and 79.5 cm^{-1} nm. The above relation is derivable solely from a careful application of elasticity theory to SWNTs [4]. The later discussion of the use of the temperature variation of the G^+ band in connection with thermal conductivity is based on a time-dependent perturbation of the G^+ Raman band.

Next, the Raman band at around 1300–1350 cm^{-1} is the D-Band, which has been shown in [9] to be associated with any defects or departures from perfect regularity in the sp^2 graphite lattice. Equation (3) shows the variation of typical crystallite sizes L_a with the Raman laser excitation energy, and ratio of the intensity of the D-band and G-band, based on a more recent version of Tuinstra and Koenig's original analysis [9]:

$$L_a(nm) = \left(\frac{560}{E_{laser}^4}\right)\left(\frac{I_D}{I_G}\right)^{-1} \tag{3}$$

The final Raman band featured in this section, the radial breathing mode is only present carbon nanotube spectra, and is therefore used as an indication of their presence in sp^2 Carbon samples. As implied by its name the intra-planar displacement of the C-atoms in this Raman active mode is effectively in the radial direction of the resultant SWNT, as if the entire SWNT is breathing. There are two significant features associated with this particular SWNT Raman band in connection with their structural and electronic properties. The first is the inverse relationship between the RBM frequency and SWNT diameter expressed in Eq. (4):

$$\omega_{RBM} = \frac{A}{d_t} \tag{4}$$

Similar to Eq. (1) this relation is also based on continuum elasticity theory. Since the original work done on this relationship was based on individual SWNTs of the kind typically produced via the super-growth method, further work by various groups led to Eq. (5), which is an extension of the prior equation that accounts for environmental perturbations of the RBM, especially on SWNTs in macroscopic bundled samples [4]:

http://dx.doi.org/10.5772/intechopen.72769

$$\omega_{RBM} = \frac{227}{d_t}\sqrt{1 + C_e d_t^2} \tag{5}$$

The overall environmental contribution to the RBM shift is contained in the fitted constant C_e, which is unique to SWNTs produced by any of the various available techniques. In the case of SWNTs produced by the HiPCO technique used in our studies, C_e was determined to have a value of 0.05 [4].

The second critical feature associated with the RBM band is its role in expressing the link between the quasi 1-dimensional electronic behavior and structure of SWNTs. The phenomenon referred to above is the resonant character of the Raman spectra of SWNTs, in which the intensity of Raman scattering from a SWNT is increased many fold when the laser excitation energy is very near that of an optical transition. **Figure 5** shows the density of states for a representative metallic and semi-conducting single-walled carbon nanotube, respectively, and

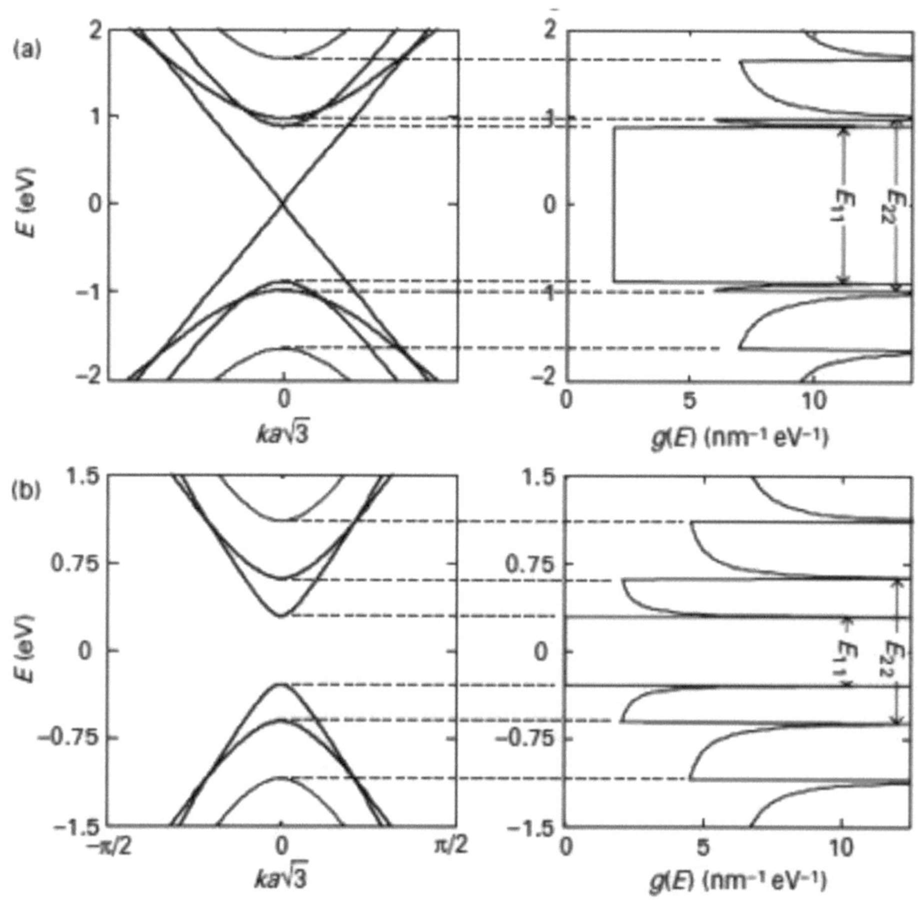

Figure 5. Electronic density of states for a metallic and semi conducting SWNT shows two transitions labeled, E_{11} and E_{22}, between the corresponding valence and conduction bands.

shows two such transitions, labeled E_{11} and E_{22}, between the corresponding valence and conduction bands (adapted from [3]).

These sharp divergences between the optical transitions are known as van Hove singularities. Such singularities arise from the form of the density of states expression for a 1-dimensional sample as indicated in Eq. (6), which is derived for the simple free electron gas model [3]:

$$g(E) = \begin{cases} \frac{1}{h}\sqrt{\frac{2m}{E - E_0}}, & E > E_0 \\ 0, & E \leq E_0 \end{cases} \tag{6}$$

The density of states of the electrons and holes in SWNTs shows the two most significant characteristics of Eq. (6), namely the inverse square root variation with energy, and the van Hove divergences at energies close to the Fermi level E_0. Kataura [4] made arguably one of the most significant contributions to both the theoretical description and practical use of these resonance Raman effects with the introduction of the so-called Kataura Plot in 1999. These plots of nanotube diameter or RBM frequency vs. optical transition energy E_{ii}, which are now in common use for the identification of the chiralities present in a SWNT sample, are possible due to the E_{ii} values' inverse dependence to SWNT diameter. The use of a theoretically derived Kataura plot for (n, m) identification begins the following section discussing our actual use of Raman spectral data in SWNT sample characterization.

5. Raman spectral SWNT characterization

5.1. Kataura plot analysis

The discussion of Section 4 means that in Raman spectra from bundled SWNTs, which is the case in this research, the majority of the signal comes from those tubes in the sample with diameters that are resonant with the excitation wavelength [4]. Therefore, this makes Kataura plots extremely useful in identifying the possible chiral indices contained in any SWNT samples one may be working with. Displayed in **Figures 6–8**, and **Table 1**, are the theoretical Kataura plots and Raman spectral data that were used to identify the chiralities present in one of the SWCNT samples used in the present study.

5.2. Thermal expansion

In this section, we discuss the effect of temperature on the Raman vibrational modes of SWNTs and subsequent use of this valuable effect in obtaining an estimate of the thermal expansion for one of the SWNT samples used in the present study [10].

Considering the phonon frequency ω(V, T) as a function of volume V, and temperature T, the derivatives of this quantity with respect to pressure and volume can be connected to each other via the equations

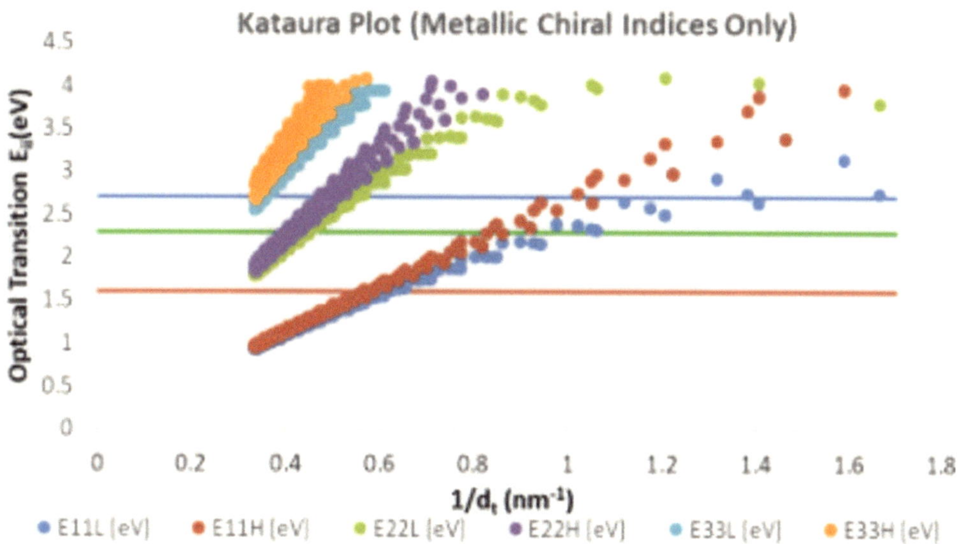

Figure 6. Theoretical Kataura plot for metallic SWNTs. The three horizontal lines from top to bottom represent respectively, the 455, 532, and 780 nm laser excitation wavelengths used on the SWNT sample. The intersection of the horizontal lines with the Kataura plot at any of the diameters obtained from the radial breathing modes help to identify possible chiralities in the sample. The Kataura plot data was obtained from K. Saito, et al. [5].

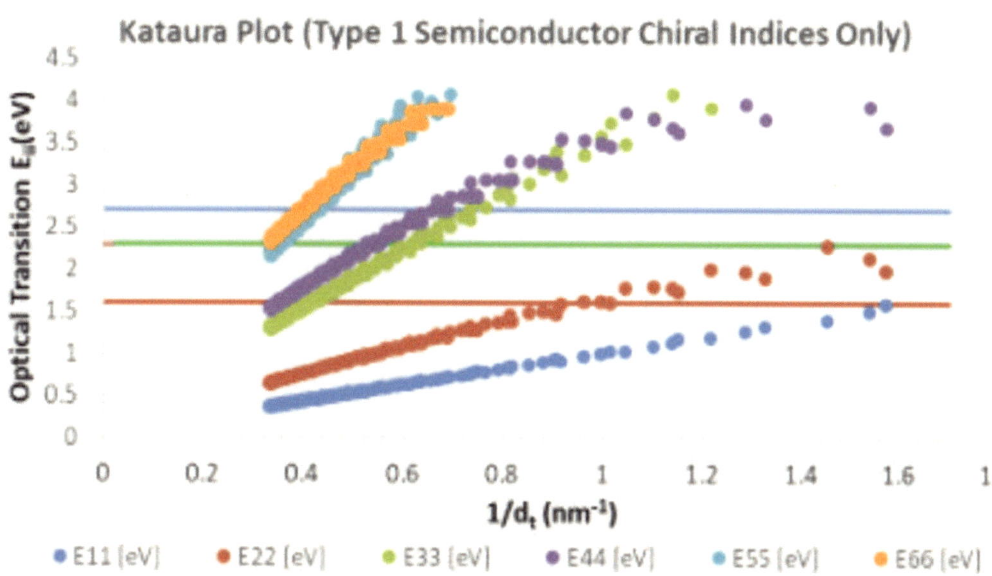

Figure 7. Theoretical Kataura plot for type 1 semi conducting SWNTs. The three horizontal lines from top to bottom represent respectively, the 455, 532, and 780 nm laser excitation wavelengths used on the SWNT sample. The intersection of the horizontal lines with the Kataura plot at any of the diameters obtained from the radial breathing modes help to identify possible chiralities in the sample. The Kataura plot data was obtained from Saito et al. [5].

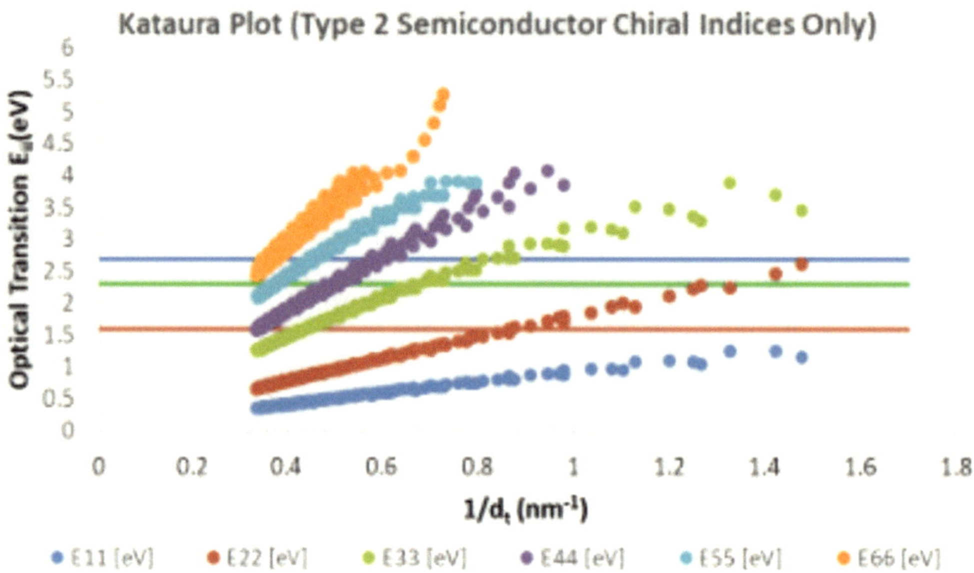

Figure 8. Theoretical Kataura plot for type 2 semi conducting SWNTs. The three horizontal lines from top to bottom represent respectively, the 455, 532, and 780 nm laser excitation wavelengths used on the SWNT sample. The intersection of the horizontal lines with the Kataura plot at any of the diameters obtained from the radial breathing modes help to identify possible chiralities in the sample. The Kataura plot data was obtained from Saito et al. [5].

(1) Laser Excitation (nm)	Radial Breathing Mode (cm-1)	1/d_t (nm-1)	Resonant Chiralities
780	150.51	0.632	(12,12), (20,2), (17,5), (16,7)
532	166.04	0.697	***†
455	171.57	0.721	***†
(2) Laser Excitation (nm)	**Radial Breathing Mode (cm-1)**	**1/dt (nm-1)**	**Resonant Chiralities**
780	150.51	0.632	***†
532	166.04	0.697	***†
455	171.57	0.721	(15,4), (17,3), (18,1)
(3) Laser Excitation (nm)	**Radial Breathing Mode (cm-1)**	**1/d_t (nm-1)**	**Resonant Chiralities**
780	150.51	0.632	***†
532	166.04	0.697	(14,7), (15,5), (16,3), (17,1)
455	171.57	0.721	***†

†No resonant SWCNTs

Table 1. Identification of metal type, type 1 semiconducting and type 2 semiconducting chiral indices in SWNT samples using Kataura plots.

$$\left(\frac{\partial\omega}{\partial V}\right)_T = \left(\frac{\partial\omega}{\partial P}\right)_T \left(\frac{\partial P}{\partial V}\right)_T \tag{7}$$

and

$$\left(\frac{\partial\omega}{\partial V}\right)_T = -\left(\frac{B_T}{V}\right)_T \left(\frac{\partial\omega}{\partial P}\right)_T \tag{8}$$

where B_T is the isothermal bulk modulus which equals $B_T = -V\left(\frac{\partial P}{\partial V}\right)_T$ [11]. Next, a connection between the temperature derivatives at fixed volume and pressure are obtained by writing

$$d\omega = \left(\frac{\partial\omega}{\partial V}\right)_T dV + \left(\frac{\partial\omega}{\partial T}\right)_V dT \tag{9}$$

which after dividing through by dT at constant pressure yields

$$\left(\frac{\partial\omega}{\partial T}\right)_P = \left(\frac{\partial\omega}{\partial T}\right)_V + \left(\frac{\partial V}{\partial T}\right)_P \left(\frac{\partial\omega}{\partial V}\right)_T \tag{10}$$

Equation (10) gives a breakdown of the measurable temperature variation of the phonon frequency on the left in terms of two contributions. The first term on the right represents the intrinsic, purely thermal contribution to the lowering of the phonon frequency caused by the anharmonic intermixing of the various phonon modes [12]. The second term, often referred to as the "pure volume" effect, represents the induced phonon shift to lower frequency with temperature due to a reduction in the bonds' force constants brought about by typical volume increases with temperature, namely thermal expansion.

The laser excitation wavelength used to obtain the Raman data for the purified Hipco produced sample was 780 nm with a 6 mW power setting. Some other collection parameters associated with this sample were an exposure time of 10.0 s during each of the 3 exposures for each recorded Raman spectrum. Lastly, the aperture setting used for the entrance slit was a 25-micron slit.

Similarly, for the second SWNT sample, the laser power setting for the heated Raman spectra was 6 mW, but this time with 532 nm excitation. A larger entrance slit aperture, 50 microns in width was used, in addition to the exposure time and number of exposures for the Raman spectra at each temperature being 10.0 s and 6 respectively. It must be noted that there is an unexpected "kink" or change in slope in **Figures 9** and **10** at approximately 350 K temperature for all of the Raman bands. This phenomenon which is more noticeable in **Figure 9** for the purified Hipco sample requires further examination.

Figures 9 and **10** show the results of our reproduction [10] of the linear downshift of the primary first order Raman frequencies with temperature for two SWNT samples, along with each sample's corresponding Scanning Electron Microscopy (SEM) images. The linearity of this trend is due to the dominance of the "purely temperature" effect's contribution over that of the contribution from thermal expansion as was noted in [13] and [14], where the latter authors were able to separate each of the two contributions' effects on graphite.

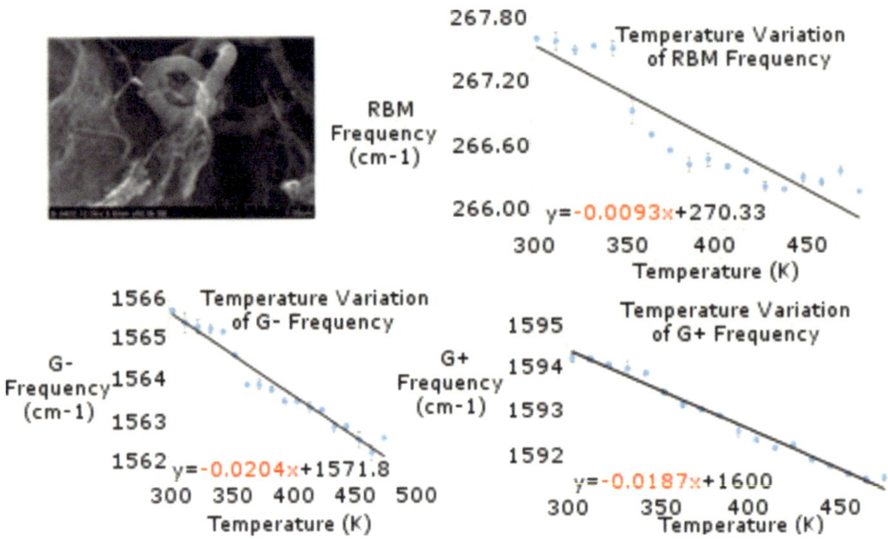

Figure 9. Top left: Scanning electron microscopy (SEM) image of HipCO produced SWNT sample; top right: Variation with temperature of the first order RBM vibration; bottom left: G^- Raman band and bottom right: G^+ Raman band of a SWNT sample heated externally with the Ventacon heat cell. A 780-nm wavelength laser excitation was used for recording the Raman spectra on this HipCO produced sample.

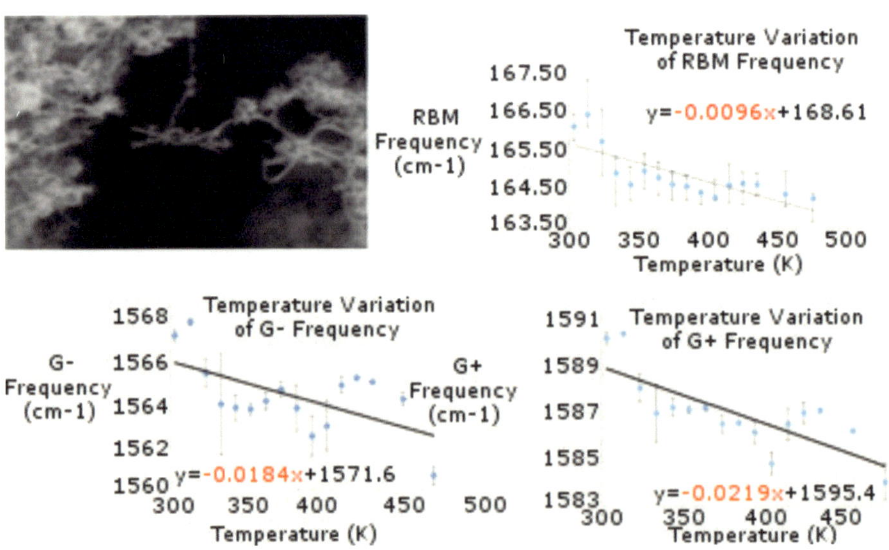

Figure 10. SEM image of SWNT sample (top left); temperature variation of the first order Raman RBM (top right), G^- band (bottom left), and G^+ band (bottom right), with temperature of a SWNT sample heated externally with the Ventacon heat cell. 532 nm laser excitation was used for recording the Raman spectra.

The temperature effect dominance of the phonon frequency downshift mentioned above also manifests itself in the empirical polynomials usually used to express experimental data associated with this phenomenon, such as the example in Eq. (11):

$$\omega(T) = \omega_0 + a_1 T + a_2 T^2 \tag{11}$$

In the reported values of the temperature redshift of Raman frequencies throughout the literature the second-order term a_2 is often negligible compared to its first order counterpart a_1, resulting in a primarily linear trend that was caused mainly by thermal effects not associated with volumetric changes as discussed above.

We now present our results of the use of Resonant Raman Spectroscopy to determine the volume coefficient of thermal expansion (CTE) behavior of the second SWNT sample obtained through the use of the temperature shifted radial breathing mode band. The technique used to determine the volume CTE β, is the same as that used by Espinosa-Vega et al. [15]. **Figure 11** shows the variation with temperature of the fractional volume change of the resonant SWNTs associated with the sole radial breathing mode band of 166.0 cm^{-1} that was present in our spectra. The resulting volume thermal expansion coefficient β, and the same results from [15] are also displayed in **Figure 11**. Based on the premise of the bundled SWNTs being arranged in the sample as circular cylinders of equal length, which was the same assumption made by Espinosa-Vega et al. [15], we obtained the volume at each temperature using the previously discussed relationship between ω_{RBM} and tube diameter:

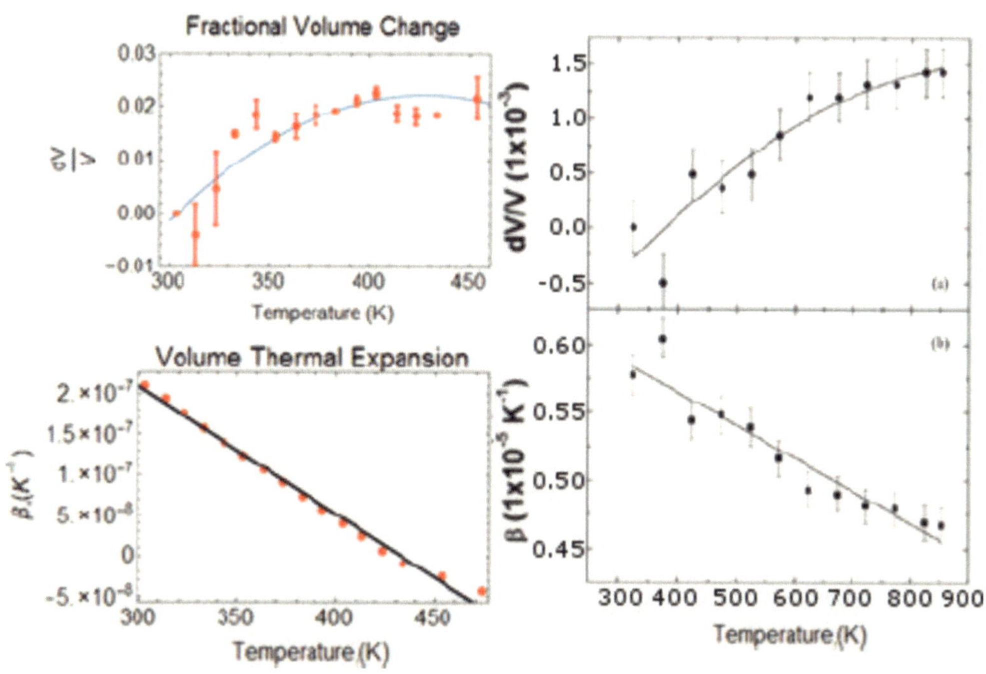

Figure 11. Temperature variation of the fractional change in volume and the volume thermal expansion obtained from the radial breathing mode Raman band (left). Corresponding data are from Espinosa-Vega et al. [15] (right).

$$\omega_{RBM} = \frac{A}{d_t} + B \tag{12}$$

The respective values for A and B of 248.0 cm^{-1}·nm and 10.0 cm^{-1} were used by Espinosa-Vega et al. [15].

The linear temperature dependence of the volume coefficient of thermal expansion (CTE) β was then obtained as shown in **Figure 11** with Espinosa-Vega's data on the right for comparison. According to [12], both the fractional volume data and subsequent linear trend in β led to the conclusion that the temperature dependence of the volume varied as $e^{c+bT+aT^2}$. The resulting values for the parameters a, b, and c from our model were -1.46×10^{-6}, 0.00125142, and -0.245 respectively. The linear downshift in the data for β goes from 0.2×10^{-6} to -0.5×10^{-7} K^{-1}, with a slope of -1.6×10^{-9} K^{-2}, over the experimental temperature range of ~300–473 K. Espinosa-Vega et al.'s data show β decreasing from 5.8×10^{-6} to 4.7×10^{-6} K^{-1} over a larger temperature range of 300–875 K [15].

Espinosa-Vega et al.'s Raman spectra were obtained in a nitrogen atmosphere, whereas ours were collected under open air ambient conditions. This is why the Espinosa-Vega experiment was allowed to operate at the very high (maximum) temperature of 875 K before the onset of any irreversible changes to the Raman spectrum frequencies, such as the loss of intensity brought about by the thermal deterioration of the SWNTs. The smaller values of the temperature slope values $\Delta\omega/\Delta T$ obtained in the nitrogen atmosphere, as opposed to in open air, also meant that Espinosa-Vega et al.'s SWCNTs had a greater thermal stability [15].

Another possible reason for Espinosa-Vega et al.'s β values being much greater than ours was that their experiment was performed on SWNT samples at a much lower density, possibly even at the individual level, due to the pre-processing they performed on their samples. They initially formed a dispersion of their SWNTs with benzene, which was then annealed on a Silicon surface before any Raman spectra were done. The Raman measurements in the present study, however, were all performed on macroscopic bundled samples where van der Waals interactions among the individual tubes were a significant factor. Espinosa-Vega et al. also provided data on the linear decrease of the volume CTE for SWNTs in ambient air, which ranged from 3.3×10^{-6} to 2.7×10^{-6} K^{-1} again over the operating temperature range of 300–875 K. Although, their volume CTE values are still larger, the decrease from their values associated with their experiments done in a nitrogen atmosphere support our earlier suggestion of our lower values of β(T) being due in part to the lower thermal stability of the SWNTs under ambient air conditions.

5.3. Thermal conductivity

We conclude this section dealing with SWNT properties of interest obtainable via Resonant Raman spectroscopy, with a discussion on obtaining an estimate of sample thermal conductivity. The method used was developed by Terekhov et al. [1], who show that there is a demonstrable correlation between the slope of the variation of the G^+ Raman band with increasing laser power at the sample spot for SWNT samples containing different percentages of true

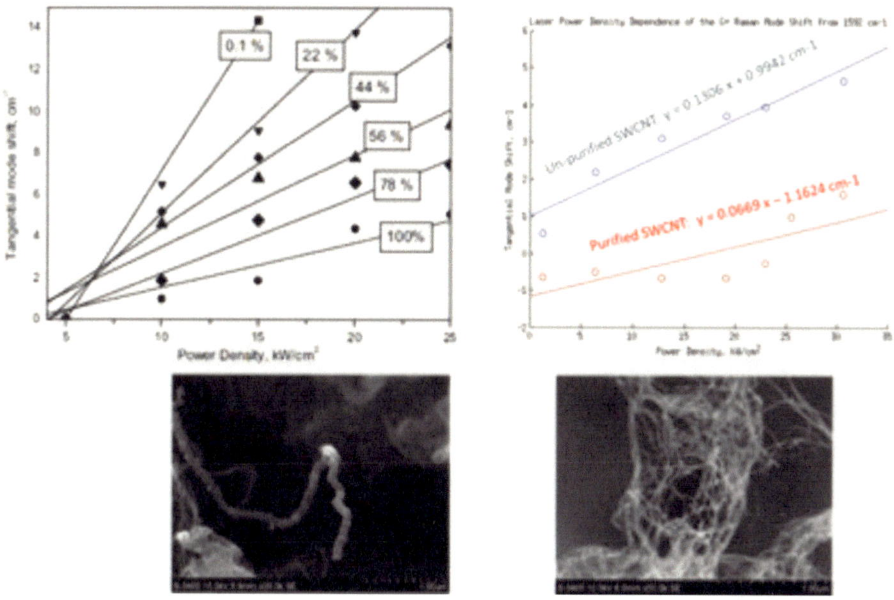

Figure 12. Variation of the G^+ Raman band of SWNT samples of differing purity levels. (top left) data from Terekhov, S.V., et al., AIP Conference Proceedings, (685), 2003, where the percentages indicate the estimated amount of actual carbon nanotubes present in the sample. (Top right) Our data for the two SWNT samples imaged below. The higher purity HipCO SWNT sample on the left contains ~8% residual catalyst material.

SWNT content. In other words, the slope depends on sample "purity," where samples with less purity exhibit larger/steeper slope variations in contrast with high purity samples composed almost entirely of SWNTs. Since additional analysis performed by Terekhov et al. [1] also show that the slope rate of change of the G^+ band with increased laser power is inversely proportional to thermal conductivity κ, using this method allows one to estimate the thermal conductivity of a SWNT sample by simply taking the ratio of the experimentally determined slope variation and an accepted literature value of the thermal conductivity of amorphous Graphite. **Figure 12** shows our reproduction of the above described effect on two SWNT samples (also pictured in the figure) of differing purity, with the less pure sample indeed exhibiting a greater rate of decline of the Raman G^+ band with increased laser power at the sample spot.

6. Temperature and gas exposure effects on graphene Raman spectra

The Raman spectra of graphene were also recorded at varying temperatures (30–200°C) using the Ventacon heated cell and the 780-nm laser with the DXR Raman spectrometer. The graphene samples were all on a silicon/SiO_2 substrate and subjected to consecutive heating/cooling cycles between 30 and 200°C in a sealed chamber. **Figure 13** is the spectrum collected at 30°C. As discussed earlier, the G-band at 1598 cm^{-1} originates from intraplanar stretching, while the peak at 2703 cm^{-1} corresponds to the 2D band. The latter band is due to a second-order two-phonon process that is highly dispersive. It was discovered that this band can be used in estimating the number of layers of a graphene sample. Based on the analysis for this study, the graphene sample(s) in this experiment proved to be largely single-layered. A third band around 3078 cm^{-1} was also present among the Raman spectra collected. It was tracked

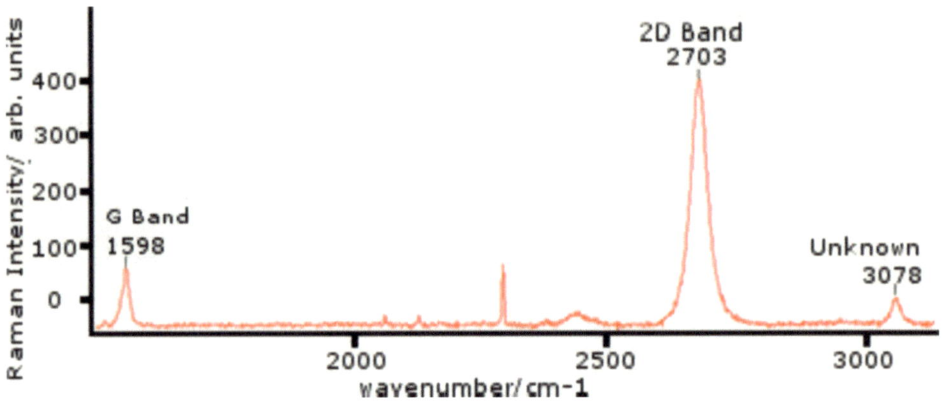

Figure 13. Raman spectrum of graphene at 30°C.

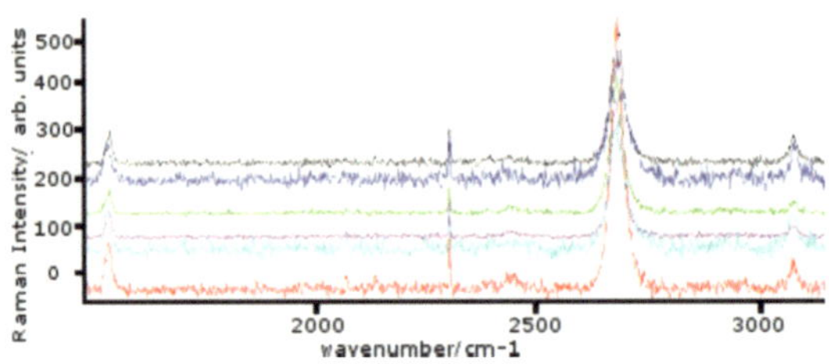

Figure 14. Raman spectra illustrating the sensing of water vapor and humidity effects on graphene as a function of temperature in the range 24.0–150.0°C [top to bottom: Sample at 24.0°C (pristine graphene before heating/pre-exposure), 150.5°C (after heating/pre-exposure), 29.3°C (after cooling/pre-exposure), 26.4°C (last of gas exposure), 149.0°C (after heating/post-exposure), and 28.5°C (after cooling/post-exposure, respectively].

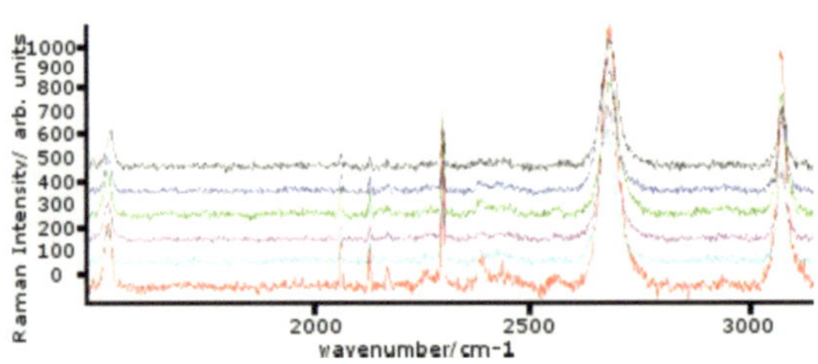

Figure 15. Raman spectra illustrating the sensing of NO gas on graphene as a function of temperature in the range 24.0–150.0°C. [top to bottom: Sample at 24.8°C (pristine graphene before heating/pre-exposure), 139.2°C (after heating/pre-exposure), 27.7°C (after cooling/pre-exposure), 27.5°C (last of gas exposure), 146.2.0°C (after heating/post-exposure), and 28.9°C (after cooling/post-exposure, respectively].

and is probably due to the substrate or glass surface of the microscope stage. Another possibility is that it is either an overtone Raman band of graphene or a C-H stretch benzene ring vibration. Further analysis is needed to determine with certainty the precise origin of the spectral feature at 3078 cm^{-1}.

Raman spectra of the graphene sample(s), under similar heating/cooling cycles was also performed simultaneously with exposure to gaseous (H_2O, NO, SO_2, NO_2). This analysis allowed us to search for any possible patterns in the response of the graphene as its temperature was increased before gas exposure. Each Raman spectral acquisition was analyzed with regards to band frequency, band intensity, and peak width.

Plots of the Raman shift, light intensity, and peak width vs. temperature (in the range 30–150°C) were recorded before exposure and after being exposed to a specific gas. These plots (**Figures 14–17**) were made for all four vapor and gases of interest (H_2O, NO, SO_2, NO_2).

The plots of the Raman shift, light intensity, and peak width over the temperature range 24.0–150°C before and after being exposed to NO are shown in **Figure 18** [16].

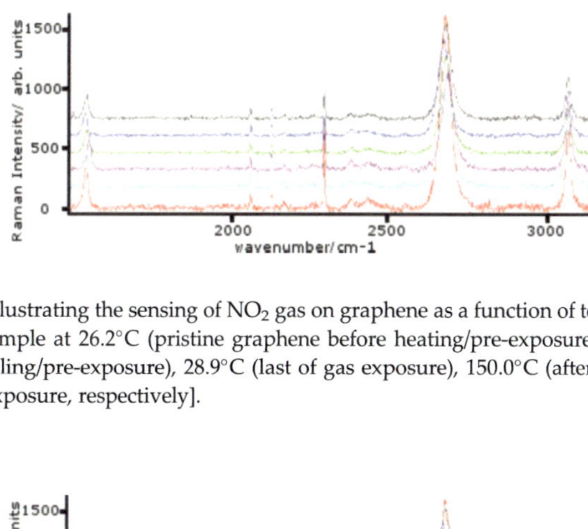

Figure 16. Raman spectra illustrating the sensing of NO_2 gas on graphene as a function of temperature in the range 26.0–150.0°C. [top to bottom: Sample at 26.2°C (pristine graphene before heating/pre-exposure), 151.2°C (after heating/pre-exposure), 31.4°C (after cooling/pre-exposure), 28.9°C (last of gas exposure), 150.0°C (after heating/post-exposure), and 32.4°C (after cooling/post-exposure, respectively].

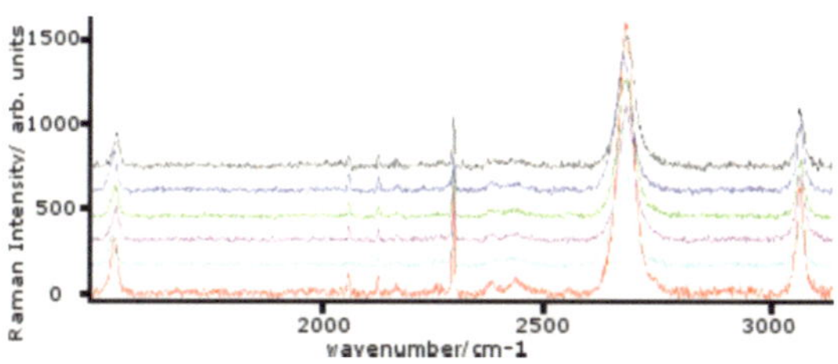

Figure 17. Raman spectra illustrating the sensing of SO_2 gas on graphene as a function of temperature in the range 24.0–137.0°C. [top to bottom: Sample at 24.3°C (pristine graphene before heating/pre-exposure), 149.0°C (after heating/pre-exposure), 29.3°C (after cooling/pre-exposure), 27.7°C (last of gas exposure), 135.6°C (after heating/post-exposure), and 31.1°C (after cooling/post-exposure, respectively].

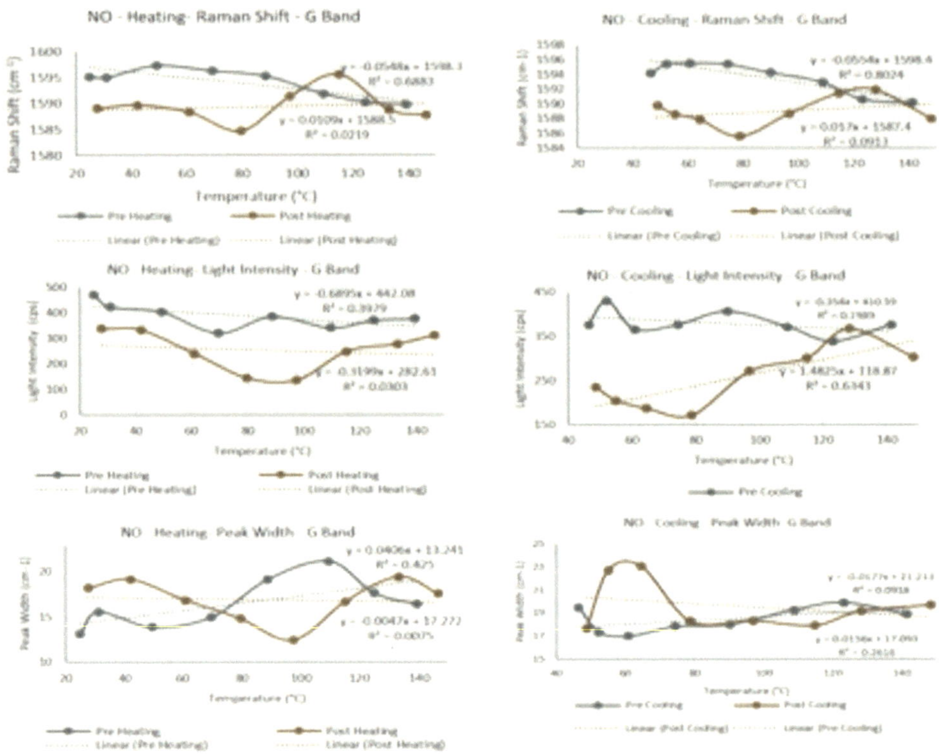

Figure 18. Plots showing the change in Raman frequency shift, light intensity, and peak width, over a temperature range (30–150°C), before and after being exposed to NO [16].

7. Physical properties and Raman spectra of graphene Nanoplatelets

All of the graphene nanoplatelet samples investigated in the current study (functionalized oxygen, nitrogen, argon, ammonia, carboxyl and fluorocarbon) have similar shapes (see **Table 2** and **Figure 19**) [17]. Graphene nanoplatelet aggregates (aggregates of sub-micron platelets with diameters of <2 microns and a thickness of a few nanometers) were identified and studied, rather than individual nanoplatelets (STREM Data Sheets) [18].

The electronic structure of graphitic nanocarbons is linked to its structure, and Raman spectroscopy is sensitive to this intimate and unique relationship, which makes it very effective at studying the various functionalized graphene nanoplatelets used in this study. Akin to pristine

Functional species	X average (μm)	Y average (μm)	Z average (μm)
Argon	4.8	3.9	0.50
Carboxyl	4.3	4.5	0.57
Oxygen	4.7	4.3	0.90
Ammonia	4.4	3.7	0.64
Fluorocarbon	5.0	3.6	0.55
Nitrogen	6.7	6.5	0.91

Table 2. Average x, y, z axis spatial measurements of functionalized graphene nanoplatelet aggregates.

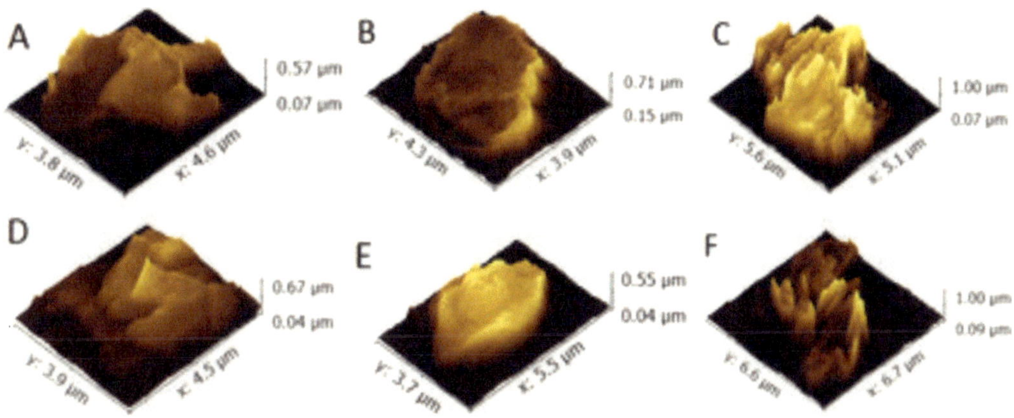

Figure 19. 3D view of SEM data of functionalized graphene nanoplatelet aggregates doped with argon (**a**), carboxyl (**b**), oxygen (**c**), ammonia (**d**), fluorocarbon (**e**), and nitrogen (**f**) via Gwyddion software [19].

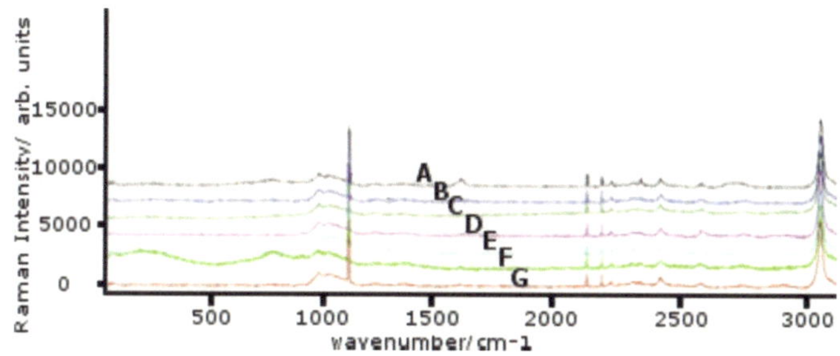

Figure 20. Offset Raman measurements of CVD graphene (**a**), functionalized graphene nanoplatelet aggregates doped with ammonia (**b**), argon (**c**), carboxyl (**d**), fluorocarbon (**e**), nitrogen (**f**), and oxygen (**g**) at room temperature (~25°C) displayed using the Renishaw's WiRE software [21].

graphene, the Raman spectra are characterized by the G and 2D bands. The D band present in the spectra signaled some defects. The behavior of the sharp Lorentzian G-band, at 1587 cm^{-1}, can also be used to verify the sample layer thickness. An increase in the number of layers lowers the frequency of this band, along with an increase in peak intensity. The 2D band, however, depends on the band position and shape, exhibiting distinct band shape differences with the numbers of layers present (AZO Materials) [20]. **Figure 20** indicates identical position and shape of G and 2D bands visible in the spectra of the CVD graphene and the functionalized graphene nanoplatelet samples at room temperature (~25°C).

8. Conclusions and outlook

The thermal characteristics of a variety of graphitic nanomaterials (single-walled and multi-walled carbon nanotubes, graphene and functionalized graphene in the form of nanoplatelets) have been investigated in the temperature range 24.0–200°C using Raman spectroscopy for enhanced gas-sensing and optoelectronic applications. A Kataura plot analysis has been presented for the Radial Breathing Mode vibrations of single-walled carbon nanotubes and possible chiralities identified that pertain to metallic, semiconductor and type 2 semiconducting SWNTs. The effect of temperature on the Raman vibrational modes (RBM, G^{+} and G^{-} bands) of SWNTs has been investigated and the thermal expansion of the SWNT sample determined. A demonstrable correlation between the slope of the variation of the G+ Raman band with laser power for varying levels of SWNT purity has been obtained showing clearly that less pure samples exhibit a steeper slope variation with enhanced laser power. We have also investigated in some detail the behavior of Raman vibrational modes of graphene as a function of temperature in the range (24–150°C), following exposure to a variety of toxic gases (NO, NO_2 and SO_2) at 500 ppm concentration in nitrogen with an eye toward developing sensitive chemical and biological sensors that are efficient, sensitive and portable.

Acknowledgements

Financial support from the National Science Foundation (Award No. PHY1659224) *REU Site in Physics at Howard University* is gratefully acknowledged.

Author details

Daniel Casimir[1], Iman Ahmed[1], Raul Garcia-Sanchez[1], Prabhakar Misra[1]* and Fabiola Diaz[2]

*Address all correspondence to: pmisra@howard.edu

1 Laser Spectroscopy Laboratory, Department of Physics and Astronomy, Howard University, Washington, DC, USA

2 Department of Physics and Astronomy, Francis Marion University, Florence, SC, USA

References

[1] Terekhov SV, Obraztsova ED, DetlaffWeglikowska U, Roth S. Calibration of Raman based method for estimation of carbon nanotube purity. In: Proceedings of the XVII International Winterschool/Euroconference on Electronic Properties of Novel Materials; 8-15 March 2003; Austria. pp. 116-119

[2] Cancado LG, Jorio A, Pimenta MA. Measuring the absolute Raman cross section of nanographites as a function of laser energy and crystallite size. Physical Review B. 2007;**76**. DOI: 10.1103/PhysRevB.76.064304

[3] Wong HSP, Akinwande D. Carbon Nanotube and Graphene Device Physics. New York: Cambridge University Press; 2011. 251 p. DOI: 10.1017/CBO9780511778124

[4] Jorio A, Dresselhaus G, Dresselhaus MS, editors. Carbon Nanotubes: Advanced Topics in the Synthesis, Structure, Properties and Applications. 1st ed. Springer; 2008. 709 p. DOI: 10.1007/978-3-540-72865-8

[5] Saito R, Dresselhaus G, Dresselhaus MS, editors. Physical Properties of Carbon Nanotubes. 1st ed. London: Imperial College Press; 1998. 259 p

[6] Misra P, Casimir D, Garcia-Sanchez R. Thermal Expansion Properties of Single Walled Carbon Nanotubes, by Raman Spectroscopy at 780nm Wavelength. Annual International Conference on Optoelectronics Photonics and Applied Physics (OPAP). 2013. pp. 52-55

[7] Garcia-Sanchez R. Characterization of Metal Oxide Gas Sensors Materials Using Raman Spectroscopy and Computer Simulations. Washington, DC: Howard University; 2015

[8] Scanning Electron Microscope JEOL JSM – 7600F [Internet]. Available from: http://www.sfr.st.keio.ac.jp/forms/FE-SEM-J_manual_in_EN.pdf [Accessed: 2017-10-06]

[9] Tuinstra F, Koenig JL. Raman Spectrum of graphite. The Journal of Chemical Physics. 1970;**53**:1126-1130. DOI: 10.1063/1.1674108

[10] Casimir D. Investigation of Thermal Expansion Properties of Single Walled Carbon Nanotubes by Raman Spectroscopy and Molecular Dynamics Simulation. Washington, DC: Howard University; 2014

[11] Wallace DC. Thermodynamics of Crystals. 1st ed. New York: Dover, 1998. 512 p

[12] Calizo I, Balandin AA, Bao W, Miao F, Lau CN. Temperature dependence of the Raman spectra of graphene and graphene multilayers. Nano Letters. 2007;7:2645-2649. DOI: 10.1021/nl071033g

[13] Dresselhaus MS, Eklund PC. Phonons in carbon nanotubes. Advances in Physics. 2000;**49**:705-814. DOI: 10.1080/000187300413184

[14] Tan P, Deng Y, Zhao Q, Cheng W. The intrinsic temperature effect of the Raman spectra of graphite. Applied Physics Letters. 1999;**74**:1818-1820. DOI: 10.1063/1.123096

[15] Espinosa-Vega LI, Rodriguez AG, Navarro-Contreras H, Vidal MA. Determination of the thermal expansion coefficient of Single-Wall carbon nanotubes by Raman spectroscopy. Spectroscopy Letters. 2013;**48**:139-143. DOI: 10.1080/00387010.2013.860172

[16] Department of Physics and Astronomy: Research Experiences for Undergraduates in Physics, Diaz REU Research paper [Internet]. 2017. Available from: http://physics1.howard.edu/REU/2017/Final/Fabiola%20Diaz%20Research%20Paper.docx [Accessed: 2017-10-06]

[17] Casimir D, Ahmed I, Garcia-Sanchez R, Misra P. Raman spectroscopy & scanning electron microscopy of carbon nanotubes, graphene & graphene nanoplatelets of relevance to energy storage & gas-sensing applications. In: Proceedings of the International Conference on Computational & Experimental Engineering and Sciences (ICCES '17); 24-30 June 2017; Funchal, Madeira Island, Portugal: ICCES; pp. 484-487

[18] Graphene Supermarket: HDPlas Graphene Nanoplatelets: Functionalized Graphene Nanoplatelet Trial Kit [Internet]. 2009. Available from: https://graphene-supermarket.com/HDPlas-Graphene-Nanoplatelet-Trial-Kit.html [Accessed: 2017-10-06]

[19] Gwyddion – Free SPM (AFM, SNOM/NSOM, STM, MFM, …) data analysis software [Internet]. 2004. Available from: http://www.trichord-inc.com/pricing/frames/content/solar_power.pdf [Accessed: 2017-10-06]

[20] Characterization of Graphene Using Raman Spectroscopy. AZO Materials. 2012. http://www.azom.com/article.aspx?ArticleID=6271#3

[21] Renishaw WiRE Raman Software. 2001. Available from: http://www.renishaw.com/en/raman-software–9450 [Accessed: 2017-10-06]

Chapter 9

Graphene Nanocomposites Studied by Raman Spectroscopy

Elena Iuliana Bîru and Horia Iovu

Additional information is available at the end of the chapter

http://dx.doi.org/10.5772/intechopen.73487

Abstract

The goal of this chapter is to provide a general introduction about graphene nanocomposites studied by Raman spectroscopy. The chapter will therefore begin with a brief description of the major Raman bands of carbon allotropes. In the following chapter a concise comparison between single walled carbon nanotubes (SWCNTs), multi-walled carbon nanotubes (MWCNTs), fullerenes and graphene is exposed. The characteristic features in Raman spectra of carbon allotropes, namely the intense signals D and G are investigated. In particular, the chapter will outline the Raman spectrum of graphene and different types of graphene oxide. The last part of the chapter is devoted to graphene nanocomposites.

Keywords: carbon allotropes, carbon nanotubes, graphene Raman bands, graphene oxide, polymeric nanocomposites

1. Introduction

As a nondestructive chemical analysis technique, Raman spectroscopy has become a powerful research tool providing detailed information about chemical structure and identity, phase and polymorphism, molecular interactions and crystallinity. Raman spectrum is a distinct chemical fingerprint for a particular molecule or material and it can be used to quickly identify the sample, or distinguish it from others. Therefore, Raman spectroscopy may be used in any application where nondestructive, microscopic chemical analysis or imaging is required. The use of Raman spectroscopy initially originating in physics and chemistry analysis has now spread to a variety of applications in materials science [1] or even in biology for ultrafast reveling of bacteria [2] and medicine [3].

Recently, carbon materials have revolutionized the field of material science. In order to illustrate the importance of Raman spectroscopy in the field of carbon nanocomposites a short description of some representative carbon allotropes will be exposed. Due to its unique electronic structure, carbon is an element available in a variety of structural forms being able to form sp^3, sp^2 and sp hybridization networks more stable than any other element. Carbon nanomaterials offer a wide range of useful properties such as large specific area [4], excellent electrical conductivity [5], high Young's modulus [6] and thermal conductivity [7]. Raman spectroscopy is an important tool for the characterization of carbon nanomaterials offering valuable information about the existence of structure defects or further functionalization.

Although carbon materials are all entirely made of C-C bonds, the orientation of these bonds is different for each type of carbon allotrope. All these materials are exclusively composed of pure carbon but are different structural forms and exhibit quite different physical properties and chemical behavior. Carbon exists in two allotropic forms: crystalline allotropes (diamond, graphite, fullerene, carbon nanotubes) and amorphous allotropes (carbon black, coke, charcoal).

Diamond is the hardest material on earth and finds applications in cutting, drilling, and jewelry. In diamond structure each carbon atom undergoes sp^3 hybridization and it is linked with four other carbon atoms in a tetrahedral structure. Diamond does not conduct electricity because it does not exhibit any delocalized electrons.

Graphite has a layered structure and all these layers are held by Van der Waals forces. In graphite structure each carbon atoms is sp^2 hybridized and each layer is composed of hexagonal rings of carbon atoms.

Fullerenes are made by heating graphite in an electric arc in the presence of inert gas. These carbon allotropes are cage like structures, with all carbon atoms sp^2 hybridized.

Carbon nanotubes (CNTs) exhibit the form of cylindrical carbon molecules and exhibit unique features that make them extremely useful in a plethora of applications especially in nanotechnology, electronics, optics and many other fields of materials science. A carbon nanotube can be defined as a tube-shaped material, entirely made from carbon, having the diameter measuring on the nanometer scale. A carbon nanotube can be as thin as a few nanometers but as long as hundreds of microns [8]. CNTs are at least 100 times stronger than steel, but only one sixth as heavy, so nanotube fibers could strengthen almost any material [9]. Nanotubes may conduct heat and electricity better than copper. CNT are already incorporated in polymer composites to control or enhance conductivity. Carbon nanotubes may be classified as single-wall (SWCNTs) or multi-wall nanotubes (MWCNTs). SWCNTs can be simply envisaged like a regular tube entirely made of carbon atoms. In contrast to single-wall carbon nanotubes, the MWCNTs are an assemblage of an outer and at least one inner carbon tube separated one from another by interatomic forces.

Graphene is a thin single layer of pure carbon. Graphene was firstly isolated in 2004 by two researchers from The University of Manchester using the "scotch-tape" technique [10]. Basically graphene consists in a monolayer of graphite. In more complex terms, graphene is a two dimension honeycomb single layer crystal lattice formed by the tightly packed sp^2 bonded

http://dx.doi.org/10.5772/intechopen.73487

carbon atoms [11]. Having excellent thermal, mechanical, electrical and barrier properties [12], graphene is recommended for many applications such as: electronics [13], antimicrobial materials [14], construction materials [15], battery [16] and supercapacitors etc. More details about graphene and other graphene derivatives will be exposed in section 2 of this chapter.

In order to differentiate these materials there is a strong demand for techniques that can be used to characterize them. As a remarkably sensitive technique, Raman spectroscopy suits perfectly on these demands, being highly responsive to symmetric covalent bonds with very little or no dipole moment. Raman spectroscopy is capable of discerning any slight changes in structure.

When monochromatic radiation is incident upon a carbon allotrope sample the light will interact with the sample in a specific way. Every bond in the Raman spectrum corresponds directly to a specific molecular bond vibration, including bonds such as C-C, C=C, C-H etc. As a chemical fingerprint of the material, the general spectrum profile (peak position and intensity) provides unique information which can be used to identify the material and distinguish it from others.

When comparing the Raman spectra of two carbon allotropes – diamond and graphite – significant differences between these two materials can be noticed even if both are entirely made of C-C bonds (**Figure 1**). The Raman spectrum of pure diamond exhibits an extremely sharp signal at ~1332 cm^{-1}. The Raman spectrum of graphite shows different features compared to diamond. Two distinguishable peaks are revealed at ~1350 cm^{-1} (D band) and ~1580 cm^{-1} (G band). The G band arises from the stretching of the C-C bond in graphitic materials and it is common to all sp^2 carbon systems. The presence of an additional band in the graphite spectrum reveals that graphite is not as uniform in structure as diamond. The D-mode in graphite is induced by disorder or defects and increases linearly with decreasing graphite crystallite size (**Figure 2**).

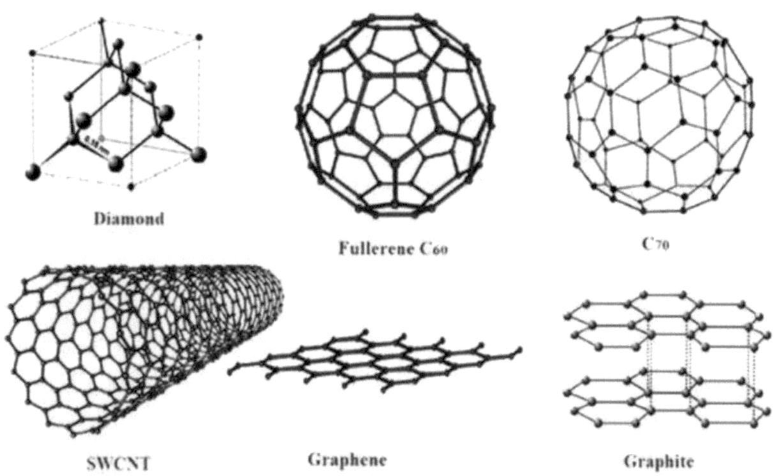

Figure 1. Structure of the most representative carbon allotropes.

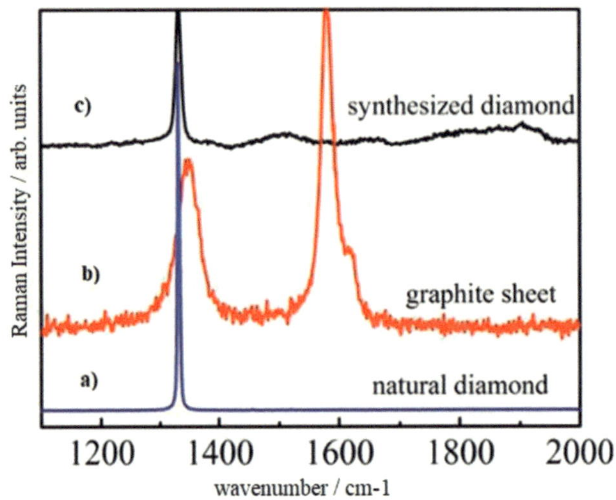

Figure 2. The Raman spectra for a) natural diamond and c) synthesized diamond compared with b) Raman spectrum of graphite (Yao K et al. (2017) copyrights) [17].

More complex structures can be also investigated by Raman spectroscopy. Fullerenes have attracted much attention for their applications in non-linear optics [18] or biomedical devices [19]. C60 (also called Buckminster fullerene) and C70 have been investigated in a large number of experiments because of their potential applications fabrication of nanodevices such as field effect transistors and flat panel display devices based on field emission [20].

Figure 3 compares the Raman spectra of C60 and C70. The Raman spectrum of C60 exhibited strong signals at 1467 cm^{-1} and 1567 cm^{-1}. This fact reveals that C60 is composed of sp^2 bonded carbon and the sharpness of the signal shows that C60 exhibits a uniform structure. On the contrary, the Raman spectrum of C70 exhibits numerous other peaks. In case of C70 film, the main peaks are located at 1564 and 1228 cm^{-1} due to a reduction in molecular symmetry which results in more Raman bands. Their relative intensities strongly depend on the excitation laser wavelength.

Until 1980 only four carbon allotropes were known: graphite, amorphous carbon, fullerenes and diamond. Since their discovery in 1991 by Dr. Sumio Ijima, carbon nanotubes (CNTs) have gained tremendous attention as a versatile nanomaterial with abundant applications. Like previously mentioned, CNTs are carbon allotropes with a cylindrical nanostructure.

CNTs are essentially rolled up graphene sheets that have been sealed to form hollow tubes (**Figure 1**). Depending on the number of concentrically rolled-up graphene sheets, CNTs are classified to single-walled (SWCNT), double-walled (DWCNT), and multi-walled CNTs (MWCNT). The structure of SWNT can be imagined by wrapping a one-atom-thick layer of graphene into a seamless cylinder. DWCNT is considered as a special type of MWCNT wherein only two rolled up graphene sheets are present. MWCNT consists of two or more numbers of rolled-up concentric graphene sheets [22]. The diameter of SWCNT is generally up to 2 nm. In case of MWCNTs the diameter varies from 5 to 20 nm, seldom exceeding 100 nm [23].

http://dx.doi.org/10.5772/intechopen.73487

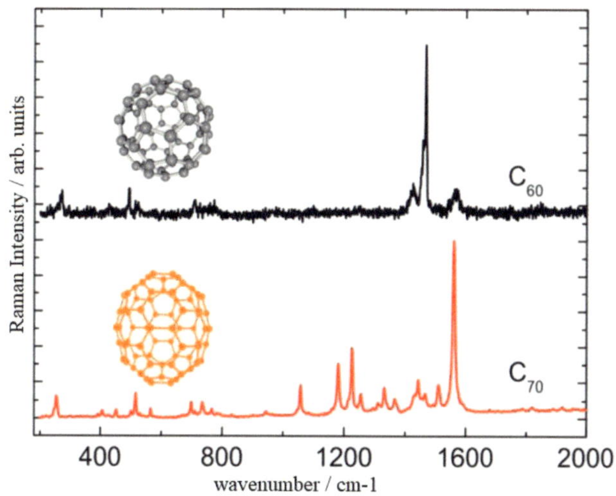

Figure 3. Raman spectra for C60 and C70 films. The insets show the molecular structures for C60 and C70 (Zhang X et al. (2016) copyrights) [21].

Having remarkable properties such as high electrical conductivity, very high tensile strength and low thermal expansion coefficient, CNTs have been investigated for many applications such as composite materials, microelectronics and electronic components, solar cells, energy storage devices [24] etc. In addition, their one-dimensional structure makes them an ideal platform for biomedical applications. Due to their important applications a significant number of methods to produce CNTs were developed: arc discharge method [25], laser method [26], chemical vapor deposition (CVD) [27], ball milling [28], etc. Final properties of CNTs are dependent on the production and purification methods.

In **Figure 4** a comparison of Raman spectra of graphite, SWCNTS and MWCNTs is depicted. The stretching of the C-C bond in graphitic materials gives rise to the so-called G-band Raman feature which is common to all sp^2 carbon systems. This spectral feature is similar for graphite and nanotubes but is not used for distinguishing one carbon nanostructure from another. The G-band is highly sensitive to strain effects in sp^2 carbon materials and can be used to investigate any modification to the structure of graphene, such as the strain induced by external forces in multiwall nanotubes, or even by the curvature of the side wall when growing a SWCNT [29]. A prominent G band can be noticed in the graphite Raman spectrum at ~1580 cm^{-1}. As one can see the G band is present also in the SWCNT and MWCNT spectra but with different width. The G-band of investigated SWCNTs splits in two band components because of large diameter nanotubes and it can be used to distinguish metallic and semiconducting nanotubes.

Another important band in the Raman spectra of the investigated nanotubes at ~1350 cm^{-1} was observed known as the D band. The D band is caused by disordered structure of graphene sheets. The presence of disorder in sp^2-hybridized carbon systems results in resonance Raman spectra as one can see in the Raman spectrum of CNTs making Raman spectroscopy one of the most sensitive techniques to characterize disorder in sp^2 carbon materials. In case of CNTs the D band is significantly increased compared to graphite. The increase of the ratio

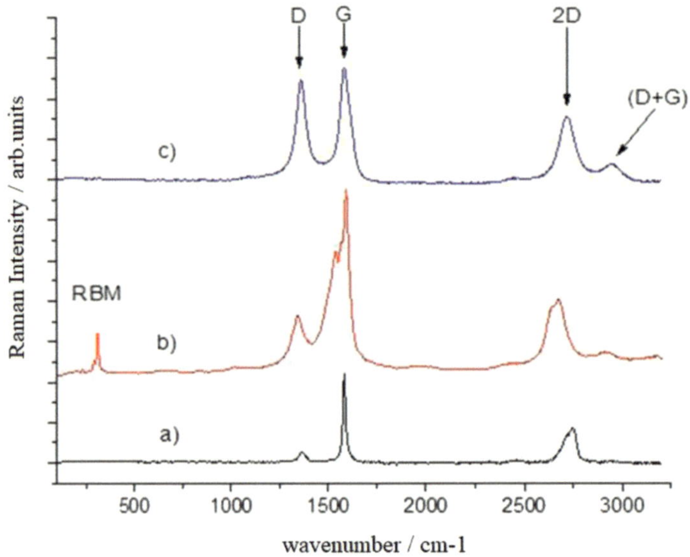

Figure 4. The Raman spectra of a) graphite, b) single wall carbon nanotubes (SWCNTs) and c) multiwall carbon nanotubes (MWCNTs).

between the intensity of D band and the intensity of G band (I_D/I_G) indicates that some disorder of the graphene sheets is induced. As previously mentioned graphite consists of stacks of planar sheets of graphene and carbon nanotubes are basically rolled-up graphene sheets. Therefore the presence of the 2D band (also named G` band) is observed in all spectra of the investigated samples which features the arrangement and the number of graphene layers [30]. The importance of the 2D band of graphene will be better explained in section 2 of this chapter. The radial breathing mode (RBM) is particularly important for the determination of the diameter of CNT, its frequency being related to the aggregation state of SWCNTs. The RBM band is unique to SWNCTs and corresponds to the expansion and contraction of the nanotubes. Comparing the Raman spectrum of MWCNTs to that of SWCNTs some important differences can be easily noticed: the absence of RBM mode in MWCNTs spectrum and much sharper D peak in MWNCTs. The RBM band is not present in case of MWNCTs due to the outer tubes that restrict the breathing mode. The presence of a more outlined D band in case of MWCNTs is observed because of the multilayer configuration of nanotubes suggesting a more disordered structure. In addition, a sharper D+G combination peak strongly supports the presence of higher disorder in the MWCNTs, compared to SWCNTs.

2. Graphene and graphene oxide

Even if for several decades the isolation of single layer graphite seemed to be impossible two researchers from Manchester University successfully managed in 2004 to isolate monolayer graphite. The "scotch-tape" technique reported by Geim and Novoselov consisted in obtaining single layer of graphene on a silicon oxide substrate by peeling the graphite by micromechanical cleavage [10]. In 2010 the two researchers won the Nobel Prize in Physics for their pioneering study [31]. Several methods have been established for graphene production, such as micromechanical or chemical exfoliation of graphite [32], graphitization of silicon carbide [33], chemical vapor deposition (CVD) growth [34], and chemical, thermal or electrochemical reduction of graphene oxide [35].

Graphene is a one-atom-thick planar sheet of sp2-bonded carbon atoms that are densely packed in a honeycomb crystal lattice [36]. Due to its 2D nature graphene exhibit a unique combination of characteristics not seen in other carbon allotropes. Graphene is the thinnest, strongest and stiffest material [37]. Graphene has extraordinary electrical properties due to the high electron mobility at room temperature [38]. In terms of mechanical properties graphene exhibit greater performances when single or few layer graphene are employed. Therefore, the superior mechanical properties of graphene and its derivatives make them the ideal candidates for incorporation into a large variety of materials in order to produce composites with enhanced properties.

Graphene also exhibits other high characteristics such as large specific surface are, high transparency and high thermal conductivity. Because of its high surface area graphene also finds potential applications as support material in catalysis field as an electrode material in electrochemical applications such as supercapacitors and batteries [39].

Due to these extraordinary properties, graphene found already a great number of important applications with potential used in touch screens displays, fuel cells, intelligent coatings, transparent conductive films and flexible electronics [40]. In addition, once functionalized with biomolecules like polysaccharides [41], proteins [42], etc. or other biological systems graphene can be integrated for developing new applications in biomedicine and bio-nanotechnology such as biosensors [43], biocatalysis [44], biofuel cells [45], etc.

It is worth noting that the electronic properties of graphene drastically depend on the number of graphene layers. For that reason, the graphene community distinguishes between monolayer graphene, bilayer or few-layer graphene. A structure composed of more than 10 graphene layers exhibits the electronic properties of graphite and therefore is considered as a thin film of graphite. Being transparent as well as a good conductor, graphene may replace the electrodes in the indium used in touchscreens [46].

In order to integrate graphene into various functional structures or other materials for making performant nanodevices one preliminary condition is required: graphene sheets have to be exfoliated into individual or few-layer sheets and stabilized. Also, unwanted by-products and structural damage can be produced while synthesizing graphene. A quick and precise method for determining the number of layers of graphene sheets is essential for speeding up the research and exploration of graphene. In sp2-bonded carbon species, as a highly sensitive and non-destructive technique, Raman spectroscopy can be used to investigate the number of layers, the type and relative quantity of defects, mechanical strain, and any further functionalization. Therefore Raman spectroscopy is one of the most powerful tools available for analysing graphene.

In specific case of graphene, Raman spectroscopy can evaluate not only the number of graphene layers, but also can provide a quick and non-destructive means to distinguish monolayer, bilayer and few layer graphene. The most prominent Raman features of graphene are the so-called G band and 2D as easily seen in **Figure 5**, which depicts a typical Raman spectrum for graphite and graphene respectively obtained using a 514 nm excitation laser.

Graphite consists of sp^2 bonded planar graphene sheets stacked through Van der Waals intermolecular forces. When comparing the Raman spectra of graphite and graphene one can observe a tremendous similarity. The reason for that is the fact that graphite consists in multiple stacked graphene layers. As in the case of graphite, the G mode from graphene occurs around 1580 cm^{-1} and the 2D signal is situated around 2700 cm^{-1}. Two further signals can be noticed: the D band that is observed at ~1350 cm^{-1} and 2D` band at ~3250 cm^{-1}.

In terms of vibrational behavior, the G band originates from in-plane stretching vibrations of sp^2 carbon atoms in both rings and chains. Even at low intensity the D mode can be observed in the Raman spectrum of graphene which occurs due to the breathing modes of sp^2 carbon atoms rings. Generally the D mode is associated with the presence of graphene structural defects. When the D band is higher it means that the sp^2 bonds are broken and new sp^3 bonds are created. Consequently the increase of the ratio between the intensity of D band and the intensity of G band (I_D/I_G) demonstrates that new defects are created during the modification of pristine graphene.

The D mode is almost absent in well-ordered structure of graphene and graphite. Despite the similarities, there are some significant differences as one can notice in **Figure 5**. In case of pure graphene the 2D band situated at ~2700 cm^{-1} is much sharper. The 2D band originates from a two-phonon double resonance process and it is interrelated to the band structure of graphene layers. **Figure 5(b)** indicates a significant change in the shape and intensity of the 2D band of graphene compared to graphite. It can be easily observed that in case of graphene the 2D peak is much narrower and its position is down-shifted. Another difference is observed in the

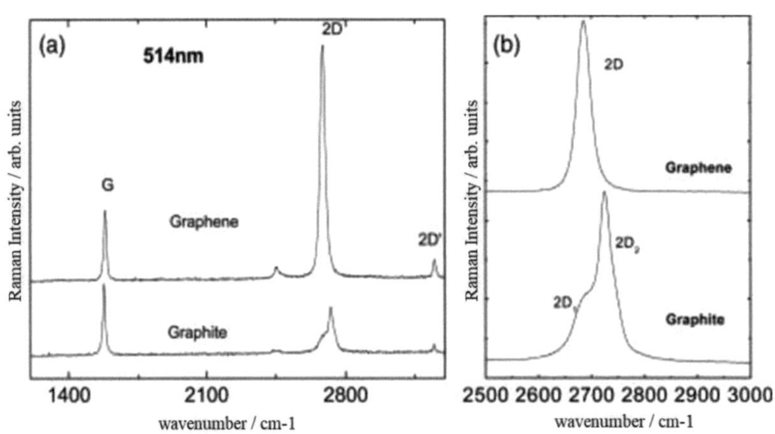

Figure 5. (a) The Raman spectra of graphene and graphite measured at 514.5 nm. (b) 2D peaks in graphene and graphite (Ferrari A (2007) copyrights).

Raman spectrum of graphite in which the 2D band is formed from two elements, namely $2D_1$ and $2D_2$ [45], which are roughly ¼ and ½ of the height of the G peak, respectively. Graphene exhibits a single sharp 2D signal, approximately four times more intense than G band [47]. When more graphene layers are present, the 2D peak is shifted to higher frequencies due to the interactions between the graphene layers.

Many studies showed that Raman spectroscopy can be used as an indicator for single or multi-layer graphene [48]. As one can see the 2D peak evolves as the number of graphene layers increases to about ten layers upon its profile resembles with that of graphite. As the number of graphene layers increases an important reduction of the relatively intensity for the $2D_1$ mode is noticed. Therefore graphene stacks that have more than five layers are more difficult to discriminate from graphite by Raman spectroscopy (**Figure 6**).

It is worth pointing out that this technique for identifying the number of graphene sheets is precisely established only for graphene with AB Bernal stacking [49]. Graphene samples that exhibit AB Bernal stacking features are graphene layers where half of their atoms lie directly over the center of a hexagon in the lower graphene sheet, and half of the atoms lie over an atom. Bernal stacked bilayer graphene exhibit much interest for functional electronic and photonic devices due to the feasibility to continuously tune its band gap with a vertical electrical field [50]. Such type of samples are obtained from highly oriented pyrolytic graphite (HOPG) produced by mechanical exfoliation. Also chemical vapor deposition (CVD) or thermal deposition of SiC can be used to synthetize bilayer graphene but these procedures do not lead to homogeneously AB stacking layers.

Since the graphene flakes have small dimensions it is important to select a Raman instrument with high microscopy performances. Consequently for more accurate results most Raman measurements are performed using an optical microscope which allows a better localization

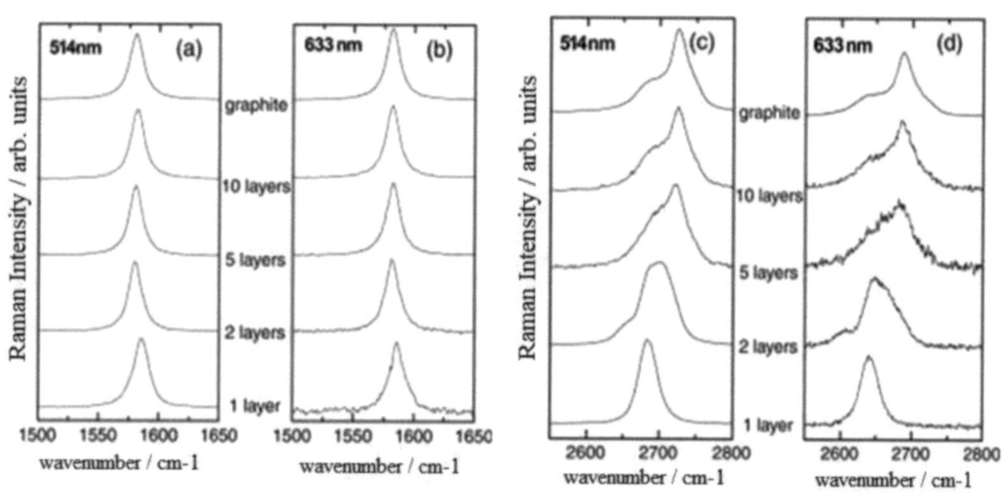

Figure 6. (a) and (b) The evolution of G band by increasing the number of layers, (c) and (d) The evolution of the 2D band by increasing the number of graphene layers using 514 and 633 nm excitation laser (Ferrari A (2007) copyrights).

of the graphene layers. Raman microscopy couples a Raman spectrometer to a standard optical microscope, allowing high magnification visualization of graphene and Raman analysis with a microscopic laser spot.

However, the addition of a microscope to a Raman spectrometer does not provide full 3D spatial resolution. More recently, confocal Raman microscopy (CRM) was used in order to investigate graphene layers [51]. As the name suggests confocal Raman microscopy refers to the ability to spatially filter the analysis volume of the sample, in the XY (lateral) and Z (depth) axes. A confocal microscope practically designs clear images of a sample by removing most of the light from the investigated sample [52]. Apart from allowing better observation of fine details of the sample CRM gives rich information concerning the distribution of individual chemical components, and variation in other effects such as phase, polymorphism, stress/strain, and crystallinity. Based on thousands of Raman spectra acquired from different positions on the sample Raman spectral mapping can be created generating detailed chemical images.

Moreover, the substrate on which the graphene samples are deposited and the Raman equipment performances has a significant role for graphene investigation. Monolayer graphene can be observed on many types of substrates like sapphire, single crystal quartz, glass, metal alloys like NiFe, or polymers like polytetrafluoroethylene (PTFE), but the most popular substrate surface used to discriminate the monolayer graphene is the silicon wafer with a silicon dioxide layer (SiO_2/Si). The Si/SiO_2 substrate usually with 300 nm thickness was reported in many studies as the most appropriate for visual detection of single layer graphene (SLG) [53]. Additionally, it is important to study the interaction between monolayer graphene and the substrate because possible interaction may appear due to defects or surface changes between the two interfaces. In case of monolayer graphene obtained by epitaxial growth from SiC substrates a strong interaction between graphene and SiC substrate appears by strong blue-shifting the G and 2D signals (~11 cm^{-1} and ~34 cm^{-1}, respectively) due to covalent bonding of the two interfaces [54]. On the other hand, the interactions between graphene produced by micro-cleaving and standard SiO_2/Si substrate do not influence the physical structure of graphene because only weak Van der Waals forces could appear [55].

It was found that the Raman spectrum depends not only on the substrate, but also on the wavelength of the excitation laser. Regarding the laser excitation energy for graphene investigation usually the visible lights laser are employed (from 633 to 473 nm). The near infrared (NIR) or ultraviolet (UV) sources are not frequently used. When analyzing graphene layers placed on silicon wafers with silicon dioxide (SiO_2/Si) strong fluorescence signals are observed using a NIR laser (780 or 785 nm). Also, graphene layers are difficult to investigate using a UV excitation laser (from 244 to 364 nm) due to the fact that the obtained Raman spectrum exhibits differences concerning relative intensities of the characteristic graphene signals. Not least the selection of the excitation source is important. In order to avoid sample damaging usually powers between 0.04 to 4 mW are employed. At higher laser power it was observed that the laser may burn the graphene sample leading to graphitization and therefore to spectral variations. Also, if lower laser power is used the ratio between the Raman signal and noise is very poor [56].

Graphene oxide (GO) is another graphene material that can be intense characterized by Raman spectroscopy. By chemical oxidation of graphite in the presence of strong oxidizing agents

http://dx.doi.org/10.5772/intechopen.73487

and ultrasonic cleavage graphite oxide is obtained, at the end presenting on its surface a significant number of functional groups such as hydroxyl, carboxyl and epoxy [57]. The introduction of oxygenated functionalities not only expands the layer separation, but also makes the material hydrophilic and relatively easy to disperse in aqueous media or other polar solvents [58]. This property enables graphite oxide to exfoliate in hydrophilic medium under sonication and finally to produce single or few layer graphene oxide (GO). Consequently the main difference between graphite oxide and graphene oxide is the number of layers.

In the last decade GO attracted the researchers attention due to its many important properties which can be used to tailor novel applications. Graphene oxide sheets exhibit high design flexibility. GO is decorated with a significant number of carboxyl groups (-COOH) and most of them are located at the GO edges. These carboxyl groups are extremely useful as they can easily react for attachment of various functionalities. Thus, reactions may be established with (a) amines and various organic molecules or polymers which exhibit in their composition amino groups, by forming an amide linkage, (b) alcohols, phenols or epoxy groups to form ester bond, (c) various other organic reactive macromolecules, resulting in the functionalization of GO.

Functionalization of GO can fundamentally change graphene oxide's properties and consequently, graphene oxide's applications. Graphene oxide flakes can be used to remove radioactive ions from water for disposal [59]. Also, graphene oxide can be used to develop sensors that can detect tumorous cells by attaching to GO surfaces molecules that contain antibodies that are further linked to the cancer cells. The cancer cells are then tagged with fluorescent molecules to make the cancer cells stand out in a microscope [60]. Withal graphene oxide is used to obtain anodes for rechargeable lithium-ion batteries. The graphene oxide is thermally treated in order to extract the oxygen form the film and driven to cause pores in the film which are rapidly filled with lithium ions, resulting in quicker charge-discharge process for batteries [61].

With respect to electrical conductivity, graphene oxide behaves as an electrical insulator, because of the disruption of its sp^2 bonding networks. It is important to reduce the graphene oxide so as to restore the honeycomb hexagonal lattice of graphene, in order to recover electrical conductivity. The product of this reduction reaction has been named in different ways, including: reduced graphene oxide (rGO), chemically-reduced graphene oxide (CRGO), and graphene. For the purposes of clarity, we will refer to the product as reduced graphene oxide (rGO). Chemical reduction of graphene oxide is mostly employed in the presence of hydrazine (N_2H_4) where the majority of the oxidized groups of GO are reduced. But the use of anhydrous N_2H_4 demands a dry environment which creates difficulties for the large-scale production. Usually chemical reduction agents are classified as toxic or corrosive. The electrochemical method to reduce graphene oxide in order to produce large rGO films is a greener, safer and more convenient procedure for reducing graphene oxide films. Also the thermal expansion of graphite oxide can be used for exfoliating graphite layers and finally to produce functionalized graphene sheets. Temperatures around 550°C or higher can break the Van der Waals forces that stack the graphene layers together and exfoliation occurs. After the reduction of graphene oxide defect sites within the lattice are produced which provide new routes for chemical functionalization. Chemical modification of the graphene oxide by functionalization with different other molecules or polymers opens new routes for the incorporation of

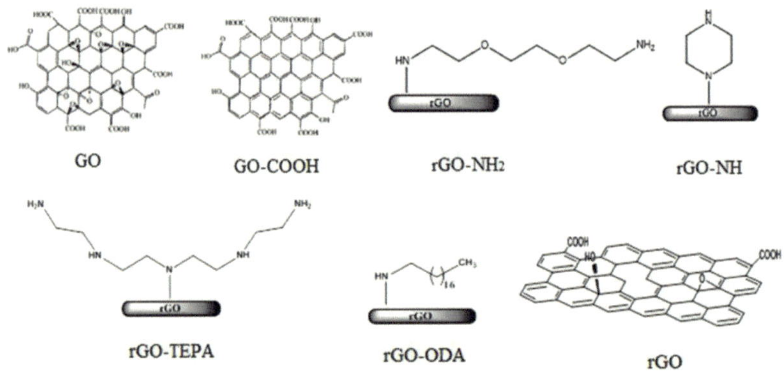

Figure 7. Chemical structures of graphene oxide investigated by Raman spectroscopy.

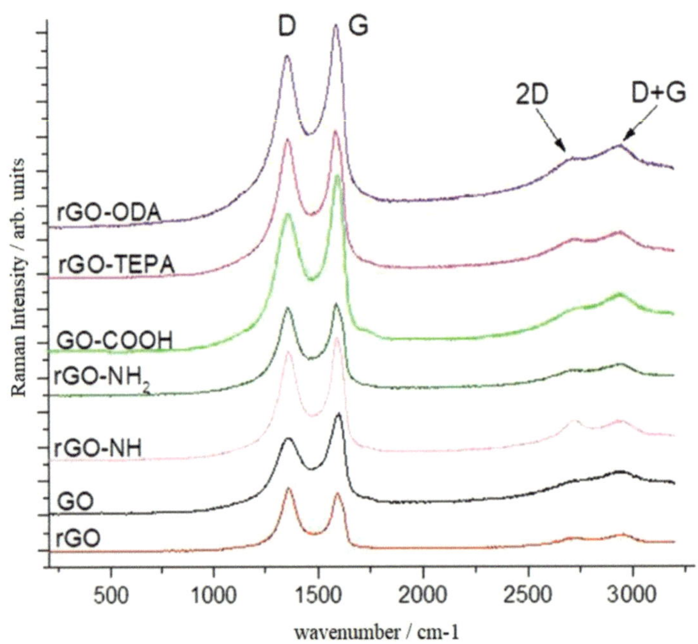

Figure 8. Raman spectra of various graphene oxides.

graphene in other matrices and surfaces to enhance its applicability, that which would otherwise be more difficult using pristine graphene.

Being a derivative of graphene, the graphene oxide structure and functionalized graphene oxide can be also successfully characterized by Raman spectroscopy. For this study different structures of commercial graphene oxide (**Figure 7**) were investigated using a Raman spectrometer equipped with confocal microscope.

In the Raman spectra of the studied graphene oxide structures one can observe that the G band is much broader than in case of graphene and also blue-shifted to ~1590 cm^{-1} (**Figure 8**). The D band from graphene oxide Raman spectra is also modified exhibiting a much higher intensity due to the disorder in the sp^2 structure induced after the oxidation of graphite and also due to the attachment of hydroxyl and epoxide groups on the planar carbon structure. Depending on the functionalization degree of graphene oxide, the Raman spectrum may exhibit sometimes even stronger D band than G band. Regarding the 2D band its intensity is very small compared to the D and G peaks, but may be enhanced by reducing the number of graphene oxide layers. The D+G combination peak is also observed which strongly supports the presence of a higher disorder structure for graphene oxide.

3. Raman investigation of graphene and graphene oxide nanocomposites

Polymeric composites are biphasic materials consisting essentially of a continuous phase, commonly referred to as polymer matrix, and a reinforcing or filler agent, which is the discrete phase. The purpose of this association is to obtain materials with enhanced properties, superior to those of individual components, capable of replacing natural materials (wood, rocks, etc.), aluminum and its alloys, and other metallic materials. Polymeric composites are obtained from a wide range of matrices (epoxy resins, polyester resins, phenol-formaldehyde resins, vinyl polymers, elastomers, polyimides, etc.) with reinforcing materials (boron fibers, glass fibers, or filler materials (wood flour, starch, silica, talc, asbestos, etc.).

Nanocomposites represent a new class of composites, characterized by the coexistence of two distinct phases (an organic one which is the polymer as the continuous phase and an inorganic phase dispersed in the continuous phase, the latter exhibiting nanometric dimensions). The advantages of these structures consist in global properties superior to the individual components such as improved optical clarity, high mechanical resistance, better conductivity, leading to important uses in electronics, optics, constructions, etc. In order to obtain nanocomposites, two important aspects should be considered: firtsly, the nanoparticle must be compatible with the polymer and to show satisfactory interfacial interaction; secondly, the most convenient way to uniformly disperse the nanoparticles in the polymer matrix should be chosen. In most cases polymeric nanomaterials exhibit multifunctionality by combining more than one properties.

When the polymer is unable to intercalate between the graphene layers, a phase separation (two distinct phases) is obtained, the properties of which resemble the microcompounds. In addition to this class, two other types of composites can be prepared: intercalated structures where most of the time a single polymer chain is interposed between layers of graphene, resulting in a multilayered structure in which the polymer-graphene layers alternate and exfoliated structures in which the graphene layers are completely dispersed in the continuous polymer matrix (**Figure 9**).

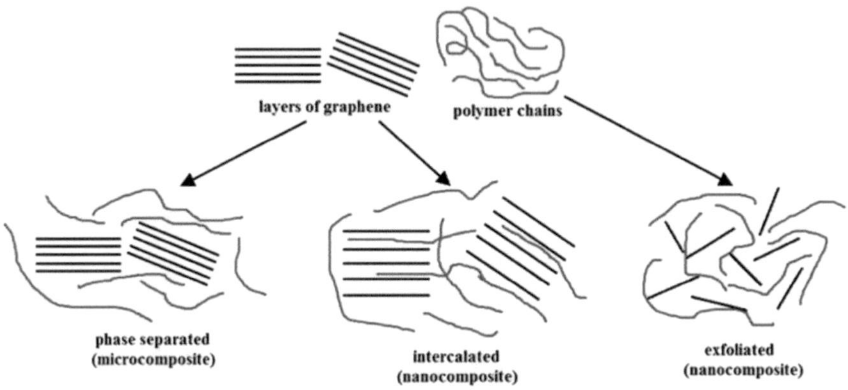

Figure 9. Schematic structure of graphene – based polymer composites.

Graphene nanosheets–polystyrene nanocomposites (GNS-PS) were prepared by in situ emulsion polymerization and reduction of graphene oxide using hydrazine hydrate [62]. The nanocomposites displayed high electrical conductivity, and a considerable increase in glass-transition temperature and good thermal stability of PS are also achieved. Raman spectroscopy was employed as an efficient tool to probe the structural characteristics and properties of graphene and graphene-based materials.

Figure 10 shows a schematic diagram for the formation of the GNS-PS nanocomposites. TEM images revealed polystyrene microspheres with diameters ranging from 90 to 150 nm attached to the graphene surface, particularly along the edges of the stacked nanosheets with a thickness of several nm. This suggests that the compatibility between PS microspheres and GNS is sufficient to obtain nanosized dispersion without an additional surface treatment.

Regarding the Raman spectrum of GNS, two intense features are assigned to the D band at 1331 cm^{-1} and the G band at 1594 cm^{-1}. The G peak was assigned to vibrations of sp^2 carbon atoms. The peak intensity ratio I_D/I_G of GNS nanocomposites was calculated as 1.156. This fact demonstrated the presence of localized sp^3 defects within the sp^2 carbon network, which shows the chemical grafting of polymers to the GNS surface.

The prominent D peak revealed some structural defects are created during the reduction process of the oxidized functional groups, while GNS synthesized through the exfoliation method is considered a more effective route for graphene sheets production. The G band from the GNS Raman spectrum was noticed at 1594 cm^{-1} and upshifted by 5 cm^{-1} in the composites of GNS–PS. By combining the XRD and Raman spectroscopy, Hu and co-workers proved that the substantial structure of the GNS has been maintained after PS microspheres were linked to the edges of the stacked graphene nanoplatelets, which is advantageous for improving the electrical and thermal properties of polymer.

Interest in graphene oxide has increased significantly due to its epoxy, hydroxyl, carbonyl and carboxyl functional groups, allowing its functionalization and the formation of various monomers on its surface. Dispersibility of graphene oxide in water and other solvents makes

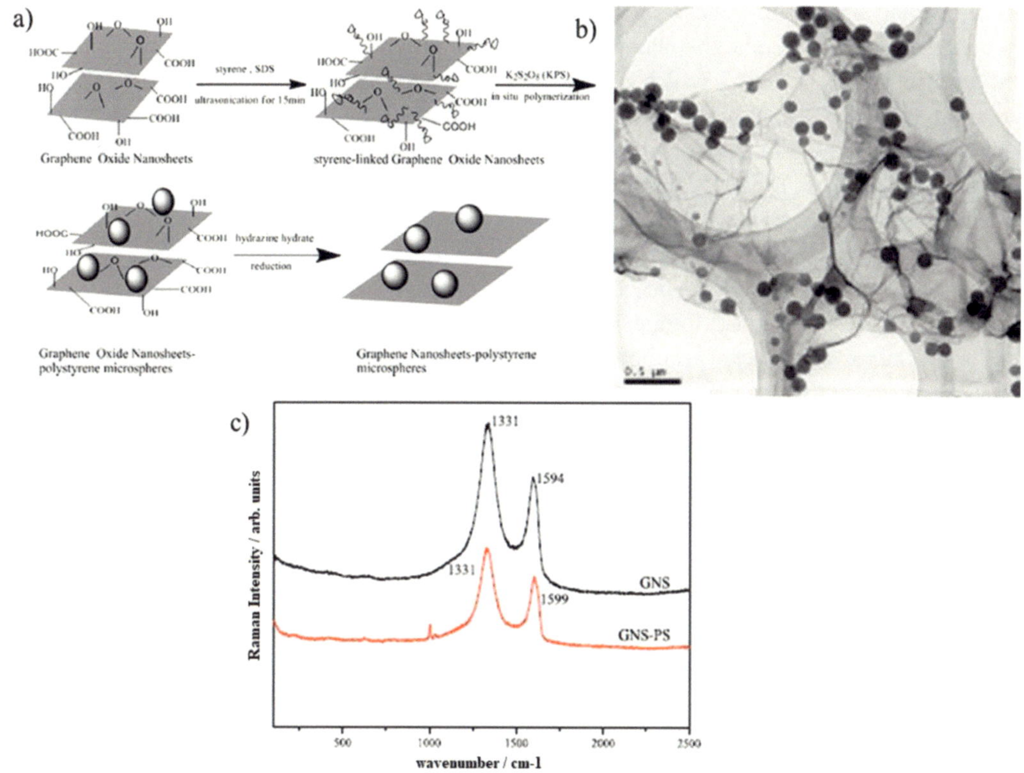

Figure 10. a) Synthesis of GNS-PS nanocomposites; (b) TEM image of CNS-PS nanocomposites; (c) Raman spectra of the pristine GNS and GNS-PS nanocomposites (Hu H et al. (2010) copyrights).

it more attractive compared to graphene for the manufacture of electronic components or in use for nanocomposites synthesis. Polymer reinforced with graphene oxide has been reported in the literature [62].

Recently, graphene oxide with numerous carboxylic groups (GO-COOH) was modified with benzoxazine rings in order to produce exfoliated graphene oxide – polybenzoxazine [63]. The carboxylic groups from GO surface were treated with tyramine (TYR) in order to synthesize a lot of phenolic groups using the activation of the carboxylic groups from GO surface by 1-ethyl-3-(3-dimethylaminopropyl)-carbodiimide/N-Hydroxysuccinimide system (EDC/NHS) and the chlorination method employing $SOCl_2$ respectively. GO-TYR previously obtained further reacted with benzylamine and formaldehyde in order to form the benzoxazine rings. Finally a nano structure with strong covalent bonds between the graphene sheets and the polybenzoxazine chains was achieved (GO-Bz). The study demonstrated that GO-COOH is a good candidate for the preparation of benzoxazine-based nanocomposites due to the abundance of oxidized functional groups on its surface. Raman spectroscopy was successfully employed to demonstrate the efoliation of the graphene sheets through the polybenzoxazine matrix.

In the Raman spectra for the raw material (GO-COOH), GO-TYR and final monomers GO-Bz obtained by both methods, the characteristics of the graphene structure are noticed, namely the intense signals D and G, which proves the presence of the graphene structure in the final compound (**Figure 11**). At the same time, it is worth mentioning the appearance of the 2D band that characterizes the arrangement and the number of graphene plans. Graphene, the two-dimensional form of graphite, consisting of sp^2 hybridized carbon atoms has attracted the attention of researchers in recent years due to its excellent thermal, mechanical, electrical and barrier. All these excellent properties have been shown to the monolayer graphene, the increase of the number of layers leading to the decrease of its properties. For this reason graphene structure has been extensively studied. Raman spectroscopy allows the investigation and determination of the number of layers of the graphene, this information being extracted from the 2D spectrum band. Thus, for products obtained by the EDC/NHS method, the band is wider even in the final benzoxazine product, indicating aggregation in the form of multiple layers of the graphene plans, provided that a part of the benzoxazine monomer did not polymerize and therefore, there was no driving force needed to move the graphene aggregates into independent layers.

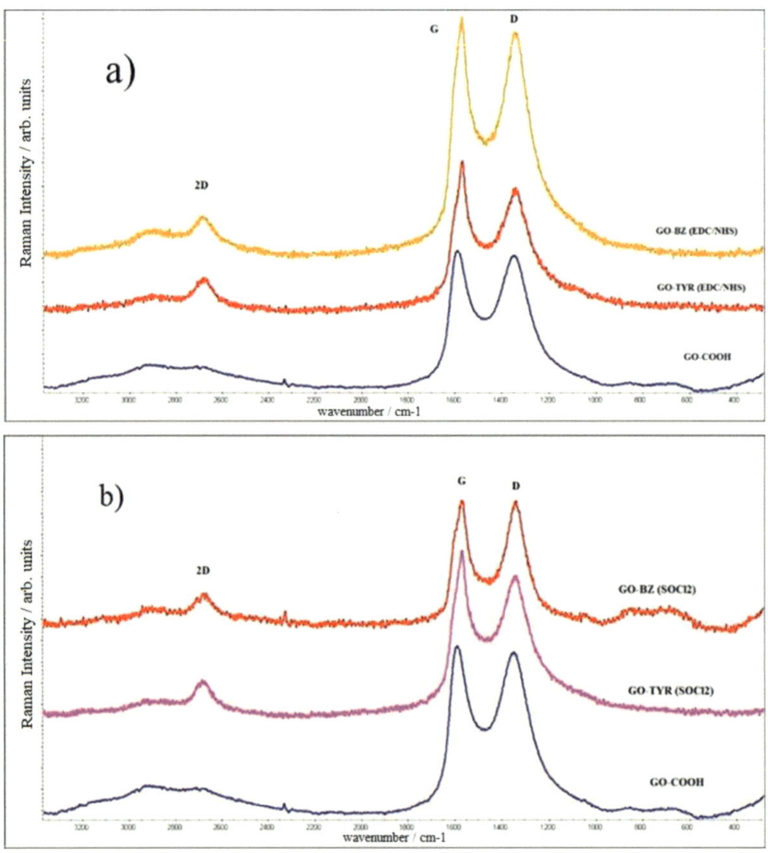

Figure 11. Raman spectra of GO-COOH, GO-TYR, GO-Bz obtained by: a) EDC/NHS activation method; b) chlorination with $SOCl_2$ (Biru I et al. (2016) copyrights).

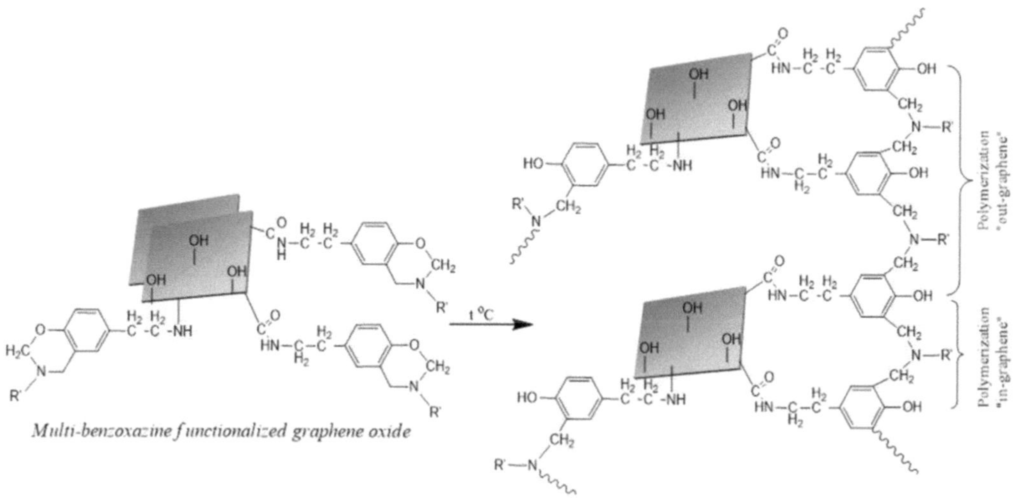

Figure 12. The nanostructure synthesized by polymerization of benzoxazine - functionalized graphene oxide (Biru I et al. (2016) copyrights).

In the case of thionyl chloride products, the 2D band is sharper, which proves that most of the graphene aggregates have disintegrated due to the polymerization of the benzoxazine rings, which has led to the cancellation of the attractions between the graphene plans.

The benzoxazine polymerization may take place either between the rings of the same GO layer (*"in-graphene"* polymerization) or between the rings of two neighbors of GO layers (*"outgraphene"* polymerization). The final balance between these two types of structure will give the ratio between intercalated and exfoliated structures. Consequently the *"in-graphene"* polymerization will lead to more exfoliated structures of GO-Bz (**Figure 12**).

4. Conclusions and outlook

Raman spectroscopy is a powerful instrument for investigating carbon nanomaterials. As a highly sensitive technique Raman spectroscopy is recommended for detection of small changes in structural morphology of various carbon nanomaterials playing an important role as a direct or complementary tool in any laboratory working with carbon allotropes. Raman spectrum shows specific signals for each carbon allotrope when the monochromatic radiation interacts with the sample. Diamond and graphite exhibit significant differences in the Raman spectrum even if both are entirely made of C-C bonds. The Raman spectrum of pure diamond exhibits an extremely sharp signal at ~1332 cm^{-1} which arises from the stretching of the C-C bond. Instead, in the Raman spectrum of graphite two distinguishable peaks are revealed at ~1350 cm^{-1} (D band) and ~1580 cm^{-1} (G band) revealing that the graphite is not as uniform in structure as diamond.

Also more complex structures can be investigated by Raman spectroscopy. The Raman spectrum of C60 fullerene exhibits strong signals at 1467 cm^{-1} and 1567 cm^{-1} revealing that C60 is composed of sp^2 bonded carbon. The sharpness of the signal shows that C60 exhibits a uniform structure. On the contrary, the Raman spectrum of C70 exhibits numerous other peaks. In case of C70 sample, the main peaks are located at 1564 $cm^{-1}$and 1228 cm^{-1} due to a reduction in molecular symmetry which results in more Raman bands.

With this technique it is easily possible to distinguish graphite from diamond, SWCNTs from MWCNTs or graphene form bulk graphite. The stretching of the C-C bond in graphitic materials gives rise to the G-band Raman feature which is common to all sp^2 carbon systems. The G-band of SWCNTs splits in two band components because of large diameter nanotubes and it can be used to distinguish metallic and semiconducting nanotubes. In case of CNTs the D band is significantly increased compared to graphite indicating that some disorder of the graphene sheets is induced. The radial breathing mode (RBM) is particularly important for the determination of the diameter of CNT, its frequency being related to the aggregation state of SWCNTs. The RBM band is unique to SWNCTs and corresponds to the expansion and contraction of the nanotubes.

Raman spectroscopy could be even used to discriminate single layer graphene from multilayer graphene and to determine number of graphene sheets. In case of pure graphene a sharper 2D band situated at ~2700 cm^{-1} is observed. The 2D band originates from a two-phonon double resonance process and it is interrelated to the band structure of graphene layers. The 2D peak evolves as the number of graphene layers increases to about ten layers upon its profile resembles with that of graphite. Moreover, as a non-destructive characterization technique Raman spectroscopy is frequently used to investigate graphene nanocomposites in order to prove successfully graphene functionalization.

Acknowledgements

This research study was supported by a grant of Ministery of Research and Innovation, CNCS - UEFISCDI, project number PN-III-P4-ID-PCE-2016-0818, within PNCDI III. The authors gratefully acknowledge financial support of the GEX project no. 75/25.09.2017, CH 38-17-05. All data presented were recorded using the Renishaw confocal spectrometer with 473 nm laser excitation wavelength and 1.25 laser power. For this study graphene oxides with different types of functionalities were used as received from NanoInnova Technologies. Graphite, SWCNTs and MWCNTs were purchased from Aldrich.

Author details

Elena Iuliana Bîru[1] and Horia Iovu[1,2*]

*Address all correspondence to: horia.iovu@upb.ro

1 Advanced Polymer Materials Group, University Politehnica of Bucharest, Romania

2 Academy of Romanian Scientists, Romania

References

[1] Cantarero A. Raman scattering applied to materials science. Procedia Material Science. 2015;**9**:113-122. DOI: 10.1016/j.mspro.2015.04.014

[2] Toniazzo V, Mustin C, Benoit R, Humbert B, Berthelin J. Superficial compounds produced by Fe(III) mineral oxidation as essential reactants for bio-oxidation of pyrite by Thiobacillus ferrooxidans. Process Metallurgy. 1999;**9**:177-186. DOI: 10.1016/S1572-4409(99)80017-0

[3] Dégardin K, Desponds A, Roggo Y. Protein-based medicines analysis by Raman spectroscopy for the detection of counterfeits. Forensic Science International. 2017;**278**:313-325. DOI: 10.1016/j.forsciint.2017.07.012

[4] Zhang Z, Wang H, Zhang Y, Mu X, Huang B, Du J, Zhou J, Pan X, Xie E. Carbon nanotube/hematite core/shell nanowires on carbon cloth for supercapacitor anode with ultrahigh specific capacitance and superb cycling stability. Chemical Engineering Journal. 2017;**325**:221-228. DOI: 10.1016/j.cej.2017.05.045

[5] Mousaviac H, Jalilvanda S, Kurdestany JM, Grabowskic M. Electron doping effects on the electrical conductivity of zigzag carbon nanotubes and corresponding unzipped armchair graphene nanoribbons. Physica E: Low-dimensional Systems and Nanostructures. 2017;**94**:87-91. DOI: 10.1016/j.physe.2017.07.026

[6] Bora C, Bharali P, Baglari S, Dolui SK, Konwar BK. Strong and conductive reduced graphene oxide/polyester resin composite films with improved mechanical strength, thermal stability and its antibacterial activity. Composites Science and Technology. 2013; **87**:1-7. DOI: 10.1016/j.compscitech.2013.07.025

[7] Han C, Gu A, Liang G, Yuan L. Carbon nanotubes/cyanate ester composites with low percolation threshold, high dielectric constant and outstanding thermal property. Composites Part A Applied Science and Manufacturing. 2017;**41**:1321-1328. DOI: 10.1016/j.compositesa.2010.05.016

[8] Zhang R, Zhang Y, Zhang Q, Xie H, Qian W, Wei F. Growth of half-meter long carbon nanotubes based on Schulz-Flory distribution. ACS Nano. 2013;**7**:6156-6161. DOI: 10.1021/nn401995z

[9] Chang C, Hsu I, Aykol M, Hung WH, Chen CC, Cronin SB. A new lower limit for the ultimate breaking strain of carbon nanotubes. ACS Nano. 2010;**4**:5095-5100. DOI: 10.1021/nn100946q

[10] Geim AK, Novoselov KS. The rise of graphene. Nature Materials. 2007;**6**:183-191. DOI: 10.1038/nmat1849

[11] Neto AC, Geim A. Graphene: Graphene's properties. New Scientist. 2012;**214**:4-5. DOI: 10.1016/S0262-4079(12)61116-6

[12] Tong Y, Bohm S, Song M. The capability of graphene on improving the electrical conductivity and anti-corrosion properties of polyurethane coatings. Applied Surface Science. 2017;**424**:72-81. DOI: 10.1016/j.apsusc.2017.02.081

[13] Kim H, Ahn J-H. Graphene for flexible and wearable device applications. Carbon. 2017;**120**:244-257. DOI: 10.1016/j.carbon.2017.05.041

[14] Lukowiak A, Kedziora A, Strek W. Antimicrobial graphene family materials: Progress, advances, hopes and fears. Advances in Colloid and Interface Science. 2016;**263**:101-112. DOI: 10.1016/j.cis.2016.08.002

[15] PapageorgiouIan DG, Kinloch A, Young RJ. Mechanical properties of graphene and graphene-based nanocomposites. Progress in Materials Science. 2017;**90**:75-127. DOI: 10.1016/j.pmatsci.2017.07.004

[16] Dobrota AS, Pašti IA, Mentusa SV, Johansson B, Skorodumova NV. Functionalized graphene for sodium battery applications: The DFT insights. Electrochimica Acta. 2017; **250**:185-195. DOI: 10.1016/j.electacta.2017.07.186

[17] Yao K, Dai B, Zhu J, Ralchenko V, Shu G, Zhao J, Wang P, Liu B, Gao G, Sun M, Liu K, Lv Z, Yang L, Han J. Diamond micropowder synthesis via graphite etching in a microwave hydrogen plasma. Powder Technology. 2017;**322**:124-130. DOI: 10.1016/j.powtec.2017.09.021

[18] Amendolaa V, Matteib G, Cusanc C, Pratoc M, Meneghetti M. Fullerene non-linear excited state absorption induced by gold nanoparticles light harvesting. Synthetic Metals. 2005;**155**:283-286. DOI: 10.1016/j.synthmet.2005.01.032

[19] Papavasileiou KD, Avramopoulos A, Leonis G, Papadopoulos MG. Computational investigation of fullerene-DNA interactions: Implications of fullerene's size and functionalization on DNA structure and binding energetics. Journal of Molecular Graphics and Modelling. 2017;**74**:177-192. DOI: 10.1016/j.jmgm.2017.02.015

[20] Tran CM, Sakai H, Kawashima Y, Ohkubo K, Fukuzumi S, Murata H. Multi-level non-volatile organic transistor-based memory using lithium-ion-encapsulated fullerene as a charge trapping layer. Organic Electronics. 2017;**45**:234-239. DOI: 10.1016/j.orgel.2017.03.018

[21] Zhang X, Ai X, Zhang R, Ma Q, Wang Z, Qin G, Wang J, Wang S, Suzuki K, Miyazaki T, Mizukami S. Spin conserved electron transport behaviors in fullerenes (C60 and C70) spin valves. Carbon. 2016;**106**:202-207. DOI: 10.1016/j.carbon.2016.05.011

[22] Xia K, Zhan H, Gu Y. Graphene and carbon nanotube hybrid structure: A review. Procedia IUTAM. 2017;**21**:94-101. DOI: 10.1016/j.piutam.2017.03.042

[23] Hedman D, Larsson JA. Length dependent stability of single-walled carbon nanotubes and how it affects their growth. Carbon. 2017;**116**:443-447. DOI: 10.1016/j.carbon.2017.02.007

[24] Suna L, Wanga X, Wanga Y, Zhang Q. Roles of carbon nanotubes in novel energy storage devices. Carbon. 2017;**122**:462-474. DOI: 10.1016/j.carbon.2017.07.006

[25] Arora N, Sharma NN. Arc discharge synthesis of carbon nanotubes: Comprehensive review. Diamond and Related Materials. 2017;**50**:135-150. DOI: 10.1016/j.diamond.2014.10.001

[26] Santiago EV, López SH, Camacho López MA, Contreras DR, Mancilla RF, Flores-Gallardo SG, Hernández-Escobar CA, Zaragoza-Contreras EA. Optical properties of carbon nanostructures produced by laser irradiation on chemically modified multi-walled carbon nanotubes. Optics & Laser Technology. 2016;**84**:53-58. DOI: 10.1016/j.optlastec.2016.05.002

[27] Lopez D, Abe IY, Pereyra I. Temperature effect on the synthesis of carbon nanotubes and core–shell Ni nanoparticle by thermal CVD. Diamond and Related Materials. 2015;**52**: 59-65. DOI: 10.1016/j.diamond.2014.12.006

[28] Soares OSGP, Rocha RP, Gonçalves AG, Figueiredo JL, Órfão JJM, Pereira MFR. Easy method to prepare N-doped carbon nanotubes by ball milling. Carbon. 2015;**91**:114-121. DOI: 10.1016/j.carbon.2015.04.050

[29] Anoshkin IV, Nefedova II, Lioubtchenko DV, Nefedov IS, Räisänen AV. Single walled carbon nanotube quantification method employing the Raman signal intensity. Carbon. 2017;**116**:547-552. DOI: 10.1016/j.carbon.2017.02.019

[30] Varga M, Izak T, Vretenar V, Kozak H, Holovsky J, Artemenko A, Hulman M, Skakalova V, Lee DS, Kromka A. Diamond/carbon nanotube composites: Raman, FTIR and XPS spectroscopic studies. Carbon. 2017;**111**:54-61. DOI: 10.1016/j.carbon.2016.09.064

[31] Powell C, Beall GW. Graphene oxide and graphene from low grade coal: Synthesis, characterization and applications. Current Opinion in Colloid & Interface Science. 2015; **20**:362-366. DOI: 10.1016/j.cocis.2015.11.003

[32] Guo L, Yin X, Wu W, Meng H. Preparation of graphene via liquid-phase exfoliation with high gravity technology from edge-oxidized graphite. Colloids and Surfaces A: Physicochemical and Engineering Aspects. 2017;**531**:25-31. DOI: 10.1016/j.colsurfa.2017.07.074

[33] Konishi K, Cui Z, Hiraki T, Yoh K. Spin-injection into epitaxial graphene on silicon carbide. Journal of Crystal Growth. 2013;**378**:385-387. DOI: 10.1016/j.jcrysgro.2012.12.168

[34] Liu Z, Lin L, Ren H, Sun X. CVD synthesis of Graphene. A volume in Micro and Nano Technologies (Elsevier). 2017:19-56. DOI: 10.1016/B978-0-32-346240-2.00002-9

[35] Toh SY, Loh KS, Kamarudin SK, Daud WRW. Graphene production via electrochemical reduction of graphene oxide: Synthesis and characterisation. Chemical Engineering Journal. 2014;**251**:422-434. DOI: 10.1016/j.cej.2014.04.004

[36] Zhong Y, Zhen Z, Zhu H. Graphene: Fundamental research and potential applications. FlatChem. 2017;**4**:20-32. DOI: 10.1016/j.flatc.2017.06.008

[37] Vasić B, Matković A, Ralević U, Belić M, Gajić R. Nanoscale wear of graphene and wear protection by graphene. Carbon. 2017;**120**:137-144. DOI: 10.1016/j.carbon.2017.05.036

[38] Piattia E, Galassoa S, Tortelloa M, Naira JR, Gerbaldia C, Brunab M, Borinib S, Dagheroa D, Gonnelli RS. Carrier mobility and scattering lifetime in electric double-layer gated few-layer graphene. Applied Surface Science. 2017;**395**:37-41. DOI: 10.1016/j.apsusc.2016.06.192

[39] Yang W, Ni M, Ren X, YafenTian NL, Sub Y, Zhang X. Graphene in Supercapacitor applications. Current Opinion in Colloid & Interface Science. 2015;**20**:416-428. DOI: 10.1016/j.cocis.2015.10.009

[40] Han T, Kim H, Kwon SJ, Lee TW. Graphene-based flexible electronic devices. Materials Science & Engineering R: Reports. 2017;**118**:1-43. DOI: 10.1016/j.mser.2017.05.001

[41] Cao L, Zhang F, Wang Q, Fabrication XW. Of chitosan/graphene oxide polymer nanofiber and its biocompatibility for cartilage tissue engineering. Materials Science and Engineering: C. 2017;**79**:697-701. DOI: 10.1016/j.msec.2017.05.056

[42] Naskar A, Khan H, Bera S, Jana S. Soft chemical synthesis, characterization and interaction of ZnO graphene nanocomposite with bovine serum albumin protein. Journal of Molecular Liquids. 2017;**237**:113-119. DOI: 10.1016/j.molliq.2017.04.074

[43] Justino CIL, Gomes AR, Freitas AC, Duarte AC, Rocha-Santos TAP. Graphene based sensors and biosensors. TrAC Trends in Analytical Chemistry. 2017;**91**:53-66. DOI: 10.1016/j.trac.2017.04.003

[44] Mishra S, Ashaduzzamana M, Mishra P, Swart HC, Turner APF, Tiwari A. Stimuli-enabled zipper-like graphene interface for auto-switchable bioelectronics. Biosensors and Bioelectronics. 2017;**89**:305-311. DOI: 10.1016/j.bios.2016.03.052

[45] Gokhale AA, Lu J, Lee I. Immobilization of cellulase on magnetoresponsive graphene nano-supports. Journal of Molecular Catalysis B: Enzymatic. 2013;**90**:76-86. DOI: 10.1016/j.molcatb.2013.01.025

[46] Ghany NA, Elsherif SA, Handal HT. Revolution of Graphene for different applications: State-of-the-art. Surfaces and Interfaces. 2017;**9**:93-106. DOI: 10.1016/j.surfin.2017.08.004

[47] Ferrari AC. Raman spectroscopy of graphene and graphite: Disorder, electron–phonon coupling, doping and nonadiabatic effects. Solid State Communications. 2007;**143**:47-57. DOI: 10.1016/j.ssc.2007.03.052

[48] Ferrari AC, Meyer JC, Scardaci V, Casiraghi C, Lazzeri M, Mauri F, Piscanec S, Jiang D, Novoselov KS, Roth S, Geim AK. Raman Spectrum of Graphene and Graphene layers. Physical Review Letters. 2006;**97**:187401. DOI: 10.1103/PhysRevLett.97.187401

[49] Malard LM, Pimenta MA, Dresselhaus G, Dresselhaus MS. Raman spectroscopy in graphene. Physics Reports. 2009;**473**:51-87. DOI: 10.1016/j.physrep.2009.02.003

[50] Liu L, Zhou H, Cheng R, Yu WJ, Liu Y, Chen Y, Shaw J, Zhong X, Huang Y, Duan X. High yield chemical vapor deposition growth of high quality large-area AB stacked bilayer graphene. ACS Nano. 2012;**6**:8241-8249. DOI: 10.1021/nn302918x

[51] Peng H, Sun M, Zhang D, et al. Raman spectroscopy of graphene irradiated with highly charged ions. Surface and Coatings Technology. 2016;**306**:171-175. DOI: 10.1016/j.surfcoat.2016.05.064

[52] Semwogerere D, Weeks ER. Confocal Microscpy. Encyclopedia of Biomaterials and Biomedical Engineering. London: Taylor & Francis; 2005. pp. 1-10. DOI: 10.1081/E-EBBE-120024153

[53] Gui Y, Sun H. Effects of substrates on Raman spectroscopy in chemical vapor deposition grown graphene transferred with poly (methyl methacrylate). Solid State Communications. 2017;**264**:31-34. DOI: 10.1016/j.ssc.2017.04.010

[54] Ni Z, Wang Y, Yu T, Shen Z. Raman spectroscopy and imaging of graphene. Nano Research. 2008;**1**:273-291. DOI: 10.1007/s12274-008-8036-1

[55] Calizo I, Ghosh S, Bao W, Miao F, Lau CN, Balandin AA. Raman nanometrology of graphene: Temperature and substrate effects. Solid State Communications. 2009;**149**:1132-1135. DOI: 10.1016/j.ssc.2009.01.036

[56] Ferrari AC, Robertson J. Raman spectroscopy in carbons: From nanotubes to diamond. Philosophical Transactions of the Royal Society A. 2008;**362**:2267-2565

[57] Yuan R, Yuan J, Wu Y, Chen L, Zhou H. Jianmin Chen. Efficient synthesis of graphene oxide and the mechanisms of oxidation and exfoliation. Applied Surface Science. 2017; **416**:868-877. DOI: 10.1016/j.apsusc.2017.04.181

[58] Liu Y, Zhang Y, Duan L, Zhang W, Su M, Sun Z, He P. Polystyrene/graphene oxide nanocomposites synthesized via Pickering polymerization. Progress in Organic Coatings. 2016;**99**:23-31. DOI: 10.1016/j.porgcoat.2016.04.034

[59] Zong P, Cao D, Cheng Y, Zhang H, Shaoc D, Wangad S, Hee C, Zhao Y. Functionally reduced graphene oxide supported iron oxides composites as an adsorbent for the immobilization of uranium ions from aqueous. Journal of Molecular Liquids. 2017;**240**:578-588. DOI: 10.1016/j.molliq.2017.05.101

[60] Bahreyni A, Yazdian-Robati R, Hashemitabar S, Ramezani M, Ramezani P, Abnous K, Taghdis SM. A new chemotherapy agent-free theranostic system composed of graphene oxide nano-complex and aptamers for treatment of cancer cells. International Journal of Pharmaceutics. 2017;**526**:391-399. DOI: 10.1016/j.ijpharm.2017.05.014

[61] Yan L, Yu J, Luo H. Ultrafine TiO_2 nanoparticles on reduced graphene oxide as anode materials for lithium ion batteries. Applied Materials Today. 2017;**8**:31-34. DOI: 10.1016/j.apmt.2017.02.001

[62] Hu H, Wang X, Wanga J, Wana L, Liu F, Zhen H, Chen R, Xu C. Preparation and properties of graphene nanosheets–polystyrene nanocomposites via in situ emulsion polymerization. Chemical Physics Letters. 2010;**484**:247-253. DOI: 10.1016/j.cplett.2009.11.024

[63] Biru I, Damian CM, Gârea SA, Iovu H. Benzoxazine - functionalized graphene oxide for synthesis of new nanocomposites. European Polymer Journal. 2016;**83**:244-255. DOI: 10.1016/j.eurpolymj.2016.08.024

Chapter 10

Surface-Enhanced Raman Spectroscopy Characterization of Pristine and Functionalized Carbon Nanotubes and Graphene

Sabina Botti, Alessandro Rufoloni,
Tomas Rindzevicius and Michael Stenbæk Schmidt

Additional information is available at the end of the chapter

http://dx.doi.org/10.5772/intechopen.74065

Abstract

Carbon nanotubes (CNTs) and graphene are at the center of a significant research effort due to their unique physical and chemical properties, which promise high technological impact. For the future development of all the foreseen applications, it is of particular interest the study of binding interactions between carbon nanostructures and functional groups. An appropriate method is the surface-enhanced Raman spectroscopy (SERS), which provides a large amplification of Raman signals when the probed molecule is adsorbed on a nanosized metallic surface. In this chapter, we present a review of principal results obtained applying SERS for the characterization of pristine and functionalized CNTs and graphene. The obtained results encourage us to consider SERS as a powerful method to obtain a rapid monitor of the procedures used to interface graphene and nanotubes.

Keywords: carbon nanotubes, graphene, functionalization process, SERS

1. Introduction

Carbon nanotubes (CNTs) and graphene are at the center of a significant research effort due to their unique physical and chemical properties, which promise high technological impact. For the future development of all the foreseen applications, it is of particular interest the study of binding interactions between carbon nanostructures and functional groups [1–3]. Indeed, the ability to engineer the electrical, optical, sensing, and dispersion properties of graphene and CNTs by chemical functionalization is widening considerably their potential applications generating a vast and yet largely unexplored family of carbon nanostructures for the realization of devices with novel functionalities. This task is highly challenging and the achievement of a procedure that enables a high density of functional groups, with little or no damage to the carbon structure, is of prime importance as the development of a nondestructive technique that can fully characterize the structure of functionalized graphene and CNTs.

Since long time, Raman spectroscopy has been considered as one of the most powerful tool for the characterization of carbon-based materials (see Section 2); however, Raman features associated with functional groups are usually often not observed in the Raman spectra of nanotubes and graphene due to the small quantity of the molecules attached to the carbon lattice.

In our studies, we have used the surface-enhanced Raman spectroscopy (SERS), which provides a large amplification of Raman signal when the probed molecule is adsorbed on nanosized metallic surface [4, 5]. Since the SERS effect increases by decreasing the distance between the nanostructure and the adsorbed molecule, by depositing CNTs/graphene as dilute dispersions on SERS active substrates, the Raman signal from the molecular groups deriving from functionalization/synthesis process and bound to the carbon surface, can be amplified. As a consequence, the spectral features of functional groups, otherwise very difficult to see, were recorded in the SERS spectra.

Here, we present a review of principal results obtained by applying SERS for the characterization of pristine and functionalized graphene and multiwalled nanotubes (MWNTs). The obtained results encourage us to consider SERS as a powerful method to obtain a rapid monitor of the procedures used to interface graphene and nanotubes with functionalizing groups.

2. Raman spectroscopy

In **Figure 1**, the Raman spectra of graphite (HOPG, Cree Corporation, USA), multilayer graphene, single-walled carbon nanotubes (SWNTs), double-walled nanotubes (DWNTs), and MWNTs in the range 100–2100 cm^{-1} are reported. All the samples that we used were commercially available.

The pristine SWNTs, DWNTs, and MWNTs were grown by catalytic carbon vapor deposition process, with the following characteristic: SWNTs (Nanointegris, USA, selected semiconductor) average diameter: 1.2–1.7 nm, length: 300 nm to 5 μm; DWCNTs (Nanocyl, Belgium) average outer diameter of 3.5 nm, length between 1 and 10 μm, and specific surface area of 500 m^2/g; MWNTs (Nanocyl, Belgium) average diameter: 9.5 nm, average length: 1.5 μm. We used graphene powders with particles consisting of aggregates of platelets with diameter of about 2 μm, thickness less than 5 nm, and average surface area 750 m^2/g (grade C particles, XG Sciences Inc., USA).

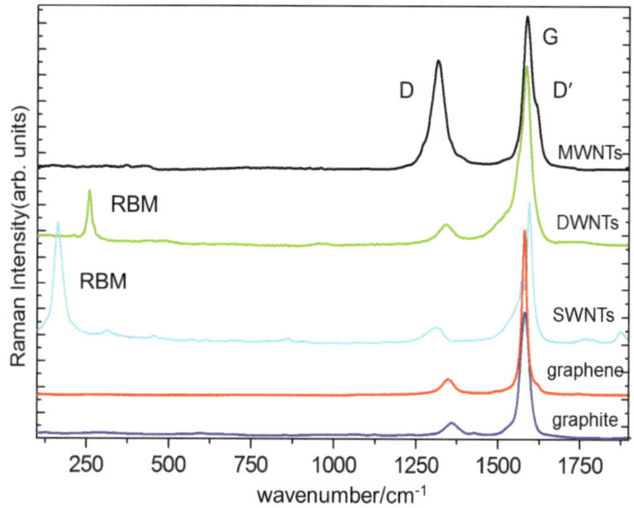

Figure 1. Raman spectra of different carbon forms: graphite, graphene, SWNTs, DWNTs, and MWNTs. The spectra were excited at 785 nm.

The Raman spectrum of graphite and its derivatives has two main peaks assigned to the first order D (1320 cm^{-1}) and G modes (1580 cm^{-1}) [6–12]. Second-order Raman spectra of carbon-based materials, and particularly of graphene, are dominated by 2D (G′) band at ~2700 cm^{-1}, caused by the existence of double electron-phonon resonance mechanism, thus making it possible to study the structure of electron bands from the analysis of resonance Raman spectra. The band has a double peak in graphite and a single peak in graphene. Both the intensity ratio I_{2D}/I_G and the shape analysis of 2D peak have been used to distinguish mono from bi or multilayer graphene [12, 13].

The D mode (A_{1g} symmetry) can be described as an in-plane breathing vibration of aromatic ring in which the six atoms move away from the center of the Brillouin zone. This band is forbidden in perfect graphitic lattice and it becomes active in the presence of disorder and deviations from an ideal structure. The G mode (E_{2g} symmetry) is due to the in-plane stretching vibration of carbon atoms pairs in which each atom moves vibrating tangentially against the other. This mode is always allowed and it is characteristic not only of graphitic rings but also of all the sp^2 structures. The ratio of D and G band intensities provides a way to estimate the amount of defects in graphitic structure, being inversely proportional to the size of ordered sp^2 domain size through the relation: $L = (2.4 \times 10^{-10}) \cdot \lambda_L^{\,4} \cdot (I_D/I_G)^{-1}$, where λ_L is the wavelength of laser that excites the Raman spectra [9, 11, 12].

The Raman spectra of SWNTs display peculiar features as the radial breathing mode (RBM), and the double peak splitting of the G band. The frequency ω of RBMs is inversely proportional to the value of tube diameter d following the Bandow relation: [13, 14]

$$\omega = \alpha / d \tag{1}$$

where the parameter α is experimentally determined.

The optical transitions of SWNTs can be identified as inter-band transitions between the Van Hove singularities in the 1D electronic density of states (DOS) [14, 15]. The singularities are spikes in the DOS, which occur at specific energies, depending on the nanotube diameter and chirality. From band structure calculations, approximate relations were deduced between the energy *E* of inter-band transitions between DOS singularities and the nanotube diameter, suggesting that the optical absorption peaks should move to higher energy at decreasing nanotube diameter. As the nanotube diameter increases, the singularities move closer together, whereas for small diameter nanotubes, the "spikes" in the DOS are well-separated, especially near the Fermi energy and can provide initial or final states or both for a highly resonant Raman scattering process. The energy of these optical transitions are in the visible range, near to the wavelength of the most commonly used laser for excite Raman scattering. When the laser energy matches the energy difference between spikes for a particular nanotube diameter, the corresponding RBM will dominate the Raman trace. Therefore, the Raman scattering in small diameter SWNTs is diameter selective as direct consequence of the 1D electronic quantum-confinement in small diameter SWNTs.

The nanotube curvature results in different force constants for atomic displacements along the nanotube axis compared to those in the circumferential direction, causing the separation of G peak in two components, G^{+} and G^{-} [13–17]. Differently, in the Raman spectra of MWNTs, which can be considered as an ensemble of concentric SWNTs with increasing diameters, the G peak splitting is both small and smeared out, leading to a broad G band line shape. The higher energy mode that appears as a shoulder in MWNT Raman trace is the D′ band, which is due to the same tangential vibrations of the G mode, but involving external layers that are not sandwiched between two other layers. As the D band, D′ is a double resonance Raman mode, induced by disorder and/or defects on the side walls of CNTs or ion intercalation between graphitic sheets. Indeed, this band is not observed in graphite, but it has a high intensity in intercalated graphite compounds [18].

As it will be discussed in Section 6, since less information is available on the structure of functional groups attached to the tube walls or basal plane of graphene, in our studies, we have used the SERS technique to record the Raman signal of functional groups and prove experimentally their linkage to the CNT walls.

3. Surface-enhanced Raman spectroscopy

The surface-enhanced Raman scattering (SERS), discovered three decades ago, is used to provide a drastic amplification of the Raman signal [4, 5]. The Raman signal amplification is observed when the molecule to be studied is adsorbed on a nanostructured metallic surface and the frequency of incident radiation overlaps the resonant frequency of conduction electrons in the metals. In this case, the incident electric field produces collective oscillations of metal electrons, which in turn generate a large electromagnetic field. This field is superimposed to the incoming field giving rise to its enhancement. Theoretical studies demonstrate that the induced field is particularly intense near sharp tips, interstitial crevices, and more generally, in between adjacent metal nanostructure if the distance is of few nanometers [4, 5, 19].

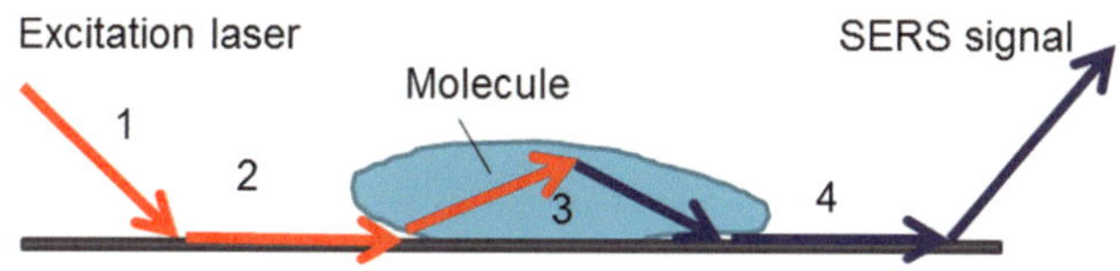

Figure 2. Scheme of SERS process steps.

However, the electromagnetic mechanism does not explain why the enhancement factor of a surface depends on the chemical nature of the adsorbed molecule [19, 20], therefore, it was hypothesized that an additional enhancement is provided by an increase of molecule polarizability due to a deformation of the distribution of the electron cloud or to the formation of resonant charge transfer complex between the metal and the adsorbed molecule.

We can then summarize the SERS process steps, depicted in **Figure 2**: (1) the incident laser impinges on nanostructured metallic surface, (2) plasmons excitation with electric field amplification, (3) Raman scattered light emission, and (4) Raman scattered light transferred back to plasmons and scattered in air. The surface enhancement effect is so pronounced because the Raman signal enhancement occurs twice.

4. SERS substrates

To be used in a sensor system, a SERS substrate should enhance the Raman effect sufficiently to enable consistent and uniform chemical detection sensitivity across the surface, maintaining its properties as long as possible in the time and to provide a high number of sites for molecular adsorption.

In principle, all systems possessing free carriers show the surface enhancement effect. However, the plasmon properties—such a wavelength and width of its resonance—depend on the nature of the metal surface and on its geometry and affect the enhancement factor (EF) of the surface [4, 5, 19]. The width of plasmon resonance resulted to be: $w = \gamma(\varepsilon_b + 3)$, where γ is the electron scattering rate and ε_b is the contribution to the inter-band transitions to the dielectric constant. Smaller conductivity and a large number of inter-band transitions in the region of plasmon resonance give resonance peak with large width and hence smaller amplification of electric field [19]. To this respect, the coin metals (Ag, Au, and Cu) resulted to be the most appropriate to be used for SERS with their amplification factors much larger than unity. Differently, the enhancement factor of good conductors as Al, Pt, and In is larger but not much larger than unity and is only slightly greater than unity for most other metals. The other advantage of using coin metals is that their plasmon resonance wavelength is in the visible–near infrared.

Early SERS experiments used gold colloids in solution. Nowadays, by exploiting semiconductor lithographic fabrication technology, periodic patterns on Si surface can be reproducibly fabricated over large areas. Ordered geometry provides uniform SERS signals from anywhere on the active surface, avoiding that only small uncontrolled areas of the total metal surface

Substrate	EF
Klarite™	1.1×10^6
Nanova	8.0×10^5
Technical University of Denmark	8.0×10^8

Table 1. Enhancement factors of gold coated substrates with different geometry.

have the correct geometry for surface enhancement, as occurred by using substrates with random nanosized roughening or nanoparticle separation and sharp metallic features, produced with the previous techniques. Usually, the SERS substrates are gold coated because, although the gold has a smaller EF than silver, it is more resistant against oxidation in air.

We used three variations of SERS active gold substrates: an ordered array of inverted square-based pyramids, (Klarite™, D3 Technologies Ltd3), an electrochemically nano-roughened surface (Nanova), and an array of nanopillars, 50–80 nm in width, 600 nm in height, fabricated by the Department of Micro and Nanotechnology of the University of Denmark [21–23].

We measured the EFs of all the above cited substrates by measuring the Raman signal of the same molecule, in this way only electromagnetic enhancement due to the geometry of the used structure was considered, avoiding the effect of chemical enhancement [20, 24]. The obtained results are reported in **Table 1**. The regularity of nanostructure guaranteed a uniform enhancement factor across the surface and a high reproducibility of Raman signal in spite of the differences among the EFs.

5. Sample preparation

As discussed in the previous paragraph, the field enhancement strongly decreases with the distance from the metallic surface, as consequence delivering the substance under study in close proximity of the metal surface is a prerequisite for measuring SERS signals. Therefore, the samples were analyzed as evaporated films prepared by dropping on the SERS active surface a controlled volume of a solution of CNTs or graphene in ultra-pure water (1 mg/mL). The solvent spreads across the surface selectively evaporates leaving a dried film, which is clearly visible under the optical microscope coupled with the spectrometer (see **Figure 3**).

In our procedure, the substrates were used as-received without any pre-treatment and the solvent evaporated in air without any heating.

We examined the dried film morphology under high-resolution scanning electron microscope (HR-SEM), using a Leo 1525 hot cathode field emission microscope, with a resolution of 1.5 nm at 20 kV.

The SERS activity depends either from the substrate morphology either from the number of sites in the surface that are accessible to the molecular adsorption. In fact, the SERS signal from molecules on the second monolayer and beyond is reduced because the SERS effect depends on the distance between the nanostructure and adsorbed molecule, as shown in **Figure 4**.

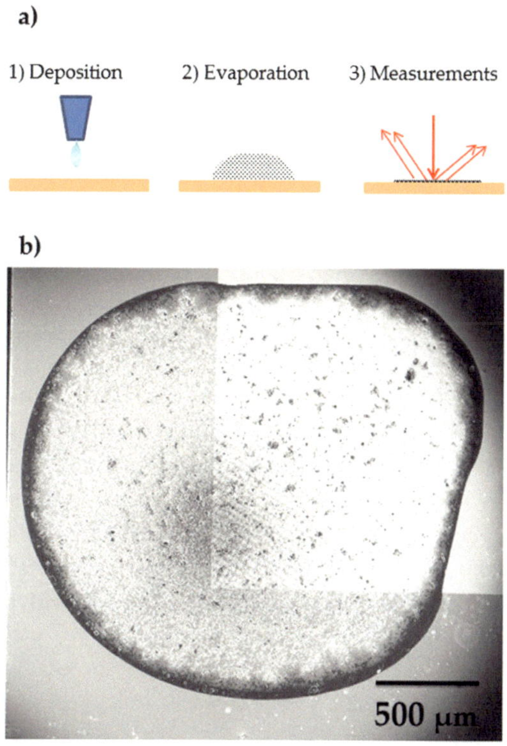

Figure 3. (a) Sequential steps of the drop and dry technique for films deposition and (b) top view of dried film.

When the number of analyte molecules is too large, a sample layer is formed covering the molecules with surface-enhanced Raman signal. To this respect, it is of prime importance that the substrate geometry provides enhancement region accessible for molecular adsorption.

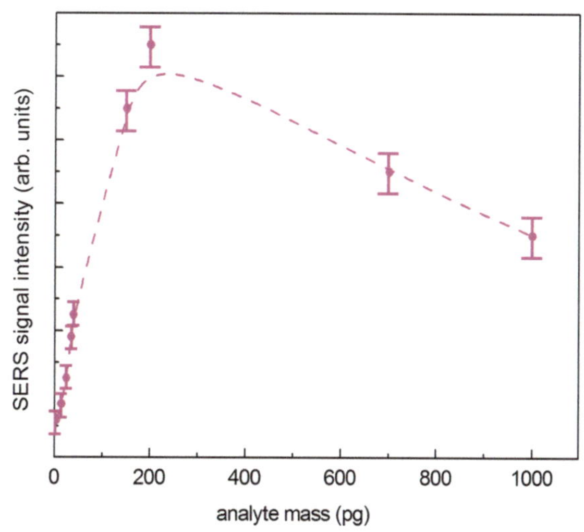

Figure 4. Variation in the intensity of surface-enhanced Raman signal with the amount of analyte mass probed by the laser.

6. Raman spectrometers

In the first steps of our works, we acquired SERS and Raman spectra using a Raman spectrometer BwTek (i-Raman 785.), equipped with a micro-positioning system for fine xyz adjustments and a video camera for sampling viewing. The system uses an air-cooled charge coupled device (CCD) detector. The 785 nm laser light, to guarantee the coupling with gold plasmons, was focused onto the sample using a 20× objective (corresponding to a laser beam diameter of 90 μm, as already mentioned). Samples were moved into position using the xyz translational stage. The Raman spectra were acquired in the wavelength range 789–1048 nm corresponding to Raman shifts of 75–3200 cm^{-1} (resolution better than 3 cm^{-1}).

Raman mapping was performed with a high-resolution micro-Raman spectrometer (Horiba Xplora) with 785 nm excitation wavelength. The Raman signal was collected through a 100× objective in the range 100–2000 cm^{-1} with accumulation time of 10 s per spectrum. Areas of the substrates up to 10 × 10 μm were scanned with an acquisition grid of 0.6 and 0.8 μm step size in x and y, respectively. All the spectra were obtained with 100 μW laser power, to avoid sample heating and damaging. The SERS maps were obtained analyzing the peak intensity of every spectrum. Raman spectra were also acquired with 532 nm excitation (range 400–3000 cm^{-1}, acquisition time 10 s).

7. Results and discussion

Controlled volumes of pristine MWNT solutions were dropped on SERS active substrates, the SERS spectra acquired in different spatial positions, maintaining all the experimental parameters unchanged are reported in **Figure 5**.

The SERS trace (a) has no significant differences in comparison of that reported in **Figure 1**, if one excepts the amplification of the signal whereas curves (b) and (c) show variations in some Raman line intensities, line shape and appearance of several sharp peaks, in the frequency range 400–1100 cm^{-1}, characteristics of C_{60}-like molecules [25].

We observed a gradual increase of the D band, which indicates an enhanced degree of disorder, with a concomitant modification of the G band profile, consisting in a peak splitting due to the increase of D′ band at its high energy side. The peak at about 250 cm^{-1} in the SERS curves (b) and (c) is similar to that at 171 cm^{-1} in Raman spectrum of SWNTs, reported in **Figure 1**.

The RBMs are not often observed in MWNTs, its observation in SERS trace is consistent with the fact that for a molecule adsorbed on a nano-structured surface, the modes which involve atoms vibrating perpendicularly to the surface are more enhanced than the others [26]. However, the observed RBM width (45 cm^{-1}) is larger than that measured in the SWNT Raman spectra [16, 17].

This finding can be explained considering that in a MWNT also the radial vibrations of outer shells contribute to this mode, producing a number of overlapping RBM peaks larger than those expected for SWNTs with the same distribution of diameter [18].

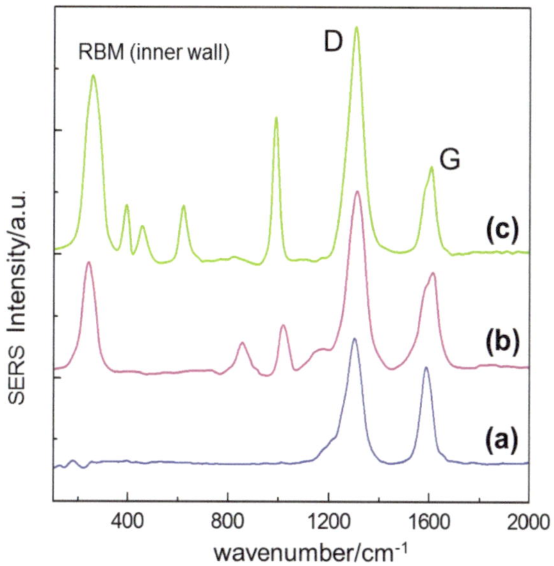

Figure 5. SERS spectra of pristine MWNT dried film. The MWNT layer thickness is decreasing from spectrum (a) to (c).

The diameter of SWNTs can be calculated from the frequency of corresponding RBM, through the Bandow relation [13–15]. Since the RBM intensity strongly decreases when the tube diameter

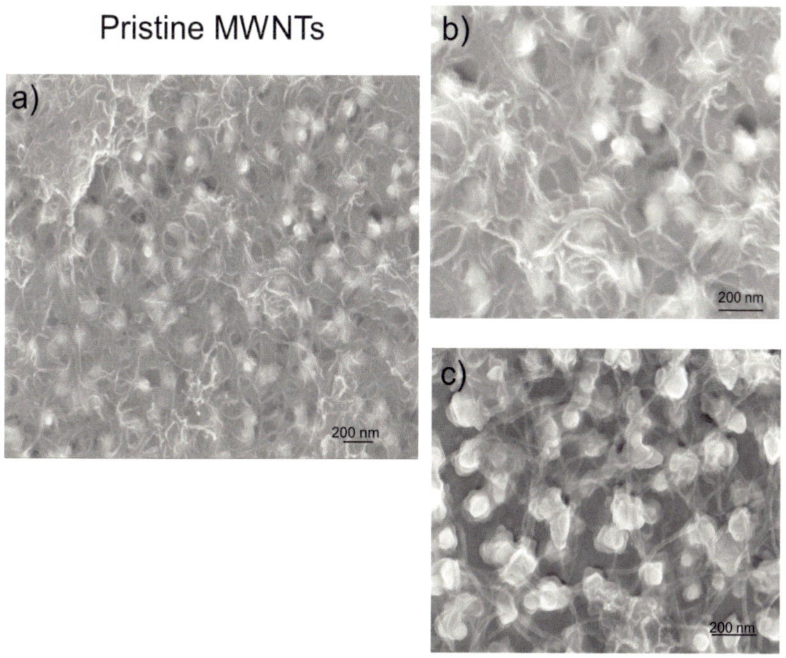

Figure 6. (a) HR-SEM image of pristine MWNT film deposited on SERS substrate. (b) Magnified view of region with larger thickness layer (c) Magnified view of region with thinner layer. The leaning effect of nanopillars is evidenced.

increases, we can roughly estimate the inner diameter of MWNTs from their RBMs in the SERS spectra by using the same rule. The RBM at 250 cm^{-1} corresponds to an inner tube diameter of 0.8 nm. These values are compatible with those reported in the data sheet of MWNTs from the manufacturer.

Figure 6 presents a HRSEM view, taken at different magnification of pristine MWNT films. The MWNTs are tightly interlaced forming a web of filaments that are distributed across the SERS active surface, occupying the regions with highest Raman signal amplification as the space between the nano-pillars. However, the layer thickness is not uniform across the deposited patch and this is reflected in the observed differences between the SERS spectra reported in **Figure 5**.

In fact, these variations can be interpreted by considering the breaking of the nanotubes into species such as amorphous carbon, tubular fragments, and closed shell fullerenes. Such reactions are of chemical nature occurring at the nanotube metal substrate interface.

Figure 7 illustrates the comparison between the Raman spectra of pristine and covalently functionalized MWNTs. The carboxyl functionalization was performed by a reflux in sulfuric/nitric acid (Nanolab, average diameter = 30 nm, length 1–5 μm, from now on denoted as COOH-MWNTs). This process leads to the concentration of COOH groups on the nanotube surface. By reacting the carboxylate CNTs with ethylene diamine, nanotubes with amide linkage and a primary amine group at the end of carbon chain are obtained (Nanolab, average diameter = 15 nm, length 1–5 μm, from now on denoted as NH_2-MWNTs). However, in the Raman spectra of covalently functionalized vibrational modes that can be directly related to molecular groups are not observed.

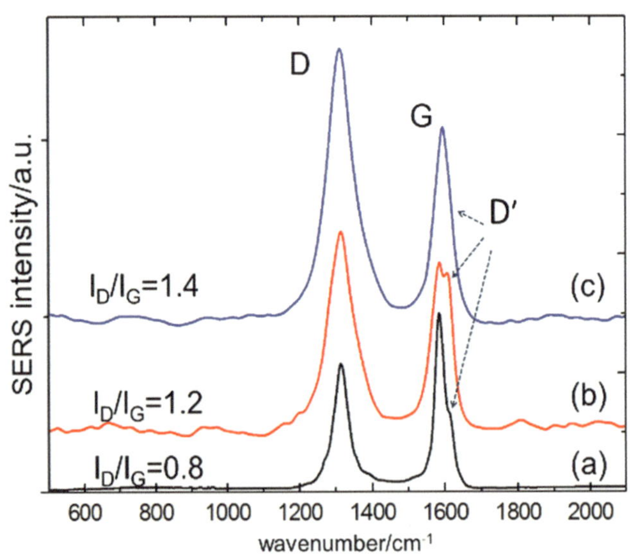

Figure 7. Raman spectra of MWNT mat. (a) Raman spectrum acquired from pristine MWNT film, I_D/I_G = 0.8; (b) Raman spectrum acquired from COOH-functionalized MWNTs, I_D/I_G = 1.2; (c) Raman spectrum acquired from NH_2-functionalized MWNTs, I_D/I_G = 1.4.

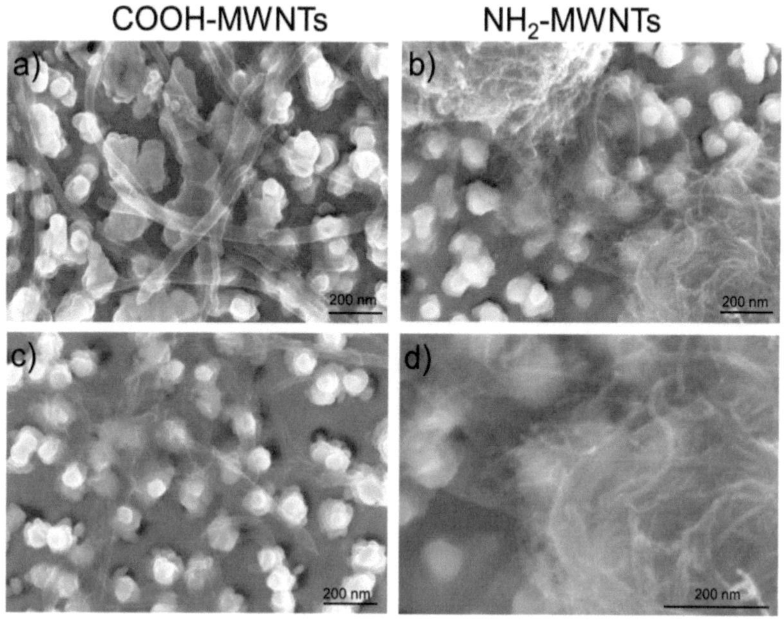

Figure 8. HR-SEM images of (a) COOH-MWNT and (b) NH_2-MWNT films deposited on SERS substrate; (c) magnified view of region (a); (d) magnified view of region (b) evidencing the formation of a dense layer.

As a consequence, the use of Raman spectroscopy to study the structure of functionalized CNTs is routinely limited to the evaluation of the I_D/I_G ratio, since the interaction between carbon nanotube wall and functional groups is expected to produce a higher density of sp^3 hybridized carbon sites, with consequent increase of I_D/I_G ratio [27–29].

However, when a set of spectra is recorded from the same pristine CNTs sample, we frequently observed a variation in the I_D/I_G ratio, up to the 15% of its average value, which is sometimes apparently of the same order as that recorded after the chemical functionalization. Although a more efficient discrimination can be reached by applying the principal component analysis [30], the observation of Raman spectral changes induced by the chemical functionalization is challenging.

In **Figure 8** are illustrated the different morphologies of functionalized MWNT-dried films. Due to the small outer diameter, the NH_2-MWNTs are tightly interlaced forming a dense web of filaments. Differently, the drying of COOH-MWNT solution forms a very thin layer with some groups of few isolated nano-tubes.

We acquired the SERS spectra from functionalized MWNT films in different spatial positions, maintaining all the experimental parameters unchanged obtaining reproducible spectra, shown in **Figure 9**.

In addition to the signal enhancement, compared to conventional Raman spectra, the SERS traces show sharp peaks in wavenumber region ascribable to vibrations of molecular groups;

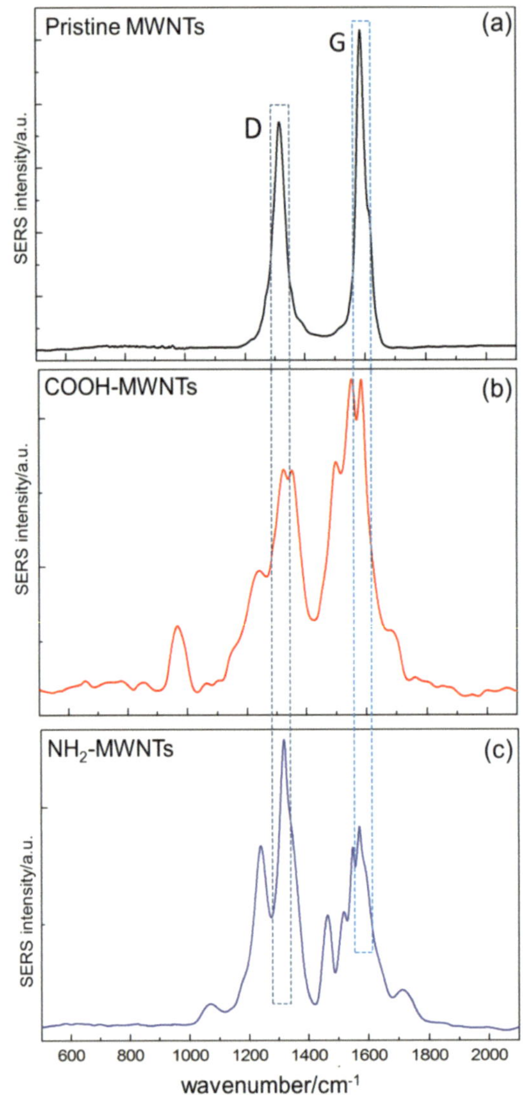

Figure 9. SERS spectra of pristine (a), COOH-MWNTs (b), and NH_2-MWNTs (c).

namely the C—O—C stretching (shoulder at around 1260 cm^{-1}), CH_3 deformation (1320 cm^{-1}), C—N stretching (1240 cm^{-1}, 1460 cm^{-1}), and C=O stretching at 1730 cm^{-1} [31, 32]. The presence of these bands proves experimentally the linkage of desired molecular groups to the nanotube walls. It is worth to note that the C=O stretching appears in both covalently functionalized samples owing to the functionalization procedure, which introduces the linkage of amides after that of carboxylate.

In the SERS spectrum of COOH-MWNTs, the D′ band is not distinguishable as shoulder next to the G band, indicating that this sample has a lower amount of defects, as expected for carbon nanotubes with large diameter, which exhibit reduced tip curvature and in turn fewer structural defects. It is worth to note that because there is no shift between SERS bands and the corresponding ones in the Raman spectrum, the interaction between MWNTs and the active surface is physical.

We can note also that in the SERS spectra of functionalized MWNTs, shoulders of G band at 1510 and 1544 cm^{-1} are clearly distinguishable. Although in the SERS trace of NH_2-MWNTs, the presence of the latter peak could be ascribed to the combination of N-H bend and C-N stretch, since these bands appear in both functionalized samples, they were assigned to disordered carbon with sp^3 bonds. In fact, the peak at 1510 cm^{-1} was observed in carbon with sixfold rings and rings with other orders, whereas the higher frequency peak, was often recorded in amorphous hydrogenated carbon, due to the C—C bond strain caused by the formation of C—H covalent bonds [6, 7, 9, 11, 33].

Surprisingly, also in the SERS spectra of pristine graphene, we recorded a large number of sharp peaks belonging to molecular group vibrations, as illustrated in **Figure 10**: namely C—C=O bending (520–570 cm^{-1}), C—O—C deformation, (850 cm^{-1}), O—H (900–960 cm^{-1}) and CH_2 (1040 cm^{-1}) bending, and C—C(O)—C stretching 1180–1270 cm^{-1}) [31, 34].

Functional groups as ethers, carbonyls, carboxyls, and hydroxyls are likely retained in the graphene structure, as a consequence of chemical exfoliation process and the presence of oxygen and hydrogen atoms in few at percentage was confirmed by XPS measurements of manufacturer [35].

We can also note that close to the G band are present the peaks of disordered carbon already observed in functionalized MWNTs. The presence of disordered carbon peaks in the SERS

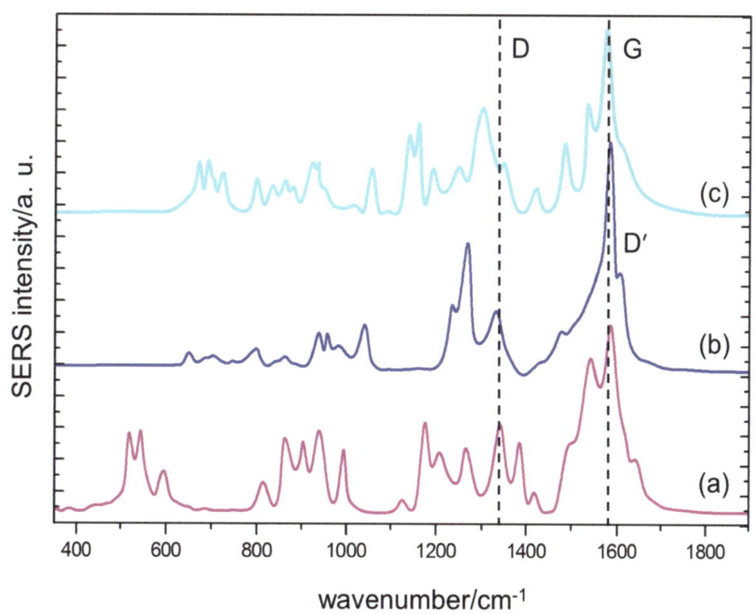

Figure 10. SERS spectra of pristine graphene. The spectra (a), (b), and (c) were acquired at different scanning positions across the dried film.

spectra of graphene and functionalized MWNTs is a further proof of evidence of the formation of tetrahedrally coordinated bonds between carbon atoms and molecular groups.

We acquired SERS spectra for each point of a selected area of pristine graphene films with a high-resolution micro-Raman spectrometer (Horiba Xplora) (see **Figure 11**). The areas were scanned with an acquisition grid of 0.6 and 0.8 μm in x and y, respectively. We selected the wavenumber region 500–1280 cm^{-1} to represent the functional groups concentration and we constructed the SERS images, illustrated in **Figure 11(b)** and **(d)**, plotting the intensity of the peaks located in this region.

Optical images show significant variation in the dried film thickness along the scanned area, due to a poor dispersibility of pristine graphene in the distilled water. However, the SERS signal is always detectable with good spectral reproducibility and an intensity variation much smaller than that expected, considering that the SERS signal from the second monolayer and beyond is strongly reduced. This finding further confirms that the functional groups are distributed at the edges of graphene structure: the larger thickness and hence the weaker enhancement experienced by the molecular groups, can be compensated by a higher concentration.

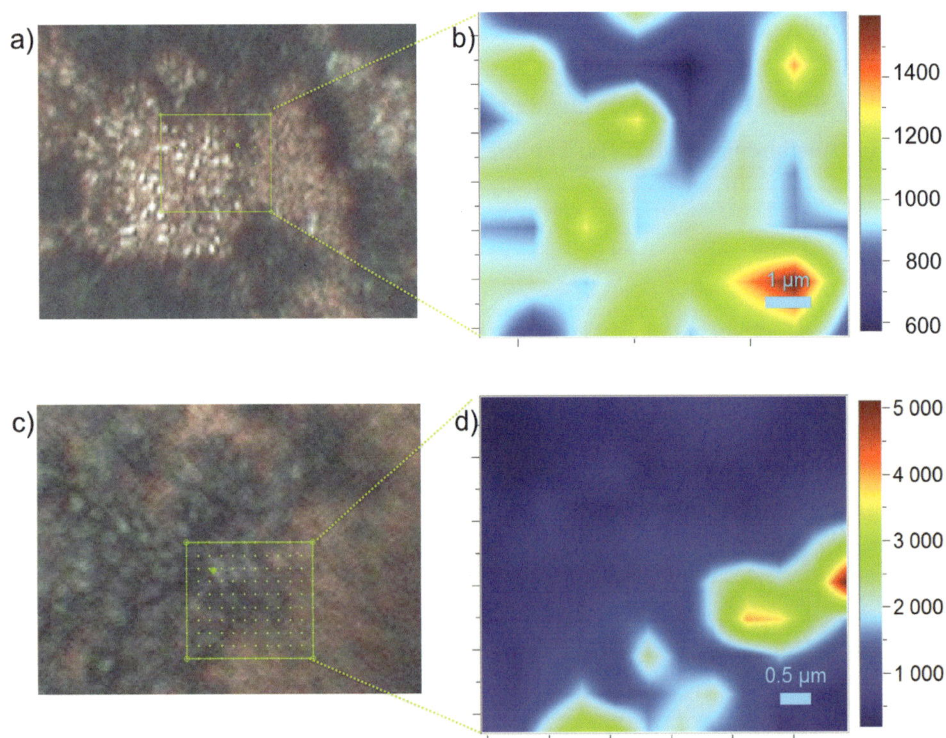

Figure 11. (a) Image of pristine graphene dried film, acquired with the optical microscope coupled with the Raman spectrometer; the green rectangle indicates the region selected to acquire the map (b). (c) and (d) represent the same for another region of the substrate. The visualization of Raman maps is obtained by using as contrast parameters the intensity of functional group peaks.

http://dx.doi.org/10.5772/intechopen.74065

8. Conclusions and outlook

In summary, we performed SERS measurements on graphene pristine and functionalized MWNTS and graphene nano-structured film deposited on gold-coated Si nano-pillar substrates. The strong surface enhancement effect allowed us to record the Raman signal from functional molecules that are not recorded in conventional Raman spectra proving experimentally their linkages to the CNT walls and graphene edges. By using the relative intensities of specific SERS features as contrast parameters to obtain SERS maps, it is possible to study the distribution of functional groups across the scanned area.

The obtained results encourage us to consider SERS as a powerful method to obtain a rapid monitor either of the procedures used to interface graphene and nanotubes with functionalizing groups either of synthesis process of graphene as chemical exfoliation.

Author details

Sabina Botti[1]*, Alessandro Rufoloni[1], Tomas Rindzevicius[2] and Michael Stenbæk Schmidt[2]

*Address all correspondence to: sabina.botti@enea.it

1 Fusion and Nuclear Security Department, ENEA, Frascati, Italy

2 Department of Micro and Nanotechnology, Technical University of Denmark, Lyngby, Denmark

References

[1] Hersam MC. Progress towards monodisperse single-walled carbon nanotubes. Nature Nanotechnology. 2008;**3**:387-394. DOI: 10.1038/nano.2008.135

[2] Byrne MT, Gun'ko YK. Recent advances in research on carbon nanotube-polymer composites. Advanced Materials. 2010;**22**:1672-1688

[3] Kim SW, Kim T, Kim YS, Choi HS, Lim HJ, Yang SJ, Park CR. Surface modifications for the effective dispersion of carbon nanotubes in solvents and polymers. Carbon. 2012;**50**:3-33

[4] Jeanmaire DL, Van Duyne RP. Surface Raman spectro-electrochemistry. Journal of Electroanalytical Chemistry. 1977;**84**:1-20

[5] Haynes CL, McFarland AD, Van Duyne RP, Surface-enhanced Raman spectroscopy. Analytical Chemistry. 2005;**77**:338A-346A and references therein

[6] Ferrari AC. Raman spectroscopy of graphene and graphite: Disorder, electron-phonon coupling, doping and non-adiabatic effects. Solid State Communications. 2007;**143**:47-57

[7] Ferrari AC, Basko DM. Raman spectroscopy as a versatile tool for studying the properties of graphene. Nature Nanotechnology. 2013;**8**:235-246

[8] Malard LM, Pimenta MA, Dresselhaus G, Dresselhaus MS. Raman spectroscopy in graphene. Physics Reports. 2009;**473**:51-87

[9] Casiraghi C, Hartschuh A, Qian H, Piscanec S, Georgi C, Fasoli A, Novoselov KS, Basko DM,Ferrari AC. Raman spectroscopy of graphene edges. Nano Letters. 2009;**9**:1433-1441

[10] Tuinstra F, Raman KJL. Spectra of graphite. The Journal of Physical Chemistry. 1970;**53**(3): 1126-1310

[11] Jiang J, Pachter R, Mehmood F, Islam AE, Maruyama B, Boeckl JJ. A Raman spectroscopy signature for characterizing defective single-layer graphene: Defect-induced I(D)/I(D′) intensity ratio by theoretical analysis. Carbon. 2015;**90**:53-62

[12] Lucchese MM, Stavale F, Ferreira EHM, Vilani C, Moutinho MVO, Kapaz RB, Achete CA. Quantifying ion-induced defects and Raman relaxation length in graphene. Carbon. 2010;**48**:1592-1597

[13] Rao AM, Richter E, Bandow S, Chase B, Eklund PC, Williams KA, et al. Diameter-selective Raman scattering from vibrational modes in carbon nanotubes. Science. 1997;**275**:187-191

[14] Bandow S, Asaka S, Saito Y, Rao AM, Grigorian L, Richter E, Eklund PC. Effect of the growth temperature on the diameter distribution and chirality of single-wall carbon nanotubes. Physical Review Letters. 1998;**80**:3779-3782. DOI: 10.1103/PhysRevLett.80.3779

[15] Dresselhaus MS, Dresselhaus G, Jorio A, Souza Filho AG, Saito R. Raman spectroscopy on isolated single wall carbon nanotubes. Carbon. 2002;**40**:2043-2061

[16] Botti S, Ciardi R, Terranova ML, Piccirillo S, Sessa V, Rossi M. Carbon nanotubes and nanowires grown from spherical carbon nano-particles. Chemical Physics Letters. 2002;**355**:395-399

[17] Botti S, Ciardi R, Terranova ML, Piccirillo S, Sessa V, Rossi V-AM. Self-assembled carbon nanotubes grown without catalyst from nanosized carbon particles adsorbed on silicon. Applied Physics Letters. 2002;**80**:1441-1443

[18] Lehman JH, Terrones M, Mansfield E, Hurst KE, Meunier V. Evaluating the characteristics of multiwall carbon nanotubes. Carbon. 2011;**49**:2581-2602

[19] Le Ru E, Etchgoin P. Principles of Surface-Enhanced Raman Spectroscopy and Related Plasmonic Effects. Elsevier Science; 2008 ebook ISBN: 9780080931555 and references therein

[20] Botti S, Cantarini L, Almaviva S, Puiu A, Rufoloni A. Assessment of SERS activity and enhancement factors for highly sensitive gold coated substrates probed with explosive molecules. Chemical Physics Letters. 2014;**592**:277-281

[21] Schmidt MS, Hübner J, Boisen A. Large area fabrication of leaning silicon nano-pillars for surface enhanced Raman spectroscopy. Advanced Materials. 2012;**24**(10):OP11-OP18

[22] Yang J, Palla M, Bosco FG, Rindzevicius T, Alstrøm TS, Stenbæk Schmidt M, Boisen A, Ju J, Lin Q. Surface-enhanced Raman spectroscopy based quantitative bioassay on aptamer-functionalized nanopillars using large-area Raman mapping. ACS Nano. 2013; **7**(6):5350-5359

[23] Wu K, Rindzevicius T, Stenbæk Schmidt M, Bo Mogensen K, Hakonen A, Boisen A. Wafer-scale leaning silver nanopillars for molecular detection at ultra-low concentrations. Journal of Physical Chemistry C. 2015;**119**(4):2053-2062

[24] Botti S, Cantarini L. Unpublished data

[25] Dresselhaus MS, Dresselhaus G, Eklund PC. Raman scattering in fullerenes. Journal of Raman Specroscopy. 1996;**27**:351-371

[26] Moskovits M. Surface selection rules. The Journal of Chemical Physics. 1982;**77**:4408-4416

[27] Osswald S, Havel M, Gogotsi Y. Monitoring oxidation of multiwalled carbon nanotubes by Raman spectroscopy. Journal of Raman Specroscopy. 2007;**38**:728-736

[28] Lefrant S, Baltog I, Baibarac M, Mevellec JY, Chauvet O. SERS studies on single-walled carbon nanotubes submitted to chemical transformation with sulfuric acid. Carbon. 2002;**40**(12):2201-2211

[29] Graupner R. Raman spectroscopy of covalently functionalized single-wall carbon nanotubes. Journal of Raman Specroscopy. 2007;**38**:673-683

[30] Sato-Berrù RY, Basiuk EV, Saniger JM. Application of principal component analysis to discriminate the Raman spectra of functionalized multiwalled carbon nanotubes. Journal of Raman Specroscopy. 2006;**37**:1302-1306

[31] Lin-Vien D, Colthup N, Fateley W, Grasselli J. The Handbook of Infrared and Raman Characteristic Frequencies of Organic Molecules. 1st ed. London: Academic Press, Elsevier; 1991. ISBN: 9780080571164

[32] Ramanathan T, Fisher FT, Suoff RS, Brinson LC. Amino-functionalized carbon nanotubes for binding to polymers and biological systems. Chemistry of Materials. 2005;**17**: 1290-1295

[33] Ferrari AC, Robertson J. Interpretation of Raman spectra of disordered and amorphous carbon. Physical Review B. 2000;**61**(20):14095-14107

[34] Botti S, Rufoloni A, Laurenzi S, Gay S, Rindzevicius T, Schmidt MS, Santonicola MG. DNA self-assembly on graphene surface studied by SERS mapping. Carbon. 2016;**109**: 363-372

[35] XG Sciences grade C particle data sheets

Section 4

Raman Spectroscopy of Ferrite Nanomaterials

Chapter 11

Raman Spectroscopy in Zinc Ferrites Nanoparticles

Pietro Galinetto, Benedetta Albini, Marcella Bini and
Maria Cristina Mozzati

Additional information is available at the end of the chapter

http://dx.doi.org/10.5772/intechopen.72864

Abstract

$ZnFe_2O_4$ ferrite nanoparticles are arousing a great interest in the biomedical field, thanks to their superparamagnetic behavior at room temperature. Functional properties depend on composition, size, nanoparticle architecture and, in turn, on the synthesis methods. Bulk $ZnFe_2O_4$ has the normal spinel structure (all Zn^{2+} ions in tetrahedral and all Fe^{3+} ions in octahedral positions), but at the nanometric size inversion takes place with a cationic mixing on divalent and trivalent sites. The sensitivity of the Raman probe to cation disorder favored the appearance of several works on a rich variety of nanosized zinc ferrites. An overview on these results is reported and discussed at variance with synthesis methods, grain dimensions, and dopants. We add to this landscape our results from new nanosized powder samples made by microwave-assisted combustion, with different dopants (Ca, Sr on Zn site and Al, Gd on Fe site). A detailed analysis of A_{1g}, E_g, $3F_{2g}$ Raman modes has been performed and Raman band parameters have been derived from best-fitting procedures and carefully compared to literature data. The vibrational results are discussed taking into account the characterization from X-ray powder diffraction raction, SEM-EDS probe, EPR spectroscopy and, of course, the magnetic responses.

Keywords: zinc ferrites, nanostructures, Raman spectroscopy, cation disorder

1. Introduction

Zinc ferrite is a very popular material widely used at the nanometric size in different applicative fields [1–4]. $ZnFe_2O_4$ (ZFO) has a spinel structure and similarly to the whole ensemble of materials with the general formula MFe_2O_4 displays unique physical and chemical properties. In particular, zinc ferrite exhibits a peculiar mixing of high-quality functional properties. Indeed its magnetic, thermal, electrical, and mechanical properties coupled to a high chemical stability allowed its usage in magnetic storage, ferrofluids, catalysis, and biomedical applications, as for instance, theranostics and hyperthermia.

The spinel ferrite system can be written as $(Me_{1-x}Fe_x)^{2+}[Me_xFe_{2-x}]^{3+}O_4$ and contains two Fe^{3+} cations and a single divalent cation. The round and square brackets denote sites of tetrahedral (A) and octahedral [B] coordination, respectively, while x represents the degree of inversion defined as the fraction of the (A) sites occupied by Fe^{3+}. A completely inverse spinel is determined when the trivalent cations are equally distributed on tetrahedral and octahedral sites, while the divalent cations fill up the remaining half of the octahedral positions.

A very important characteristic of the spinel system is that it admits an extremely large variety of *total solid solutions*. The divalent cations are usually Zn^{2+}, Ni^{2+}, Mn^{2+}, Ba^{2+}, Cu^{2+}, but partial substitutions of divalent or trivalent Fe ions are possible, preserving the spinel crystal structure. When zinc enters to form bulk zinc ferrite the system assumes a normal spinel structure AB_2O_4: the oxygen atoms are arranged a network of close-packed face-centered cubic units; the divalent Zn cations (A) occupy tetrahedral sites and trivalent Fe cations (B) octahedral sites. Spinel unit cell is composed of 8 units builders. In the full unit cell, there are 8 Me^{+2}, 16 Fe^{+3} cations, and 32 anions oxide O^{-2} [1–4]. In **Figure 1**, a scheme of the direct spinel structure is reported.

The magnetic behavior of bulk ferrite system is strongly influenced by the nature, type, amount, and distribution of cations. Let us consider a deeply studied system, the mixed Zn-Ni ferrites ($Zn_{1-x}Ni_xFe_2O_4$, with $0 \le x \le 1$). The end compositions are zinc ferrite, $ZnFe_2O_4$ (for $x = 0$), i.e., a direct spinel, and nickel ferrite, $NiFe_2O_4$ ($x = 1$), an inverse spinel. In direct spinel $ZnFe_2O_4$, the behavior is antiferromagnetic with a Néel temperature about 9 K, while for nickel ferrite is ferrimagnetic with a Curie point about T = 860 K [5]. The possibility to control the chemical composition allows to finely tailor the magnetic features. These different behaviors can be understood looking at the magnetic interactions between B and A sites surrounding an oxygen [5].

In spinel ferrites, there are two sublattices: the tetrahedral one with one cation per formula and the octahedral one with two cations per formula. The magnetic behavior is determined by superexchange interactions between two transition cations separated by an oxygen. The A-O-B interactions have an axial symmetry, and thus are highly more efficient than the B-O-B

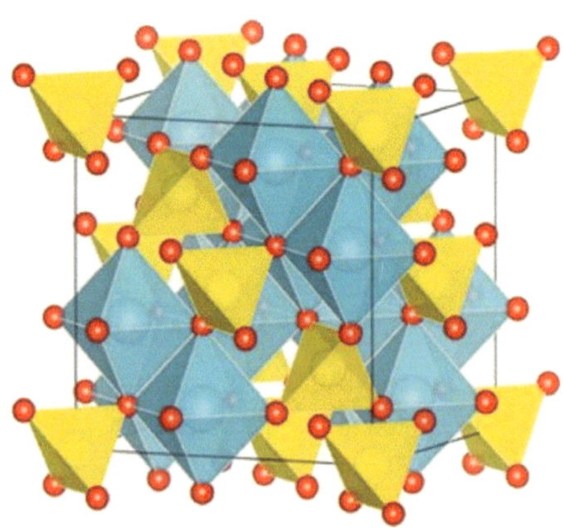

Figure 1. The direct spinel-type structure; the red circles represent oxygen ions while the ochre and cyan solids represent tetrahedral and octahedral units containing at the center Zn and Fe cations, respectively.

interactions, characterized by a 90° angle. As a result, both spins on B sites are antiparallel to the A spin. The magnetic response of spinel is thus determined by the cations distribution on A and B sites. In inverse $NiFe_2O_4$, B sites are occupied by both Ni^{2+} and Fe^{3+} ions and the remnant Fe^{3+} is placed on the A site. The magnetic moment of Fe^{3+} on A site cancels with the magnetic moment of Fe^{3+} on B site, and the resulting magnetic moment is the spin value for nickel Ni^{2+} (2.3 Bohr magnetons). This is a strong interaction with a resulting high-temperature Curie transition. In the case of zinc ferrite, iron ions occupy the B sites and zinc is located on the A site. Since Zn^{2+} is a $3d^{10}$ ion, it has no magnetic moment, and the only interaction is B-O-B, leading to an antiparallel arrangement of spins. The spins are identical, and the material becomes antiferromagnetic. The low Néel temperature clearly illustrates the weakness of this superexchange interaction, as compared with A-O-B of the nickel ferrite case [5].

The behavior of bulk spinel-type ferrites change markedly moving to the nanoscale regime. In particular, the insurgence of a superparamagnetic (SPM) behavior at room temperature (RT) has been observed [6]. In systems that are ferro- and ferrimagnetic in bulk, SPM state can appear when the grain size is reduced to 50 nm or less. In SPM phase, the thermal energy k_BT is greater than the magnetic anisotropy energy, and therefore random fluctuations of the magnetization are possible. The magnetization of a SPM material, above the so-called blocking temperature (T_B), is equal to zero in the absence of an external field, and it rapidly increases under application of an external field. This fact implies a closed sigmoidal shape of the M-H curve without appreciable hysteresis. Experimentally, the value of T_B typically corresponds to the "merging point" of the zero-field cooled (ZFC) and field-cooled (FC) magnetization curves [5].

SPM is not the only magnetic phenomenon caused by the finite size effect of nanoparticles. The reduction in size and the increase of the surface/volume ratio can, for example, produce randomly oriented uncompensated surface spins, canted spins, and magnetically dead layer at the surface [7].

Anyways, superparamagnetism is especially important in applications such as drug delivery or MRI, where the nanoparticles exhibit no magnetic properties upon removal of the external field and therefore have no attraction for each other, eliminating the major driving force for aggregation. More importantly, superparamagnetic nanoparticles allow better control over the application of their magnetic properties because they provide a strong response to an external magnetic field.

For biomedical application, it is thus essential to finely control the functional parameters: saturation magnetization (M_s), coercivity (H_c), and blocking temperature (T_B). This control can be made by tailoring the material properties in terms of size, shape, composition (substitutions, doping, cation distribution in the crystal structure), and shell-core design, with possible different chemical and physical structures of internal-core and surface-shell parts of the nanoparticle [7].

A large number of synthesis methods were reported, including conventional ceramic solid state synthesis, high-energy ball milling, microwave-assisted combustion, sol-gel, hydrothermal, co-precipitation, ultrasonic cavitation, and thermal plasma, with the peculiar aim to tune electric, catalytic, and magnetic properties, these last ones particularly appealing for biomedical application [8]. In the following, the most widespread synthesis methods for ferrites will be briefly described.

1.1. Conventional ceramic method

The conventional solid-state reaction, starting from metal oxides or carbonates, is the oldest method for the synthesis of ceramics, since it is economic, efficient, and easily scalable. In fact, despite of some disadvantages, such as particle agglomeration and growth, it is still regularly used to synthesize novel materials for the first time, also for the widespread diffusion of the necessary equipment in academic laboratories.

1.2. High energy ball milling

The use of mechanical milling in the preparation of inorganic materials (ceramics, metals, and alloys) has a long tradition. It is a cheap and relatively rapid method that, in some cases, guarantees the formation of the desired compounds without the need of other thermal treatment. The tuning of ferrites nanoparticles could be obtained by changing milling container, speed, time, ball-to-powder weight ratio, extent of filling the vial, milling atmosphere, presence of a process control agent, temperature of milling, etc. [9].

1.3. Microwave combustion method

This method is actually one of the most diffused due to its numerous advantages: is a wet chemical technique, enables fast reaction rate, chemical homogeneity, and high reactivity due to the uniform heating produced by microwaves. Some drawbacks are represented by the difficulty to vary the spherical form of particles and to control the valence states of the elements [10].

1.4. Sol-gel method

The sol-gel technique is a wet low-temperature method that provides products with a perfect chemical composition and takes advantages over the others, owing to good stoichiometric control and homogeneity, short preparation time and inexpensive precursors. Furthermore, it provides nanoparticles with controlled sizes and defined morphology. Generally, for the production of multicomponent oxides, alkoxides (substituted in some cases by acetates) are put together in alcohol. In the sol-gel method water, alcohol, pH, concentration of alkoxides, and controlled temperature are required for proper hydrolysis [11].

1.5. Hydrothermal method

The hydrothermal route is a convenient wet way to produce well crystalline materials, with particles with suitably tuned size and shape. The main advantage of hydrothermal synthesis over conventional wet-chemistry methods is that it occurs under non-standard conditions, and that non-classical crystallization pathways can be explored. Many factors can be varied such as temperature, reagents ratio, solvents, and kind of salts to customize the ferrite nanoparticles [12].

1.6. Co-precipitation method

This method is probably the most favorable to produce nanoparticles, as well as simple and very productive. A high homogeneity and uniformity can be easily obtained at moderate experimental conditions. In these last years, this kind of synthesis has been mainly used to produce nanoparticles with peculiar magnetic properties, in particular superparamagnetism for biomedical application. The particle growth is controlled only by kinetic factors, so by controlling pH, temperature, ionic strength, salts nature or the Fe(II)/Fe(III) concentration ratio, the shape and size of the nanoparticles can be customized [13].

1.7. Green synthesis

Recently, plant-based green syntheses of nanoparticles have attracted the attention of researchers. Various green resources, like micro-organisms and plants, are available for these syntheses. The extract of plants not only acts as reducing agent toward the metal ions in a short time with respect to micro-organisms or the classical organic/inorganic substances, but also provides high-yield nanosized particles. The preparation time of nanoparticles depends upon the variety of plants and the concentration of phyto-chemicals. Plant-based green synthesis of magnetic nanoparticles is still under research [14, 15].

Among the so-called physical methods, the most used to produce zinc ferrites are the following.

1.8. Ultrasonic cavitation method

Ultrasonic cavitation chemistry is useful for synthesizing a variety of compounds at milder conditions. The major advantage of this method is that it affords a reliable and facile route for the control of both the synthetic process and nanostructure of advanced materials. This process also provides chemical homogeneity and reactivity through atomic level mixing and phase pure crystalline materials can be prepared by annealing at reduced temperatures. In addition, the use of additive components (stabilizers or surfactants, precipitants) and particular calcination requirements are avoided [16].

1.9. Thermal plasma

Thermal plasmas, such as direct current arcs and radio frequency (RF) plasmas, offer unique advantages for the synthesis of ceramic powders due to the easily achievable high temperatures and energy densities. In a RF thermal plasma flame, the gas temperatures may exceed 10^4 K independently of the gas composition. In addition, a high temperature gradient exists between the hot plasma flame and the surrounding gas phase. The resulting rapid quenching rate is favorable for producing fine particles with unstable structures in thermodynamic terms [17].

The large variety of methods used to obtain zinc ferrites nanoparticles reflects on a very scattered scenery in terms of properties of the obtained materials. This is not simply due to the above-mentioned close correlation among functional parameters and morphology, gran size, intentional doping, cation distribution, and nanoparticle architecture. Indeed, it is extremely important to verify and control the magnetic, structural and chemical purity, and homogeneity of nanoparticles because even low amount of extrinsic phases or elements can alter the functional properties. In particular, it is very important to monitor the presence of unwanted iron oxides phases, possible source of extrinsic contribution to the functional properties of zinc ferrites. In addition, several factors such as nucleation, growth, aggregation, and adsorption of impurities can affect the morphology of prepared particles [18]. It is thus evident the need to deeply characterize the materials with a multi-technique approach to reveal even subtle effects due to unwanted alterations of the designed nanoparticles.

Usually, all the researchers give evidence of the magnetic properties measuring by SQUID magnetometer or Vibrating Sample Magnetometry, Saturation magnetization (M_s), Coercivity (H_c), and Blocking temperature (T_B). The local inspection about the source of magnetic behavior is usually accomplished by combining ^{57}Fe Mossbauer, XANES, XPS, and EPR investigations. Morphology and chemical compositions are usually investigated by TEM and SEM techniques, the latter with the possibility to perform EDX analyses for elemental check. Sometimes, LA-ICP is used to validate the composition. Crystalline structure quality is usually assessed by XRD analyses and micro-Raman measurements, that also allowed to estimate the inversion degree.

Raman spectroscopy (RS) has been widely used to study spinel ferrites and in particular zinc ferrites both for routine identification of materials [19] and for very fine investigations concerning basic phenomena. This is due to an interesting overlapping between characteristics of RS and properties of nanosized zinc ferrites: (i) RS sensitivity to cations distribution [20], (ii) RS ability to reveal the presence of extrinsic iron oxides phases like hematite or magnetite [21], (iii) possibility to use RS to monitor stability of spinel ferrites vs. light exposure or thermal treatment [22], (iv) influence of nanometric scaling on Raman lines features—peak energies, widths and shape [23], (v) possibility to evaluate the presence of unwanted impurities. In addition, it is important to note that Raman spectroscopy has been used to study Surface-enhanced Raman scattering (SERS) when ferrites nanoparticles are functionalized for specific biomedical applications [24].

In this chapter, we combine a brief overview on the literature about Raman data from different zinc ferrites nanoparticles to new Raman results by our group on pure and doped zinc ferrites nanoparticles prepared by using the microwave-assisted combustion method. In particular, we present data obtained from undoped $ZnFe_2O_4$, $Zn_{1-x}Ca_xFe_2O_4$ (x = 0.05 and 0.25), $ZnFe_{1.9}Gd_{0.1}O_4$, $Zn_{0.95}Sr_{0.05}Fe_2O_4$, and $ZnFe_{1.9}Al_{0.1}O_4$ doped ferrites. The Ca substitution is particularly interesting due to the low toxicity of the substituent; on the other side, Gd ions are commonly used as contrast agents for MRI. In all the cases, the Raman data are accompanied by recalling functional parameters and synthesis procedures. For our samples, a more complete description about synthesis, morphology, and magnetic behavior are outlined allowing a better comprehension about the powerful of Raman spectroscopy in this kind of system.

2. Raman effect in $ZnFe_2O_4$ nanoparticles

$ZnFe_2O_4$ spinel has a cubic structure that belongs to the space group Fd3m (O^7_h) consisting of 8 molecules within the unit cell, for a total of 56 atoms; nevertheless the smallest Bravais cell contains 2 formula units for a total of 14 atoms. By the factor group analysis, it is possible to derive the following 42 modes in $ZnFe_2O_4$ spinel, 3 acoustic modes, and 39 optical modes [19]:

$$A_{1g}(R) + E_g(R) + F_{1g} + 3\,F_{2g}(R) + 2\,A_{2u} + 2\,E_u + 4\,F_{1u}(\mathrm{IR}) + 2\,F_{2u} \quad (1)$$

where the (R) and (IR) identify Raman-active and infrared-active vibrational species, respectively, and the rest of the modes are silent modes. The $E_{g,u}$ and $F_{1g,2g,1u,2u}$ modes are doubly and triply degenerate, respectively. The three acoustic modes belong to the F_{1u} species.

Thus in the Raman spectra of zinc ferrites, only five modes $A_{1g} + E_g + 3F_{2g}$ should be observed [25]. The three Raman-active F_{2g} modes are labeled $F_{2g}(1)$, $F_{2g}(2)$, and $F_{2g}(3)$, where $F_{2g}(1)$ is the lowest frequency F_{2g} mode and $F_{2g}(3)$ is the highest frequency mode of this vibrational species.

Figure 2 shows the spectra from two different ZFO prepared by our group. These spectra are reported as an example of the features detected in the Raman spectrum of ZFO. The red spectrum is obtained from pure ZFO, synthesized by solid state reaction (see Section 2.2.), while the black curve is the Raman signal of an Al-doped ZFO sample synthesized by co-precipitation method. In the region 100–800 cm^{-1}, three different spectral intervals can be recognized for both samples: (i) 600–800 cm^{-1} is the region of A_{1g} modes; (ii) 410–550 cm^{-1} is the region of $F_{2g}(3)$ modes, and (iii) 260–380 cm^{-1} should be the region of $F_{2g}(2)$ modes. At lower frequencies, E_g and $F_{2g}(1)$ are sometimes detected. Furthermore in the spectrum from Al-doped sample, narrow extra peaks are detected at about 218, 285, and 395 cm^{-1}. These

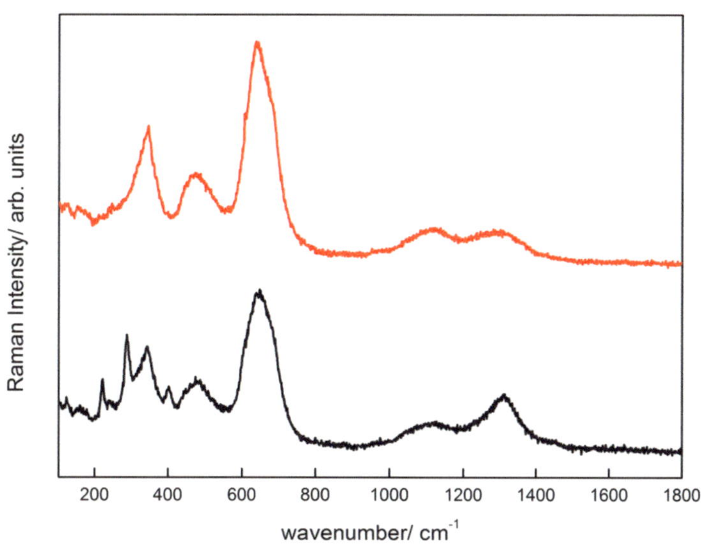

Figure 2. Raman spectra from two different ZFO prepared by our group: Pure ZFO synthesized by solid state reaction (red curve) and Al-doped ZFO synthesized by conventional co-precipitation method (black curve). The latter will not be further considered in the chapter.

Raman lines are associated to the presence of hematite inside the irradiated volume [21]. At higher energies, in the region 1000–1800 cm^{-1} only second-order features can be revealed, with a more intense feature just above 1300 cm^{-1} in the Al-doped sample due to second-order signal from hematite.

Briefly, we recall that micro-Raman measurements were carried out at RT by using a Labram Dilor spectrometer equipped with an Olympus microscope HS BX40. The 632.8 nm light from He-Ne laser was employed as excitation radiation. The samples, mounted on a motorized *xy* stage, were tested with a 100× objective and with a laser spot of 1 μm of diameter. The spectral resolution was about 1 cm^{-1}. Neutral filters with different optical density were used to irradiate the samples at different light intensities leading to power density values from 5×10^3 to 5×10^5 W/cm^2. A cooled CCD camera was used as a detector and the typical integration times were about 2 min. The sample phase homogeneity was verified by mapping the Raman spectra from different regions of each sample. The parameters of the Raman spectra were extracted by using best fitting procedures based on Lorentzian functions. In this way, the frequency, full width at half maximum, intensity, and integrated intensity of the peaks were determined.

2.1. Brief overview from the literature

Even if in the literature the assignment of the specific atomic motions within the spinel lattice during the Raman-active vibrations is controversial, it is common to explain the vibrational dynamic in term of normal modes inside the two sub-units within the spinel unit cell: the tetrahedral unit, AO_4 is comprised of the cation at the center of a cube and four oxygen atoms in the nonadjacent corners; the octahedral unit consists of a cation surrounded by six oxygen atoms, two along each dimensional axis, to form a BO_6 octahedron. It is accepted that the highest-frequency A_{1g} mode is assigned to the symmetric breathing mode of the AO_4 unit within the spinel lattice [26]. The oxygen atoms move away from the tetrahedral cations along the direction of the bonds, with all the cations at rest. For all the other Raman modes, there are some controversies about their assignments. In [27], all the other low frequency modes are attributed to vibrations inside the octahedral sites (BO_6). Besides the highest-frequency Raman-active $F_{2g}(3)$ mode is alternatively attributed to the antisymmetric breathing of the AO_4 unit [28], or to the asymmetric bending motion of the oxygens bonded to the tetrahedral cations [29].

The $F_{2g}(2)$ mode should be due to the opposite motion of cation and oxygen along one direction of the lattice [30]. In [30], the E_g mode is assigned to the symmetric bending motion of the oxygen anions within the AO_4 unit, in agreement to other researchers [28]. Finally, more agreement is found for $F_{2g}(1)$ Raman mode, the lowest frequency one, due to the complete translation of the AO_4 unit within the spinel lattice [31–34].

The controversial landscape of Raman modes in ZFO is summarized in **Table 1**, where we report a brief overview on Raman results from nanosized pure zinc ferrites obtained by different methods.

Method	Average size (nm)	Main Raman modes peak energy (cm^{-1})						References
		$F_{2g}(1)$	E_g	$F_{2g}(2)$	$F_{2g}(3)$	A_{1g} (1)	Other signals	
S MC	3–25	150	265	350–400	450–500	630 (Zn-O) 720 (Fe-O)	—	[35, 36]
Honey-mediated SG C	10–20	241	351	498	438	605 (Zn-O) 653 (Fe-O)	221, 286, 403 α-Fe_2O_3	[37]
S MC	20	160	265	370	450	637(Zn-O) 670(Fe-O)	721 A_{1g} maghemite	[38]
TD	4.7	159	254	364	459	665	—	[39]
CG	13	—	351	452	490	650 (Zn-O) 700 (Fe-O)	—	[40]
CP	5	—	330	480	—	660	—	[41]
SG AC	21	262	357	497	—	682	—	[42]
SG C	19.6	174.9	340.3	491	—	652.7 688.6	—	[43]
HT F	—	221	246	355	451	647	800 noise	[27]
CP	7	235.56	339.61	487.4	—	664.97	—	[23]

The data are listed at variance with synthesis method and grain size. For syntheses, we used the following acronyms: S MC for soft mechanochemical, SG C for sol-gel combustion, TD for thermal decomposition, CG for citrate-gel, CP for co-precipitation, SG AC for sol-gel autocombustion and HT F for high temperature flux.

Table 1. Overview on data from literature on Raman modes in pure ZFO nanoparticles.

At first, one can notice the absence of a clear correlation between grain size and peak positions of A_{1g} modes. Looking at the spectra, the effect of grain size reduction causes as expected an asymmetric broadening of Raman bands and a decreasing of Raman intensities [44, 45]. A_{1g} mode is mainly the most intense Raman feature; while in [27, 37, 46], the highest Raman signal is the Eg mode. Besides, in ZFO, the nanometric regime is primarily the cause of inversion, with marked influence on Raman spectra. Indeed in tetrahedra and octahedra a mixing between A and B cations takes place and this mixing reflects on Raman modes due to different cations involved for the same vibrational modes [35].

In **Table 2**, we summarize some results from nanocrystalline doped zinc ferrites and other common ferrites.

Both for pure and doped nanoparticles zinc ferrites, it is difficult to derive general rules for Raman modes behavior. Different parameters (synthesis, grain size, composition, and chemical purity) can give rise to competing effects with opposite effects on Raman features. In addition, it is evident from the reported data in **Tables 1** and **2** that even the attribution of a certain Raman signal to a specific mode is uncertain, in particular for F_{2g} and E_g modes. Furthermore for F_{2g} and E_g modes, the analyses are complicated by the usually lower intensities. Recently, even the attribution of A_{1g} mode has been reconsidered. Some authors [50, 51] claimed that

Sample	Method/ SynthesIs	Average size (nm)	Main Raman modes peak energy (cm^{-1})						References
			F_{2g}(1)	E_g	F_{2g}(2)	F_{2g}(3)	A_{1g} (1)	Other signals	
$ZnFe_{1.8}La_{0.2}O_4$	SG AC	10	250	343	487	—	663	—	[42]
$Ni_{0.5}Zn_{0.5}Fe_2O_4$	S MC	20	125 (Zn-O) 200 (Ni-O)	256 (Zn-O) 323 (Ni-O)	362 (Zn) 480 (Ni)	370 (Zn) 480 (Ni)	637(Zn-O) 670(Fe-O) 689(Ni-O)	721 A_{1g} maghemite	[38, 47]
$Mn_{0.45}Ni_{0.05}Zn_{0.5}Fe_2O_4$	C	9	—	336	482	540	636 (Zn-O) 680 (Fe-O)	—	[48]
$Mn_{0.05}Ni_{0.45}Zn_{0.5}Fe_2O_4$	C	12	—	339	484	550	650 (Zn-O) 695 (Fe-O)	—	[48]
$Mn_{0.65}Zn_{0.35}Fe_2O_4$	CG	9	—	350	—	540	610 (Mn-O) 650 (Zn-O) 690 (Fe-O)	—	[40]
$Mn_{0.6}Zn_{0.4}Fe_2O_4$	SG C	9.4	217.7	275.3	386.5	490.8	591.3 645.7	240, 300 α-Fe_2O_3	[43]
$Co_{0.5}Zn_{0.5}Fe_2O_4$	CP	—	207.7	314.5	468.5	551.7	632.7 A_{1g} (1) 693.2 A_{1g} (2)	—	[49]
$NiFe_2O_4$	S MC	10	188	323	480	550	670 (Fe-O) 689 (Ni-O)	721 A_{1g} maghemite	[38]
$CoFe_2O_4$	DM	6	547.7	293.2	460.7	190.2	674.8 A_{1g} (1) 604.9 A_{1g} (2)	—	[20]
$MnFe_2O_4$	SG C	8.5	220.6	277	370.9	507.4	593 638.7	240, 300 α-Fe2O3	[43]
$FeFe_2O_4$	TD	4.9	190	290	453	535	671	720, 604, 496, 386 maghemite	[39]

The data are listed at variance with synthesis method and grain size. For the acronyms, see the caption of **Table 1**. In addition, C stays for citrate and DM for double microemulsion.

Table 2. Overview on data from literature on Raman modes in doped ZFO nanoparticles.

the Raman band at around 650 cm^{-1} is associated with the presence of Zn^{2+} ions at B-sites. This conclusion seemed to be supported by the dependence of its intensity with Zn content in $Mg_{1-x}Zn_xFe_2O_4$ and by Mossbauer data. Nevertheless, some common features can be derived: the mass of the tetrahedral cations is effective in the A_{1g} lineshape with different contribution clearly detectable, as expected in a simple "mass on a spring" model; the reduced grain size causes a decrease in the total Raman yield; all the synthesis methods seem to be able to produce nanoparticles with good crystallinity even if sometimes the Raman signals are quite weak and broadened [22, 52]. Further worsening in the quality of Raman spectra has been observed when ZFO NP is functionalized for specific biomedical applications [53].

In the Raman spectra of nanocrystalline, both pure and doped zinc ferrites or other ferrites, it is common to recognize different contributions for each Raman mode, just due to different cations involved in a specific vibration [47]. This behavior, the so-called *two-mode* behavior, is particularly evident for the most intense A_{1g} mode above 600 cm^{-1}. If doping is considered, the Raman lineshape is affected both by the inversion on A and B sites and by the substitution of divalent or trivalent cations. Thus, the Raman signal in the more complex case could be the result of the overlapping of three contributions due to vibrations inside (i) ZnO_4, (ii) FeO_4, due to inversion, and (iii) the MeO_4 due to substitution. The two mode behavior is sometimes less detectable in the lower frequency modes; where in some cases, a one-mode behavior cannot be excluded [54]. The Raman studies on ZFO NP have usually the goal to characterize the structure of the samples in view of the magnetic response, being the last affected by inversion and/or cationic substitutions with magnetic ions. Raman data, and in particular the A_{1g} modes, are usually used to perform best-fitting analyses using lorenztian curves and derive the integrated intensities of the two (or three if substituents are considered) different contributions. With the empirical formula $x_{Fe}(A) = [1 - I_{Zn}/(I_{Zn} + I_{Fe})]$, one can derive the inversion degree; the lorentzian intensities

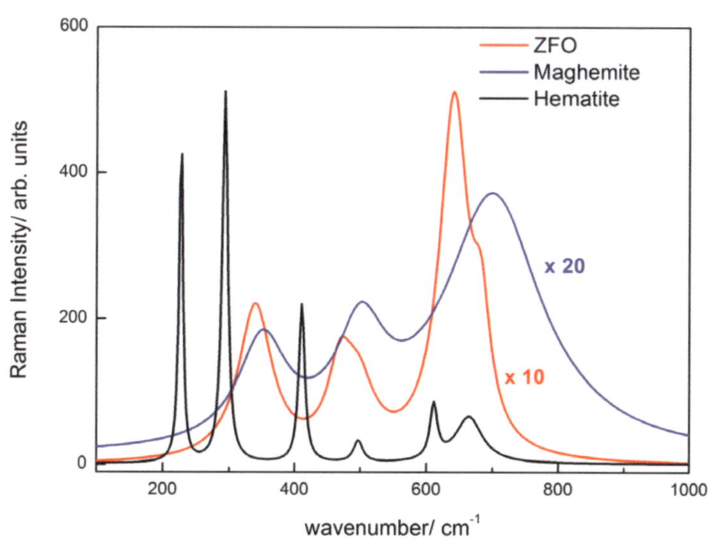

Figure 3. Simulated Raman spectra for ZFO, hematite and maghemite. The intensities are scaled to match the experimental evidences on pure micrometric powders.

can be multiplied by corresponding force constants even if it is not simply to derive a reliable value [39, 47].

For this reason, an important issue in ZFO NP studies is to reveal in synthesized samples the presence of iron oxides, like hematite and maghemite [5]. These oxides can infer an effective but extrinsic magnetic behavior with problems in terms of stability and aggregation if used in biomedical applications. To characterize the spin ordering and understand the nature of magnetic behavior, Mossbauer and neutron diffraction studies have been helpfully used [55]. Nevertheless, in this frame also the Raman spectroscopy can play an important role. Hematite (α-Fe_2O_3) belongs to the rhombohedral system while maghemite (γ-Fe_2O_3) has a cubic inverted spinel structure, thus from the point of view of Raman activity equivalent to ZFO. These differences reflect on the Raman spectra. In **Figure 3**, we report the simulated spectra for ZFO, hematite, and maghemite.

The Raman spectra of ZFO and maghemite are obviously quite similar with a total Raman yield greater for the former. The main difference is the peak position of the A_{1g}. In ZFO, the ZnO_4 vibrations are dominant and the peak is around 650 cm^{-1}, while in maghemite only the iron is present and the energy peak is usually observed around or just above 700 cm^{-1}. On the contrary, the Raman spectrum of hematite is markedly different having the main peaks in the lower energy region. It is important to note that the Raman yield of hematite is approximately 10–20 times higher than that of spinel-type ferrites, thus a small amount of hematite is enough to give clear signatures in Raman spectrum. It is difficult to quantify this small amount because it is often reported that for ZFO nanometric grains, surface shell of hematite or maghemite can be formed [37, 56]. In this case, the surface sensitivity of the Raman technique can hinder a proper estimation. A more difficult goal is to reveal the presence of few amounts of maghemite in ZFO by Raman spectroscopy. Detailed best-fitting analyses are usually performed on the A_{1g} data, and it is possible to infer about the presence of maghemite [38]; but the variability of peak energies found both for pure and doped ZFO and the weak weight of maghemite contribution in Raman signal hinder a reliable attribution simply based on Raman data. But, RS allows to monitor the stability of maghemite to laser irradiation. Indeed, it is well-known that in the oxidation route of iron ions maghemite is an intermediate metastable phase while hematite is the terminal and stable one [57]. Using this approach, the presence of maghemite in ZFO can be indirectly evaluated by the insurgence of hematite Raman signals under laser irradiation [58].

2.2. Experiments from pure and doped zinc ferrites obtained by microwave-assisted combustion method

We focused on the microwave combustion synthesis method of Ca (on Zn site) and Gd (on Fe site) substituted ferrites. Undoped $ZnFe_2O_4$ and Sr, and Al doped samples were also synthesized and used for comparison. All the samples have been investigated by X-ray powder diffraction and SEM-EDX analyses as a complement of Raman results. The magnetic behavior has been measured by SQUID magnetometry. All the experimental details are reported elsewhere [59].

We found, for all the samples, a superparamagnetic behavior with saturation magnetization between 6 and 10 emu/g at the maximum applied magnetic field of 3 T, with a more effective role played by Ca ions with respect to Gd ions substitution.

2.2.1. *Synthesis*

The samples were synthesized by microwave-assisted combustion methodology (MW), a rapid and green method as previously described, and by a conventional high energy ball milling.

For MW, the starting reagents were $Zn(NO_3)_2$ $6H_2O$ and $Fe(NO_3)_3$ $9H_2O$ taken in stoichiometric ratio to obtain $ZnFe_2O_4$ and mixed in an agate mortar with a proper amount of citric acid as fuel, as calculated from the propellant chemistry theory [60]. In brief, by taking into account the reducing and oxidizing valences of the involved elements (Zn = +2, N = 0, O = −2, Fe = +3, C = 4, H = 1), we can calculate a global valence value of −15, −10, and +18 for iron nitrate, zinc nitrate, and citric acid, respectively. We can write the balanced chemical reaction for the ferrite formation as 2·(−15) + 1·(−10) + x·(+18) = 0 and the mol amount of citric acid necessary to reduce the nitrates oxidant amount is calculated as 2.22.

The mixture was placed in a ceramic crucible in a microwave oven for 30 min at 800 W: this power ensures a temperature inside the oven between 450 and 500°C. This undoped sample will be named ZnFe. To obtain the doped samples, proper amount of $Ca(NO_3)_2$ $4H_2O$, $Sr.(NO_3)_2$, $Gd(NO_3)_3$ $6H_2O$, and $Al(NO_3)_3$ $9H_2O$ were added to the previous reagents to obtain the desired stoichiometry $Zn_{0.95}Ca_{0.05}Fe_2O_4$, $Zn_{0.75}Ca_{0.25}Fe_2O_4$, $Zn_{0.95}Sr_{0.05}Fe_2O_4$, $ZnFe_{1.9}Gd_{0.1}O_4$, and $ZnFe_{1.9}Al_{0.1}O_4$. The citric acid amount to add to the various mixtures was calculated as explained. In the following, these samples will be named Ca005, Ca025, Sr005, Gd01, and Al01, respectively.

For comparison, an undoped $ZnFe_2O_4$ sample was also synthesized by means of a classical solid-state synthesis, starting from a stoichiometric mixture of ZnO and Fe_3O_4 oxides ground by ball milling in tungsten jars for 6 h at 500 rpm with intermediate periods of last. The mixture was then treated in oven in air at 650°C for 12 h (heating rate 5°C/min, spontaneous cooling to 25°C) and ground after cooling. This sample will be named ZnFe-SS.

2.2.2. *XRD and SEM analyses*

X-ray diffraction analysis was used to determine the sample purity, in particular to control the effective cation substitution, the crystallite sizes and the eventual inversion degree of the spinel by means of the structural and profile refinement based on the Rietveld method. In **Figure 4**, the XRD patterns of the ZnFe-SS and ZnFe undoped $ZnFe_2O_4$ samples are shown together with Ca005 one. In all the cases, a pure ferrite sample is formed, whose peaks well agree with the $ZnFe_2O_4$ cubic spinel structure (PDF card N. 89-7412). It is well evident a different peaks broadening and crystallinity between MW and SS samples suggesting markedly different particle sizes, as can be expected due to the different experimental synthesis methodologies. For the doped samples, XRD patterns demonstrate that the doping successfully occurred, because neither traces of unreacted reagents nor phase impurities are present: only for Sr005, some low peaks reveal the presence of small traces of strontium nitrate, as a residual of the reagent. The structural refinement on the basis of the Rietveld method was performed on all the patterns by using the known cubic spinel model: the main refined structural parameters are reported in **Table 3**. It can be seen that the crystallite sizes are in the nanometer range: the values for the microwave synthesis are all lower than 11 nm, while for ZnFe-SS a value of 22 nm is found.

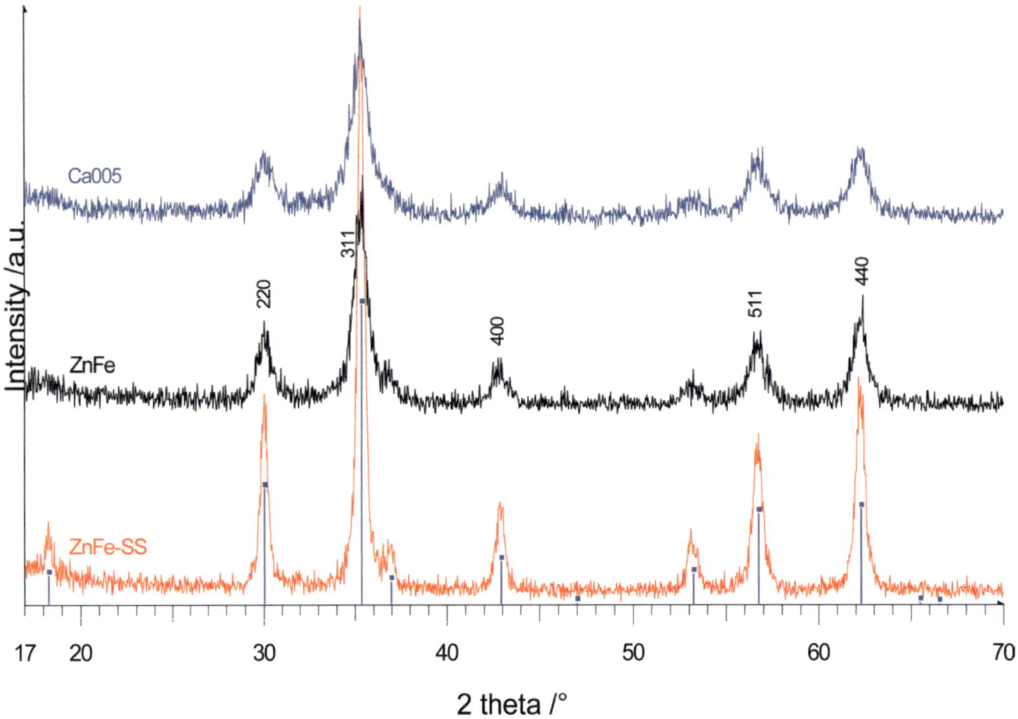

Figure 4. XRD patterns of undoped $ZnFe_2O_4$ obtained from microwave combustion and solid state methods together with the Ca005 pattern. The bars of the expected angular positions of the spinel phase (JCPDS card 89–7412) are also reported, together with the miller indices of the main peaks.

	Cations distribution	Impurities	a/Å	Size/nm
ZnFe-SS	$[Zn_{0.93}Fe_{0.07}]_T[Fe_{1.93}Zn_{0.07}]_O$	–	8.4386(6)	22.9
ZnFe	$[Zn_{0.82}Fe_{0.18}]_T[Fe_{1.82}Zn_{0.18}]_O$	–	8.4371(40)	11.1
Al01	$[Zn_{0.68}Fe_{0.32}]_T[Fe_{1.58}Al_{0.1}Zn_{0.32}]_O$	–	8.4221(25)	8.4
Ca005	$[Zn_{0.90}Ca_{0.05}Fe_{0.05}]_T[Fe_{1.95}Zn_{0.05}]_O$	–	8.4343(17)	8.7
Ca025	$[Zn_{0.21}Ca_{0.25}Fe_{0.54}]_T[Fe_{1.46}Zn_{0.54}]_O$	–	8.4331(57)	5.5
Sr005	$[Zn_{0.88}Sr_{0.05}Fe_{0.07}]_T[Fe_{1.93}Zn_{0.07}]_O$	$Sr(NO_3)_2$ 3.35	8.4380(13)	7.2
Gd01	$[Zn_{0.98}Fe_{0.02}]_T[Fe_{1.88}Gd_{0.1}Zn_{0.02}]_O$	–	8.4366(29)	8.4

Table 3. Cation distribution, impurity phases amount, lattice parameter, and crystallite size as derived from XRD analyses.

The lattice parameters although may vary due to the different ionic radii of the dopants with respect to Zn and Fe ions, are quite similar. On the other hand, a similar behavior for the lattice parameter was found for the Sr substituted samples [10]. Only for Al doping the cubic parameter decreases with respect to the undoped sample: in fact, a value of 0.53 Å is reported for Al^{3+} ions radius with respect to 0.65 Å of Fe^{3+} in high spin configuration [61]. The inversion degree, i.e., the amount of Fe ions on Zn crystallographic sites, was also determined from the structural refinement. In all the cases, the inversion takes place, although not so markedly: only for Ca025 a value of 0.54 is reached. The easily induced inversion can be a consequence of the quickness of the MW synthesis that in only about 30 min can produce a good level of crystallinity but with atomic disorder. We also verified that Ca and Sr. ions seem to prefer the A site, while Al and Gd ions the B one. This is certainly true when the substitution is about 5 atom%, while for Ca025 a different model could be hypothesized. In fact, due to the preference of calcium for octahedral coordination it is possible that these ions could be also located on Fe sites, so inducing a higher inversion degree with respect to Ca005 and a small contraction of the lattice parameter.

In **Figure 5a,b**, SEM images of Gd01 and ZnFe-SS clear up the morphological differences between the samples from MW synthesis and solid state, giving evidence of a higher surface/

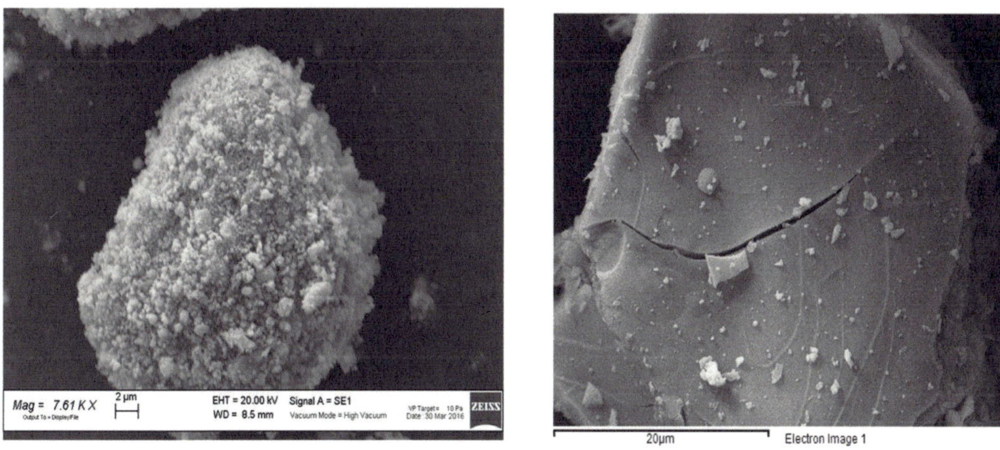

a) b)

c)

Element	App	Intensity	Weight %	Weight %	Atomic %
	Conc.	Corrn.		Sigma	
C K	5.40	0.1902	15.71	0.28	36.98
O K	18.73	0.5468	18.94	0.14	33.49
Fe K	60.44	0.9330	35.82	0.15	18.14
Zn K	37.05	0.8516	24.06	0.13	10.41
Gd L	8.60	0.8690	5.47	0.09	0.98
Totals			100.00		

d)

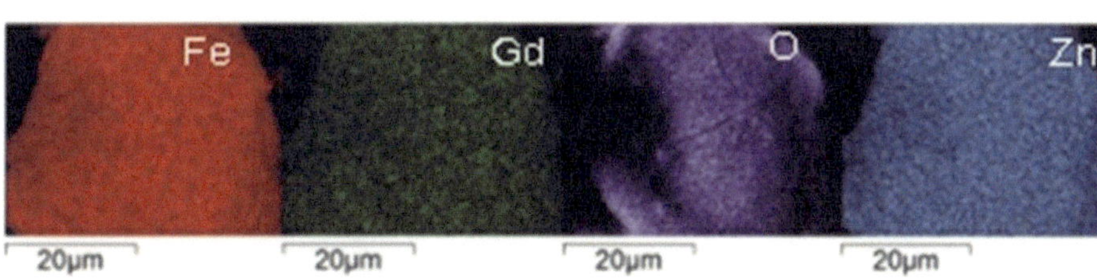

Figure 5. SEM images of Gd01 (a) and ZnFe-SS (b); data (c) and color map (d) from EDS elemental analysis for Gd01 sample.

volume ratio for MW samples with respect to solid-state one. From MW, rounded particles (lower than 100 nm) aggregates with large and open pores, regardless the doping ion, can be seen. This aspect can be due to the evolution of gases from nitrates and citric acid (such as NO_2 and CO_2) during the heating process in the microwave oven. The solid-state synthesis also leads to aggregates, but with larger rounded particles. The electronic microanalysis allowed us to verify that the stoichiometric ratio between the different elements was maintained in the final products. In **Figure 5c**, as an example, the EDS analysis of Gd01 sample is reported. The atomic percentages of all the elements are in excellent agreement with the stoichiometric values, within the EDS detection limit, suggesting that the ferrites possess the expected composition and no ions loss occurred. The maps of the different elements show good homogeneity, suggesting that Gd ions are well distributed in the sample. A similar behavior was observed for all the undoped and the other doped samples.

2.2.3. *Magnetic results*

Hysteresis cycles at RT and at 10 K have been measured as well as the zero-field and field cooled magnetization curves in temperature. The magnetic data have been carefully analyzed [59], and they provide a basic proof of the RT superparamagnetic behavior, essential requirement in magnetic hyperthermia. Examples for the hysteresis curves at RT and 10 K are reported in **Figure 6a** for the Ca005 sample, as well the relative ZFC and FC magnetization curves (**Figure 6b**).

The magnetic behavior for all the samples is summarized in **Table 4**, where all the functional parameters are reported.

We found, for all the MW samples, a superparamagnetic behavior with saturation magnetization at RT between 6 and 10 emu/g at the maximum applied magnetic field of 3 T, with a more effective role played by Ca ions with respect to Gd ions substitution. A fair agreement is found with the M_S values reported in the literature for pure and doped zinc ferrite spinels having comparable grain size and prepared by means of combustion synthesis or other methods [10, 60, 62–64]. In $ZnFe_2O_4$, the inversion promotes super exchange interactions, in turn

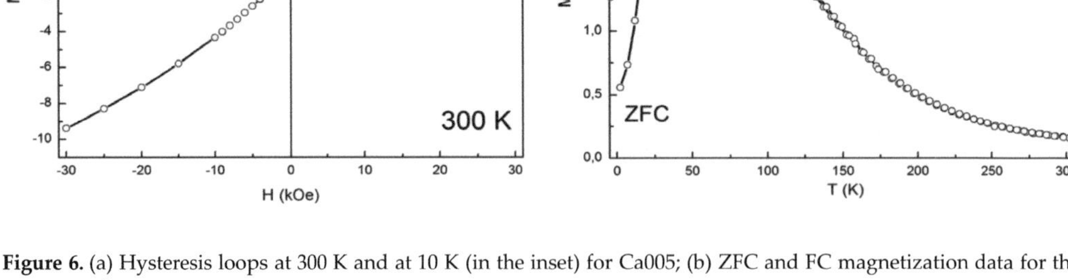

Figure 6. (a) Hysteresis loops at 300 K and at 10 K (in the inset) for Ca005; (b) ZFC and FC magnetization data for the same sample.

	T_B (K)	T_{Irr} (K)	H_C (300 K) (Oe)	M_{Rem} (300 K) (emu/g)	M_{Sat} (300 K–30 kOe) (emu/g)	H_C (10 K) (Oe)	M_{Rem} (10 K) (emu/g)	M_{Sat}(10 K–30 kOe) (emu/g)
ZnFe-SS	24	24	0	0	Linear behavior (5.3)	80	0.6	30.4
ZnFe	42	70	Negl	Negl	8.7 (S shape)	115	3.3	38.2
Al01	50	100	Negl	Negl	6.7 (S shape)	350	3.4	26.7
Ca005	45	100	Negl	Negl	10.0 (S shape)	250	5.2	45.0
Ca025	52	70	Negl	Negl	5.8 (S shape)	100	2.3	21.5
Sr005	53	85	Negl	Negl	8.1 (S shape)	90	1.6	18.2
Gd01	34	60	Negl	Negl	6.7 (S shape)	115	2.8	43.7

Table 4. Different functional parameters derived from magnetization experiments: (i) the blocking temperature, T_B, as defined in the introduction; (ii) the irreversibility temperature, T_{Irr}, corresponding to the merging of ZFC and FC curves and its distance from T_B (T_{Irr} gives suggestions about spin and structural disorder of nanoparticles also come from the temperature); (iii) coercive field H_c, remnant and saturation magnetization (M_{Rem} and M_{Sat}) both at RT and 10 K. The term negligible (Negl) means lower than 5 Oe for H_c, and lower than 10^{-2} for M_{rem}.

responsible of higher M_S values with respect to the regular spinel. Magnetization data give evidence on the effective role of inversion degree, even if M_S values correlate in a complex manner with the spinel inversion.

2.2.4. *Raman results*

In **Figure 7**, the Raman spectra of pure and doped MW ZFO samples are reported together with pure ZFO sample obtained by solid state synthesis.

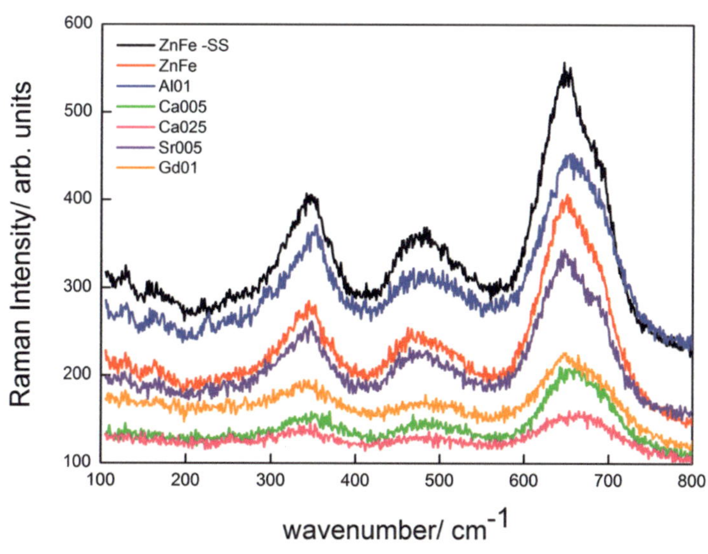

Figure 7. Raman spectra at RT in the region from 100 to 800 cm^{-1} for all the samples investigated.

For all the samples, a well-defined first order Raman scattering pertinent to the ferrite phase has been detected with the most prominent signals around 650 cm^{-1} (A_{1g} mode) accompanied by the other modes at lower energy. The substantial invariance of the Raman features for all the samples indicates a good stability of the spinel structure. The lower total Raman yield and the broadening of the signals of MW samples are consistent with the lower density and smaller crystallite size of the MW powders as derived by XRD analyses. Doping seems to play a negligible role in the peak positions, with the exception of Ca025 sample. For this heavily doped sample, the higher bands move at higher energies, while the $F_{2g}(2)$ signal further decreases in energy. For Ca and Gd samples, a weak feature at around 220 cm^{-1}, probably due to the $F_{2g}(1)$ mode, is present. By comparing the pure samples, we can notice a weak, but observable, red-shift for all the Raman features of the MW sample. This fact is compatible with the reduction of crystallite size (11 nm for ZnFe and 22.9 nm for ZnFe-SS). The peak position of A_{1g} moves from 642.5 to 638.5 cm^{-1}, while the $F_{2g}(2)$ peak is peaked at 345.0 cm^{-1} for ZnFe-SS sample and at 334.0 cm^{-1} for ZnFe one.

The lattice parameter value could play a role in this red shift. According to [44], the changes in Raman line position are related to the changes in the lattice parameter by the relationship $\Delta\omega_i = -3\ \gamma_i(q)\omega_i(q)(\Delta a/a_0)$, where γ is the Gruneisen mode parameter and $\Delta\omega_i$ is the shift of Raman line. The values of γ taken by [27] are 0.72 for A_{1g} mode and 1.88 for the F_{2g} lower energy mode. From the lattice parameters reported in **Table 3**, we derive an expected Raman shift of about 2.0 and 3.5 cm^{-1} for A_{1g} and F_{2g} modes, respectively. The measured shifts are greater and thus the lattice parameter by itself cannot account for the observed Raman softening, for which the role of inversion in the spinel structure must be considered.

As discussed in the previous section, all the Raman data in the region 200–800 cm^{-1} have been treated by best-fitting procedures. Due to the different involved masses, we used couples of oscillators for the three main Raman features with an additional curve for the low energy weaker signals. Even for higher doping levels (Al01, Gd01 and Ca025), we avoid to introduce a third oscillator. A very satisfactory agreement with experimental data has been obtained. As an example, we report the result for sample ZnFe-SS in **Figure 8**.

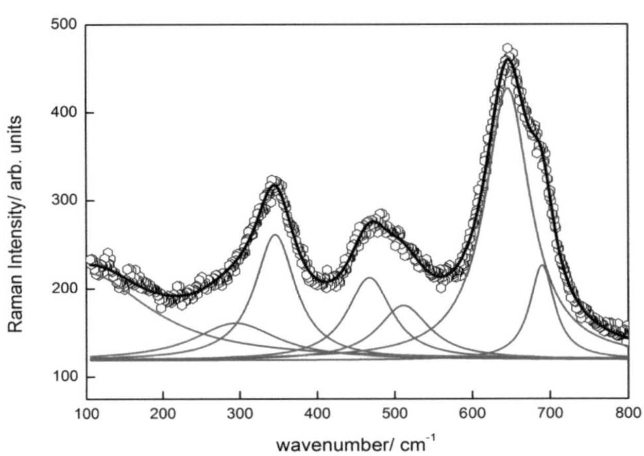

Figure 8. A representative example of best-fitting of the Raman spectrum in the range of 100–800 cm^{-1} for sample ZnFe-SS.

In particular, for the whole set of samples, the A_{1g} band is well fitted by the overlapping of two signals: the first, centered at around 641 cm^{-1}, gives the main contribution and the second, centered at around 685 cm^{-1}, results in a shoulder at higher energies. The energies are in good agreement with those expected for modes inside ZnO_4 and FeO_4 units. From the intensities of these bands, by applying the simple formula $I_{685}/(I_{685} + I_{641})$, we could estimate the inversion degree obtaining, for ZnFe-SS sample, i.e., our standard reference, a value equal to 0.095, slightly higher than 0.07, derived from XRD analysis.

The results for inversion degree are reported in **Figure 9**, where the same parameter from XRD analyses is already shown for comparison.

Even if absolute values from the two methods do not coincide, a common trend is evidenced and the highest inversion is found for Ca025 from both the techniques. Anyway, the discrepancies can be due to the different penetration depths of the two probes leading to the detection of different structural features, especially when clustered nanometric particles are involved. XRD sampling involves the bulk of the sample while Raman measurements probe mainly the surfaces, i.e., the most defective and disordered sample zones and thus it is not surprising that for undoped and low-level doped samples higher inversion degree are derived from Raman. On the contrary, for Al01 and Ca025 samples, a slightly higher inversion degree is derived from XRD analysis. In these cases, we have the higher doping level; in addition the ionic radii of Ca and Al ions are markedly higher and lower, respectively, than that of Zn and Fe ones with respect to other samples, and the differences in ionic radii with respect to the substituted ions. In fact, Ca ions are markedly greater, while Al markedly lower than Zn and Fe ones [61]. In addition, Ca025 powders exhibit the lower crystallite size. All these elements can lead to an increase of inversion degree even in the core of nanoparticles. Finally, we remark that at higher doping level, the contribution of substituents in Raman modes can give rise to a proper vibration at a specific energy. The two Lorenztian model used to interpolate A_{1g} modes remain effective even with minor physical meaning and the error in inversion degree estimation can grow.

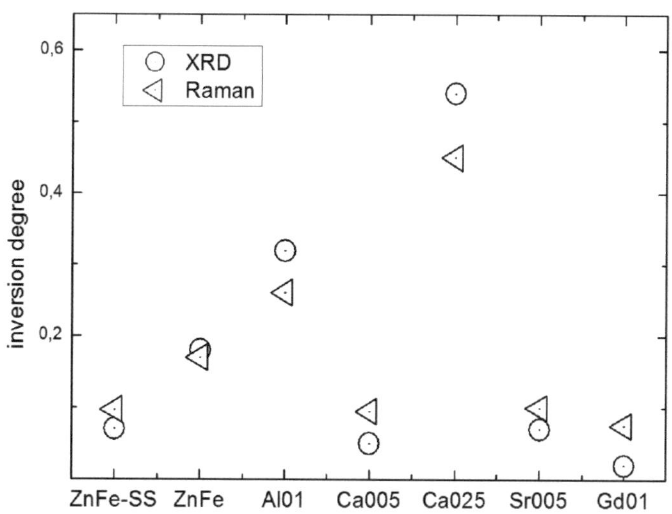

Figure 9. Inversion degree parameters obtained from Raman and XRD data as explained in the text, for all the synthesized ferrite samples. For each sample several Raman runs have been performed in different sample zones to obtain an average inversion degree with a range of variation of about 10%.

Nevertheless, it is important to underline, as evidenced in **Figure 10**, that the best-fitting procedure with two lorentzian curves allows a very good interpolation of the experimental data.

Apart from the estimation of inversion degree, from the fitting procedures the Raman band parameters have been derived for all the samples. **Figure 11** shows the integrated intensities and linewidths and allow to graphically appreciate what already observed above. A marked decrease in total Raman yield for Ca025, especially for the component peaked at 648 cm^{-1}. This component is the most affected by the substitution on A site, thus confirming the attribution for this mode.

As a final remark, one can infer that all the Raman data from MW samples indicate a good crystalline quality, absence of impurities or at least below the threshold of sensitivity, but in any case in agreement with results from XRD and morphological and chemical analyses. Thus, the magnetic functional parameters should be considered as intrinsic of the ZFO matrix, eventually doped.

Anyway, from what discussed in Section 2.1, the Raman experimental set-up allows us to evaluate the stability of the pure and doped MW ZFO powders under laser irradiation and thus to infer about the presence of cluster of maghemite. It is difficult to appreciate the presence of maghemite directly from the Raman spectra, even after detailed spectral analyses. Hence, all the samples were evaluated by means of thermal cycles. Neutral filters with optical density equal to 2, 1, 0.6, 0.3, and 0 were used leading to power density values from 5×10^3 to 5×10^5 W/cm^2. The samples were tested in different regions.

In **Figure 12a** and **b**, we report the results from the sample Al01 and the sample Sr005, respectively. These figures show two different behaviors in term of phase stability.

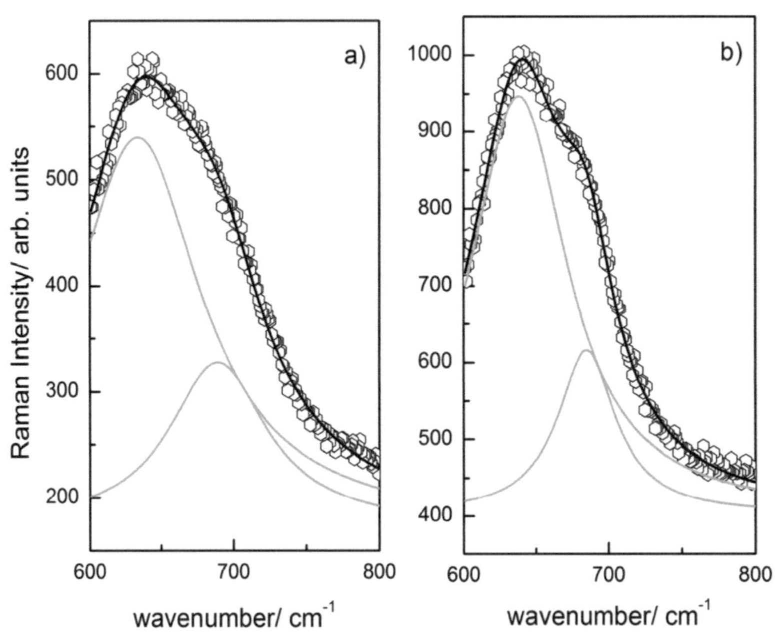

Figure 10. Raman data and curves from best-fitting runs for (a) ZnFe and (b) Sr005.

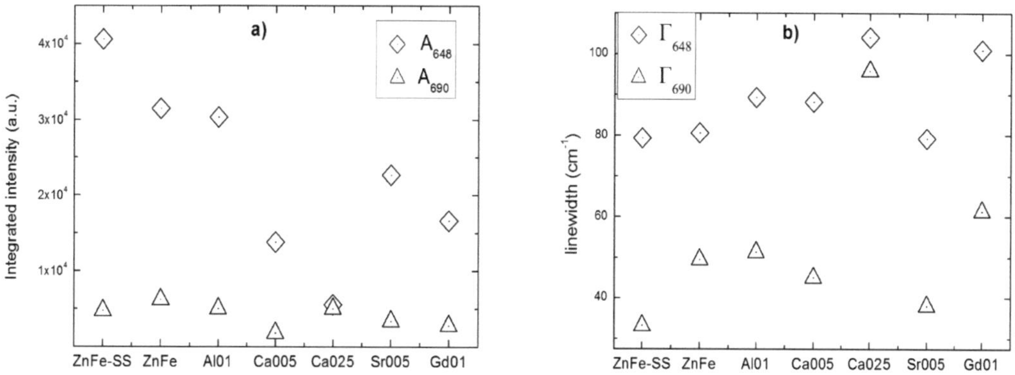

Figure 11. (a) Integrated intensities and (b) linewidths for the two lorenztian curves derived from the best fitting of A_{1g} modes for all the investigated samples.

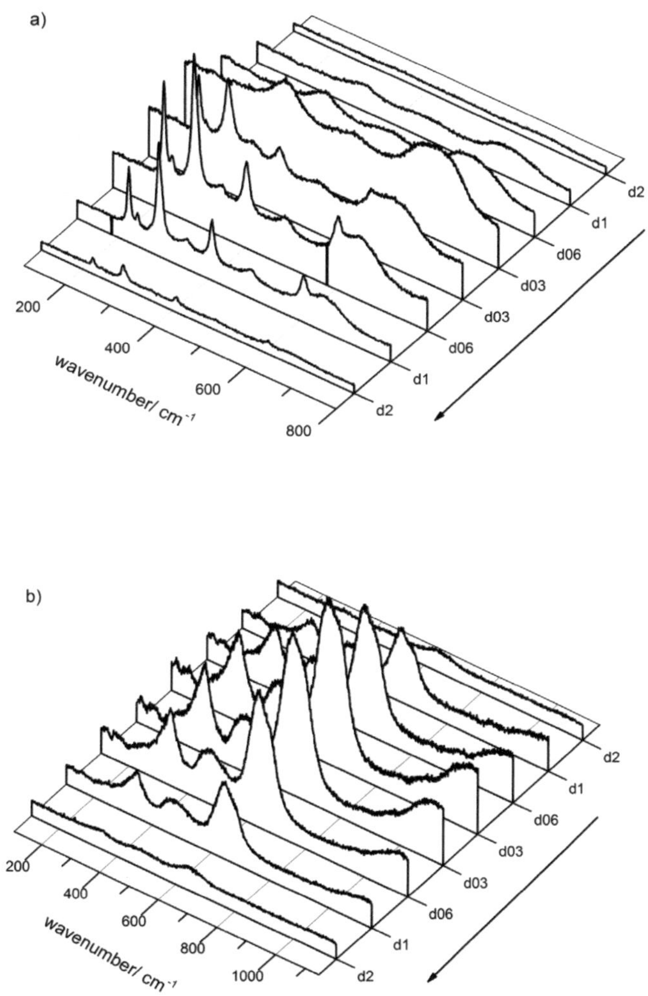

Figure 12. Raman spectra of Al01 (a) and Sr005 (b) samples collected during a thermal cycle. The samples were irradiated for 5 min and then measured at the selected power density.

The sample Sr005, even if thermally treated under laser light, remains stable showing the same Raman spectrum during the whole laser-induced thermal cycle. Small peak shifts are due to temperature effects on lattice parameters. On the contrary for Al01 sample, it is evident that upon irradiation, the Raman spectrum of the Al01 change markedly and the typical Raman features of hematite appear when the sample is maintained under laser light having a power density equal to 2×10^5 W/cm^2. When the laser light is attenuated at the end of the thermal cycle, the signal of hematite is still observable indicating a stable conversion. This is an indirect proof that at the beginning of the thermal cycle nanoregions in the irradiated volume had a structure different from the proper ZFO structure. It is likely that nanoregions of maghemite or highly defective Zn deficient zinc ferrite layers were present. We underline that the grain sizes for Al01 and Sr005 are similar.

Only three samples exhibit a stable pure zinc ferrite phase with no hematite detected at the end of the thermal cycle: ZnFe-SS, Ca025, and Sr005. The first exhibited a paramagnetic behavior, while the latter were found to be superparamagnetic. The phase stability proved by monitoring the Raman spectrum under laser heating is fundamental to validate the quality of the powders with good magnetic functional parameters, in particular in view of applications in biomedical field.

3. Conclusions and outlook

This chapter gave a look on the potential of Raman spectroscopy when applied to nanosized zinc ferrites. The nanostructured spinel $ZnFe_2O_4$ display a peculiar mixing of high-quality functional properties, and it is used in magnetic storage, ferrofluids, catalysis, and biomedical applications, as for instance theranostics and hyperthermia. For the latter applications, a strong requirement is the superparamagnetic behavior at RT, which take place in zinc ferrites thanks to the nanometric scaling. This behavior has been revealed even in pure ZFO nanoparticles due to the inversion between divalent and trivalent cations on the tetrahedral and octahedral sites of the spinel-type structure. In this frame, thanks to its intrinsic sensitivity to chemical and structural composition, Raman spectroscopy has been widely used to monitor cations distribution in pure and doped zinc ferrites. In the first part of the chapter, the state of the art concerning the results from Raman scattering in pure and doped zinc ferrites has been presented in connection to the different synthesis methods employed to obtain high quality nanostructured ferrites. It appears clear how detailed analyses on Raman data allow to evaluate the influence of nanometric scaling on Raman lines features—peak energies, widths and shape, to obtain information on the inversion degree, to reveal the presence of extrinsic iron oxides phases like hematite or magnetite and to monitor stability of spinel ferrites vs. light exposure or thermal treatment. This information is essential to assess the functional quality of the materials and it is essential to study and understand the close correlation between functional parameters and morphology, gran size, intentional doping, cation distribution, and nanoparticle architecture. Indeed magnetic, structural and chemical purity and homogeneity of nanoparticles must be thoroughly verified because even low amount of extrinsic phases can alter the functional properties. Indeed, low amount of hematite or maghemite can alter the long-term stability of nanoparticles especially in biological environment, favoring aggregation.

In the second part of the chapter, we presented original results from room temperature micro-Raman studies on nanostructured undoped $ZnFe_2O_4$, $Zn_{1-x}Ca_xFe_2O_4$ (x = 0.05 and 0.25), $ZnFe_{1.9}Gd_{0.1}O_4$, $Zn_{0.95}Sr_{0.05}Fe_2O_4$, and $ZnFe_{1.9}Al_{0.1}O_4$ doped ferrites. The samples have been prepared by using the microwave-assisted combustion method, a rapid, green, and simple synthesis route, able to ensure good physical and chemical properties. We underline the relevance of Ca substitution due to the low toxicity of the substituent; on the other side, Gd doping is rarely reported in zinc ferrites. Raman experiments have been accompanied by detailed analyses on morphological, compositional, structural, and magnetic characterization. Thus, results from XRD, SEM-EDS are briefly presented as well as the magnetic behavior. A detailed analysis of A_{1g}, E_g, $3F_{2g}$ Raman modes has been performed and Raman band parameters have been derived from best-fitting procedures and carefully compared to literature data. The results have been discussed in relation with grain size, inversion degree and doping of zinc ferrite spinel. An optimum purity level and homogeneity and crystallite sizes lower than 11 nm were determined for the doped samples. The inversion degree of the different samples was derived by both Raman and XRD data and a well agreement has been observed. The superparamagnetism seems to be favored in Ca doped samples, while in Gd doped one it is almost negligible. This is clearly associated to the inversion induced by the doping in the tetrahedral site. In addition by Raman studies we evaluated the stability of pure and doped $ZnFe_2O_4$ with laser heating just to reveal the presence of maghemite in the samples. This set of data gives evidence to the ability of Raman spectroscopy in a typical problem of solid-state material science. Finally, the present results will be the bases of further works aimed to exploit SERS effect using non-metallic nanostructured zinc ferrites eventually functionalized for specific biomedical applications.

Author details

Pietro Galinetto[1]*, Benedetta Albini[1], Marcella Bini[2] and Maria Cristina Mozzati[1]

*Address all correspondence to: pietro.galinetto@unipv.it

1 Department of Physics and CNISM, University of Pavia, Pavia, Italy

2 Department of Chemistry, University of Pavia, Pavia, Italy

References

[1] Kmita A, Pribulova A, Holtzer M, Futas P, Roczniak A. Use of specific properties of zinc ferrite in innovative technology. Archives of Metallurgy and Materials. 2016;**61**(4):2141-2146. DOI: 10.1515/amm-2016-0289

[2] Suchomski C, Breitung B, Witte R, Knapp M, Bauer S, Baumbach T, Reitz C, Brezesinski T. Microwave synthesis of high-quality and uniform 4 nm $ZnFe_2O_4$ nanocrystals for application in energy storage and nanomagnetics. Beilstein Journal of Nanotechnology. 2016; **7**:1350-1360. DOI: 10.3762/bjnano.7.126

[3] Goutham S, Kumar DS, Sadasivuni KK, Cabibihan JJ, Rao KV. Nanostructure $ZnFe_2O_4$ with *Bacillus subtilis* for detection of LPG at low temperature. Journal of Electronic Materials. 2017;**46**(4). DOI: 10.1007/s11664-016-5273-z

[4] Mady AH, Baynosa ML, Tuma D, Shim JJ. Facile microwave-assisted green synthesis of Ag-$ZnFe_2O_4$@rGO nanocomposites for efficient removal of organic dyes under UV- and visible-light irradiation. Applied Catalysis B: Environmental. 2017;**203**:416-427

[5] Díaz-Pardo R, Valenzuela R. Characterization of magnetic phases in nanostructured ferrites by electron spin resonance. In: Bashir R, editor. Advanced Electromagnetic Waves. Rijecka: InTech; 2015. DOI: 10.5772/61508. Available from: https://www.intechopen.com/books/advanced-electromagnetic-waves/characterization-of-magnetic-phases-in-nano-structured-ferrites-by-electron-spin-resonance

[6] Sheikh A, Jain P. A thorough study of zinc ferrite nanoparticles with reference to green synthesis. International Journal of Nanomedicine. 2016;**2**(3). DOI: 10.16966/2470-3206.115

[7] Issa B, Obaidat IM, Albiss BA, Haik Y. Magnetic nanoparticles: Surface effects and properties related to biomedicine applications. International Journal of Molecular Sciences. 2013;**14**:21266-21305. DOI: 10.3390/ijms141121266

[8] Kolhatkar AG, Jamison AC, Litvinov D, Willson RC, Randall Lee T. Tuning the magnetic properties of nanoparticles. International Journal of Molecular Sciences. 2013;**14**:15977-16009. DOI: 10.3390/ijms140815977

[9] Štefanić G, Krehula S, Štefanić I. Phase development during high-energy ball-milling of zinc oxide and iron – The impact of grain size on the source and the degree of contamination. Dalton Transactions. 2015;**44**:18870. DOI: 10.1039/c5dt02498f

[10] Manikandan A, Vijaya JJ, Kennedy LJ, Bououdina M. Microwave combustion synthesis, structural, optical and magnetic properties of $Zn_{1-x}Sr_xFe_2O_4$ nanoparticles. Ceramics International. 2013;**39**:5909-5917. DOI: 10.1016/j.ceramint.2013.01.012

[11] Lima-Tenório MK, Oliveira LAS, Guilherme MR, Tenório-Neto ET, Silva MF, Fernandes DM, Hechenleitner AAW, Pineda EAG. Tuning the magnetic properties of ferrite nanoparticles by Zn and Co doping. Materials Letters. 2017;**195**:151-155. DOI: 10.1016/j.matlet.2017.02.122

[12] Diodati S, Pandolfo L, Caneschi A, Gialanella S, Gross S. Green and low temperature synthesis of nanocrystalline transition metal ferrites by simple wet chemistry routes. Nano Research. 2014;**7**(7):1027-1042. DOI: 10.1007/s12274-014-0466-3

[13] Patil SB, Bhojya Naik HS, Nagaraju G, Viswanath R, Rashmi SK. Synthesis of visible light active Gd^{3+}-substituted $ZnFe_2O_4$ nanoparticles for photocatalytic and antibacterial activities. The European Physical Journal Plus. 2017;**132**:328. DOI: 10.1140/epjp/i2017-11602-x

[14] Phumying S, Labuayai S, Swatsitang E, Amornkitbamrung V, Maensiri S. Nanocrystalline spinel ferrite (MFe_2O_4, M = Ni, Co, Mn, Mg, Zn) powders prepared by a simple aloe vera plant-extracted solution hydrothermal route. Materials Research Bulletin. 2013;**48**:2060-2065. DOI: 10.1016/j.materresbull.2013.02.042

[15] Kombaiah K, Vijaya JJ, Kennedy LJ, Bououdina M. Studies on the microwave assisted and conventional combustion synthesis of Hibiscus rosasinensis plant extract based $ZnFe_2O_4$ nanoparticles and their optical and magnetic properties. Ceramics International. 2016;**42**:2741-2749. DOI: 10.1016/j.ceramint.2015.11.003

[16] Sivakumar M, Towata A, Yasui K, Tuziuti T, Iida Y. A new ultrasonic cavitation approach for the synthesis of zinc ferrite nanocrystals. Current Applied Physics. 2006;**6**:591-593. DOI: 10.1016/j.cap.2005.11.068

[17] Mohai I, Gál L, Szépvölgyi J, Gubicza J, Farkas Z. Synthesis of nanosized zinc ferrites from liquid precursors in RF thermal plasma reactor. Journal of the European Ceramic Society. 2007;**27**:941-945. DOI: 10.1016/j.jeurceramsoc.2006.04.128

[18] Ali A, Zafar H, Zia M, ul Haq I, Rehman Phull A, Sarfraz Ali J, Hussain A. Synthesis, characterization, applications, and challenges of iron oxide nanoparticles. Nanotechnology, Science and Applications. 2016;**9**:49-67. DOI: 10.2147/NSA.S99986

[19] D'Ippolito V, Andreozzi GB, Bersani D, Lottici PP. Raman fingerprint of chromate, aluminate and ferrite spinels. Journal of Raman Specroscopy. 2015;**46**:1255-1264. DOI: 10.1002/jrs.4764

[20] Chandramohan P, Srinivasan MP, Velmurugan S, Narasimhan SV. Cation distribution and particle size effect on Raman spectrum of $CoFe_2O_4$. Journal of Solid State Chemistry. 2011;**184**:89-96. DOI: 10.1016/j.jssc.2010.10.01

[21] de Faria DLA, Venaüncio Silva S, de Oliveira MT. Raman microspectroscopy of some iron oxides and oxyhydroxides. Journal of Raman Spectroscopy. 1997;**28**:873-878

[22] Singh JP, Srivastava RC, Agrawala HM, Kumarb R. Micro-Raman investigation of nanosized zinc ferrite: Effect of crystallite size and fluence of irradiation. Journal of Raman Specroscopy. 2011. DOI: 10.1002/jrs.2902

[23] Varshney D, Verma K, Kumar A. Structural and vibrational properties of $Zn_xMn_{1-x}Fe_2O_4$ (x = 0.0, 0.25, 0.50, 0.75, 1.0) mixed ferrites. Materials Chemistry and Physics. 2011;**131**:413-419. DOI: 10.1016/j.matchemphys.2011.09.066

[24] Song D, Yang R, Wang C, Xiao R, Long F. Reusable nanosilver-coated magnetic particles for ultrasensitive SERS-based detection of malachite green in water samples. Scientific Reports. 2016;**6**:22870. DOI: 10.1038/srep22870

[25] White WB, DeAngelis BA. Interpretation of the vibrational spectra of spinels. Spectrochimica Acta. 1967;**23A**:985-995. DOI: 10.1016/0584-8539(67)80023-0

[26] Hosterman BD. Raman spectroscopic study of solid solution spinel oxides [UNLV Theses, Dissertations, Professional Papers, and Capstones]. Graduate College University of Nevada: Las Vegas; 2011. p. 1087

[27] Wang Z, Schiferl D, Zhao Y, O'Neill HSC. High pressure Raman spectroscopy of spinel-type ferrite $ZnFe_2O_4$. Journal of Physics and Chemistry of Solids. 2003;**64**:2517-2523. DOI: 10.1016/j.jpcs.2003.08.005

[28] Yamanaka T, Ishii M. Raman scattering and lattice vibrations of Ni_2SiO_4 spinel at elevated temperature. Physics and Chemistry of Minerals. 1986;**13**:156-160. DOI: 10.1007/BF00308157

[29] Cynn H, Sharma SK, Cooney TF, Nicol M. High-temperature Raman investigation of order-disorder behavior in the $MgAl_2O_4$ spinel. Physical Review B. 1992;**45**(1):500-502. DOI: 10.1103/PhysRevB.45.500

[30] Larry Verble J. Temperature-dependent light-scattering studies of the Verwey transition and electronic disorder in magnetite. Physical Review B. 1974;**9**(12):5236-5248. DOI: 10.1103/PhysRevB.9.5236

[31] Julien CM, Massot M. Lattice vibrations of materials for lithium rechargeable batteries I. Lithium manganese oxide spinel. Materials Science and Engineering B. 2003;**97**:217-230. DOI: 10.1016/S0921-5107(02)00582-2

[32] Bruesch P, D'Ambrogio F. Lattice dynamics and magnetic ordering in the chalcogenide spinels $CdCr_2S_4$ and $CdCr_2Se_4$. Physica Status Solidi B. 1972;**50**:513-526. DOI: 10.1002/pssb.2220500212

[33] O'Horo MP, Frisillo AL, White WB. Lattice vibrations of $MgAl_2O_4$ spinel. Journal of Physics and Chemistry of Solids. 1973;**34**:23-28. DOI: 10.1016/0022-3697(73)90058-9

[34] Preudhomme J, Tarte P. Infrared studies of spinels III: The normal II-III spinels. Spectrochimica Acta. 1971;**27A**:1817-1835. DOI: 10.1016/0584-8539(71)80235-0

[35] Lazarević ZŽ, Jovalekić Č, Ivanovski VN, Rečnik A, Milutinović A, Cekić B, Romčević NŽ. Characterization of partially inverse spinel $ZnFe_2O_4$ with high saturation magnetization synthesized via soft mechanochemically assisted route. Journal of Physics and Chemistry of Solids. 2014;**75**:869-877. DOI: 10.1016/j.jpcs.2014.03.004

[36] Milutinović A, Lazarević Z, Jovalekić Č, Kuryliszyn-Kudelska I, Romčević M, Kostić S, Romčević N. The cation inversion and magnetization in nanopowder zinc ferrite obtained by soft mechanochemical processing. Materials Research Bulletin. 2013;**48**:4759-4768. DOI: 10.1016/j.materresbull.2013.08.020

[37] Yadav RS, Kuřitka I, Vilcakova J, Urbánek P, Machovsky M, Masař M, Holek M. Structural, magnetic, optical, dielectric, electrical and modulus spectroscopic characteristics of $ZnFe_2O_4$ spinel ferrite nanoparticles synthesized via honey-mediated sol-gel combustion method. Journal of Physics and Chemistry of Solids. 2017;**110**:87-99. DOI: 10.1016/j.jpcs.2017.05.029

[38] ZŽ Lazarević DL, Sekulić VN, Ivanovski NŽ, Romčević A. Structural and magnetic investigation of the inversion degree in spinel $NiFe_2O_4$, $ZnFe_2O_4$ and $Ni_{0.5}Zn_{0.5}Fe_2O_4$ ferrites prepared by soft mechanochemical synthesis. International Journal of Chemical, Molecular, Nuclear, Materials and Metallurgical Engineering. 2015;**9**(8)

[39] Virumbrales-del Olmo M, Delgado-Cabello A, Andrada-Chacón A, Sánchez-Benítez J, Urones-Garrote E, Blanco-Gutiérrez V, Torralvo MJ, Sáez-Puchea R. Effect of composition and coating on the interparticle interactions and magnetic hardness of MFe_2O_4

(M = Fe, co, Zn) nanoparticles. Physical Chemistry Chemical Physics. 2017;**19**:8363. DOI: 10.1039/c6cp08743d

[40] Thota S, Kashyap SC, Sharma SK, Reddy VR. MicroRaman, Mossbauer and magnetic studies of manganese substituted zinc ferrite nanoparticles: Role of Mn. Journal of Physics and Chemistry of Solids. 2016;**91**:136-144. DOI: 10.1016/j.jpcs.2015.12.013

[41] Milanović M, Stijepović I, Pavlović V, Srdić VV. Functionalization of zinc ferrite nanoparticles: Influence of modification procedure on colloidal stability. Processing and Application of Ceramics. 2016;**10**(4):287-293. DOI: 10.2298/PAC1604287M

[42] Masoudpanah SM, Hasheminisari M, Ghasemi A. Magnetic properties and photocatalytic activity of $ZnFe_{2-x}La_xO_4$ nanoparticles synthesized by sol–gel autocombustion method. Journal of Sol-Gel Science and Technology. 2016;**80**:487-494. DOI: 10.1007/s10971-016-4101-5

[43] Murugesan C, Chandrasekaran G. Structural and magnetic properties of $Mn_{1-x}Zn_xFe_2O_4$ ferrite nanoparticles. Journal of Superconductivity and Novel Magnetism. 2016;**29**:2887-2897. DOI: 10.1007/s10948-016-3604-1

[44] Spanier JE, Robinson RD, Zhang F, Chan SW, Herman IP. Size-dependent properties of CeO_2-y nanoparticles as studied by Raman scattering. Physical Review B. 2001;**64**:245407. DOI: 10.1103/PhysRevB.64.245407

[45] Arora AK, Rajalakshmi M, Ravindran TR, Sivasubramanian V. Raman spectroscopy of optical phonon confinement in nanostructured materials. Journal of Raman Spectroscopy. 2007;**38**(6). DOI: 10.1002/jrs.1684

[46] Freire RM, Ribeiro TS, Vasconcelos IF, Denardin JC, Barros EB, Mele G, Carbone L, Mazzetto SE, Fechine PBA. $MZnFe_2O_4$ (M = Ni, Mn) cubic superparamagnetic nanoparticles obtained by hydrothermal synthesis. Journal of Nanoparticle Research. 2013;**15**:1616. DOI: 10.1007/s11051-013-1616-3

[47] Lazarević ZŽ, Milutinović AN, Jovalekić ČD, Ivanovski VN, Daneu d N, Maparević I, Romĉević NŽ. Spectroscopy investigation of nanostructured nickel–zinc ferrite obtained by mechanochemical synthesis. Materials Research Bulletin. 2015;**63**:239-247. DOI: 10.1016/j.materresbull.2014.12.005

[48] Thota S, Kashyapa SC, Sharmab SK, Reddyca VR. Cation distribution in Ni-substituted $Mn_{0.5}Zn_{0.5}Fe_2O_4$ nanoparticles: A Raman, Mössbauer, X-ray diffraction and electron spectroscopy study. Materials Science and Engineering B. 2016;**206**:69-78. DOI: 10.1016/j.mseb.2016.01.002

[49] Varshney D, Verma K, Kumar A. Substitutional effect on structural and magnetic properties of $A_xCo_{1-x}Fe_2O_4$ (A = Zn, Mg and x = 0.0, 0.5) ferrites. Journal of Molecular Structure. 2011;**1006**:447-452. DOI: 10.1016/j.molstruc.2011.09.047

[50] Jovanovic S, Spreitzer M, Tramsek M, Trontelj Z, Suvorov D. Effect of oleic acid concentration on the physicochemical properties of cobalt ferrite nanoparticles. Journal of Physical Chemistry C. 2014;**18**:13844-13856. DOI: 10.1021/jp500578f

[51] da Silva SW, Nakagomi F, Silva MS, Franco A Jr, Garg VK, Oliveira AC, Morais PC. Raman study of cations' distribution in $Zn_xMg_{1-x}Fe_2O_4$ nanoparticles. Journal of Nanoparticle Research. 2012;**14**:798. DOI: 10.1007/s11051-012-0798-4

[52] Chaudhari PR, Gaikwad VM, Acharya SA. Role of mode of heating on the synthesis of nanocrystalline zinc ferrite. Applied Nanoscience. 2015;**5**:711-717. DOI: 10.1007/s13204-014-0367-5

[53] Liua J, Denga M, Huanga Z, Yina G, Liaoa X, Gub J. Preparation of $ZnFe_2O_4$ nanoparticles in the template of silk-fibroin peptide and their neuro-cytocompability in PC12 cells. Colloids and Surfaces B: Biointerfaces. 2013;**107**:19-26. DOI: 10.1016/j.colsurfb.2013.01.072

[54] Malavasi L, Galinetto P, Mozzati MC, Azzoni CB, Flor G. Raman spectroscopy of AMn_2O_4 (A = Mn, Mg and Zn) spinels. Physical Chemistry Chemical Physics. 2002;**4**:3876-3880. DOI: 10.1039/B203520K

[55] Thomas JJ, Shinde AB, Krishna PSR, Kalarikkal N. Temperature dependent neutron diffraction and Mossbauer studies in zinc ferrite nanoparticles. Materials Research Bulletin. 2013;**48**:1506-1511. DOI: 10.1016/j.materresbull.2012.12.058

[56] Gomes JA, Azevedo GM, Depeyrot J, Mestnik-Filho J, Paula FLO, Tourinho FA, Perzynski R. Structural, chemical, and magnetic investigations of core-shell zinc ferrite nanoparticles. Journal of Physical Chemistry C. 2012;**116**:24281-24291. DOI: 10.1021/jp3055069

[57] Chernyshova IV, Hochella MF, Madden AS. Size-dependent structural transformations of hematite nanoparticles.1.Phase transition. Physical Chemistry Chemical Physics. 2007;**9**:1736-1750. DOI: 10.1039/B618790K

[58] Galinetto P, Mozzati MC, Grandi MS, Bini M, Capsoni D, Ferrari S, Massarotti V. Phase stability and homogeneity in undoped and Mn-doped $LiFePO_4$ under laser heating. Journal of Raman Spectroscopy. 2010;**41**:1276-1282. DOI: 10.1002/jrs.2558

[59] Bini M, Tondo C, Capsoni D, Mozzati MC, Albini B, Galinetto P. Superparamagnetic $ZnFe_2O_4$ nanoparticles: The effect of Ca and Gd doping. Materials Chemistry and Physics. 2018;**204**:72-82. DOI: 10.1016/j.matchemphys.2017.10.033

[60] Tholkappiyan R, Vishista K. Influence of lanthanum on the optomagnetic properties of zinc ferrite prepared by combustion method. Physica B. 2014;**448**:177. DOI: 10.1016/j.physb.2014.04.022

[61] Shannon RD. Revised effective ionic radii and systematic studies of interatomic distances in halides and chalcogenides. Acta Cryst. 1976;**A32**:751-767. DOI: 10.1107/S0567739476001551

[62] Yadav RS, Havlica J, Kuritka I, Kozakova Z, Palou M, Bartonıckova E, Bohac M, Frajkorova F, Masilko J, Hajduchova M, Enev V, Wasserbauer J. Magnetic properties of $ZnFe_2O_4$ nanoparticles synthesized by starch-assisted sol-gel auto-combustion method. Journal of Superconductivity and Novel Magnetism. 2015;**28**:1417. DOI: 10.1007/s10948-014-2870-z

[63] Milanovic M, Moshopoulou EG, Stamopoulos D, Devlin E, Giannakopoulos KP, Kontos AG, Eleftheriadis K, Gini MI, Nikolic LM. Structure and magnetic properties of $Zn_{1-x}In_xFe_2O_4$ and $ZnY_xFe_{2-x}O_4$ nanoparticles prepared by coprecipitation. Ceramics International. 2013;**39**:3235. DOI: 10.1016/j.ceramint.2012.10.011

[64] Roy MK, Halder B, Verma HC. Characteristic length scales of nanosize zinc ferrite. Nanotechnology. 2006;**17**:232. DOI: 10.1088/0957-4484/17/1/039

Chapter 12

Structural Transformations in Ferroelectrics Discovered by Raman Spectroscopy

Kai Jiang, Liping Xu, Jinzhong Zhang,
Zhigao Hu and Junhao Chu

Additional information is available at the end of the chapter

http://dx.doi.org/10.5772/intechopen.72770

Abstract

Ferroelectrics systems are of great interest from the fundamental as well as applications points, such as ferroelectric random access memories, electro-optic switches and a number of electro-optic devices. Curie temperature (T_C) is one of the important parameters of ferroelectrics for high-temperature applications. Particularly, the optical modes, which are associated with the ferroelectric to paraelectric phase transition, are of great interest. Structural transformations that alter the crystal symmetry often have a significant effect on the Raman spectroscopy. This chapter systematically studies the type ferroelectric oxides and rare earth element doped ferroelectric materials such as $PbTiO_3$-$Bi(Mg_{0.5}Ti_{0.5})O_3$ (PT-BMT), $Sr_xBa_{1-x}Nb_2O_6$ (SBN), $Pb_{1-1.5x}La_xZr_{0.42}Sn_{0.4}Ti_{0.18}O_3$ (PLZST), $Bi_{1-x}La_xFe_{1-y}Ti_yO_3$ (BLFT) and $(K_{0.5}Na_{0.5})NbO_3$-$0.05LiNbO_3$ (KNN-LN) and so on synthesis of single crystal/ceramic and optical phonon vibration modes and the improvement of the Curie temperature characteristic using spectrometry measurements. The T_C, distortion degree, and phase structure of the ferroelectric materials have been investigated by temperature-dependent Raman spectroscopy. Meanwhile, the important physical parameters exhibited a strong dependence on dopants resulting in structural modifications and performance promotion.

Keywords: ferroelectrics, Raman spectra, vibration modes, phase transitions, Curie temperature

1. Introduction

As we know, ferroelectric ceramics were first found in barium titanate ceramic with the ferroelectricity in the 1940s [1]. Since that time, the ferroelectric material with high resistivity, good fatigue resistance characteristic and high dielectric constant, pyroelectric detector, uncooled infrared detectors, uncooled infrared focal plane arrays and ferroelectric memory, and other fields has great application prospect [2–5]. In recent years, complex mixed-ion ferroelectric materials have been extensively investigated in order to achieve optimum properties as well as to understand the underlying factors for property tweaking [6–9]. Therefore, the ferroelectric materials are considered to be one of the most practical materials in the future.

The ABO_3 ferroelectric materials have achieved wide usage owing to their superior electromechanical properties (**Scheme 1** shows the typical structure). Investigations of bulk ferroelectric materials have demonstrated good macroscopic homogeneity of their properties and clear ferroelectric behavior [2]. However, the development of knowledge about ferroelectric behavior at the submicrometer level is relatively slow. It has been found that the structural and chemical factors such as grain size, strain, stoichiometric and compositional homogeneity and phase structure, have great effect on optimization and reproducibility of the property coefficients in ferroelectric materials [10–12]. Therefore, a further investigation should be necessary in order to illustrate the physical mechanism in these ferroelectrics.

It is important to remember that the experimentally obtained parameters depend primarily on the spatial magnitude and time-scale of the measured physical phenomena, especially for studying the structure–property correlations in these materials. Raman spectroscopy is a sensitive technique for investigating the structure modifications and lattice vibration modes, which can give the information on the changes of lattice vibrations and the occupying positions of doping ions. Structural changes that alter the crystal symmetry often have a significant effect on the Raman spectrum. In addition, spatially resolved Raman spectroscopy can be used to probe the chemical homogeneity at sub-micrometer levels. This chapter provides a review of systematic Raman scattering study on the phase transition behavior in perovskites, tungsten bronze, Aurivillius layered, multiferroics and lead-free bulk materials. The effect of A- and B-site substitutions on the Raman spectra and phase transition behavior of these materials have been studied in detail. This chapter is arranged in the following way. In Section 1, research background; In Section 2, detailed growths of the ferroelectric materials and Raman experiment; In Section 3, results of Raman spectra in $PbTiO_3$-$Bi(Mg_{0.5}Ti_{0.5})O_3$ (PT-BMT), $Sr_xBa_{1-x}Nb_2O_6$ (SBN), $Pb_{1-1.5x}La_xZr_{0.42}Sn_{0.4}Ti_{0.18}O_3$ (PLZST), $Bi_{1-x}La_xFe_{1-y}Ti_yO_3$ (BLFT) and $(K_{0.5}Na_{0.5})NbO_3$-0.05LiN bO_3 (KNN-LN); at last, the main results and remarks are summarized.

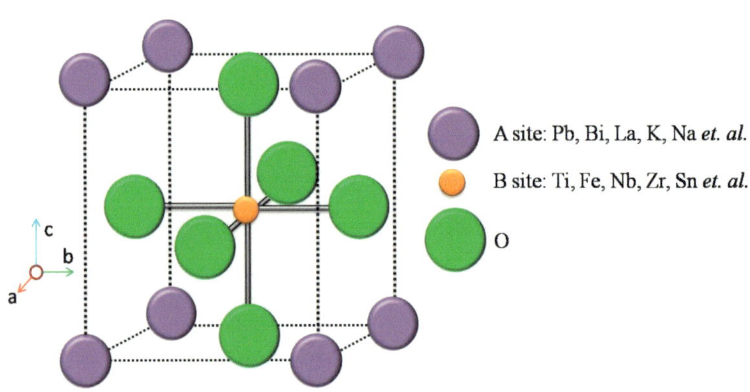

Scheme 1. Schematic representation of the typical ABO_3 ferroelectric structure.

2. Experiment details

2.1. Fabrication of ferroelectric materials

The ferroelectric single crystals have been grown by a high temperature solution method (flux method) [13, 14]. High-purity powders were selected as starting materials. The raw material powders were stoichiometrically weighed, mixed by milling with zirconia media in the ethanol as a solvent. After drying. The powders were calcined at a certain temperature for hours to form the desired perovskite phase. Details of the fabrication process for the single crystals can be found elsewhere [13].

The bulk ceramics were fabricated by a conventional solid state reaction sintering, using the appropriate amount of reagent grade raw materials [15, 16]. The samples were sintered at different temperature for several hours in air atmosphere, and then remilled for several hours to reduce the particle size for sintering. The calcined powders were added with 8 wt.% polyvinyl alcohol (PVA) as a binder. Before Raman measurements, the ceramics with the diameter of 15 mm and the thickness of 1 mm were rigorously single-side polished and cleaned in pure ethanol with an ultrasonic bath and rinsed several times by deionized water.

2.2. Raman experiment details

Raman scattering experiments were carried out using a Jobin-Yvon LabRAM HR 800 UV micro-Raman spectrometer, excited by 632.8 nm He-Ne laser or 488 nm Ar laser and recorded in the frequency range of 10–1000 cm^{-1} with a spectral resolution of 0.5 cm^{-1}. For the different temperature Raman spectra, we choose a 50× microscope with a long working distance of 18 mm. The spectrometer grating can be choosed by 600, 1800 or 2400 grooves/mm grating which is depending on the different excitation wavelength. In order to learn more about the variation trend of vibration modes, all of the experimental spectra were fitted with independent damped harmonic oscillators. The polarized Raman spectra were recorded in backscattering geometry in parallel $<x|zx|y>$ (VH) and perpendicular $<x|zz|y>$ (VV) polarization configurations. Temperature dependent Raman spectra were collected with a THMSE 600 heating/cooling stage (Linkam Scientific Instruments) in the temperature range from 77 to 800 K with a resolution of 0.1 K.

3. Results and discussion

3.1. $PbTiO_3$-based single crystals

$PbTiO_3$ (PT)-based perovskite compounds are important multifunctional materials, which have been investigated in the last half century due to their controllable physical properties. Most recently, the research hotspot for $PbTiO_3$-$BiMeO_3$ ferroelectrics have stimulated much interest [14]. A range of compelling information on thermal expansion behavior and lattice dynamics of novel ferroelectric perovskite-type $0.62PbTiO_3$-$0.38Bi(Mg_{0.5}Ti_{0.5})O_3$ (PT-BMT) single crystal has been revealed by means of temperature-dependent X-ray diffraction and polarized Raman scattering. **Figure 1** shows the polarized Raman spectra of 0.62PT-0.38BMT single

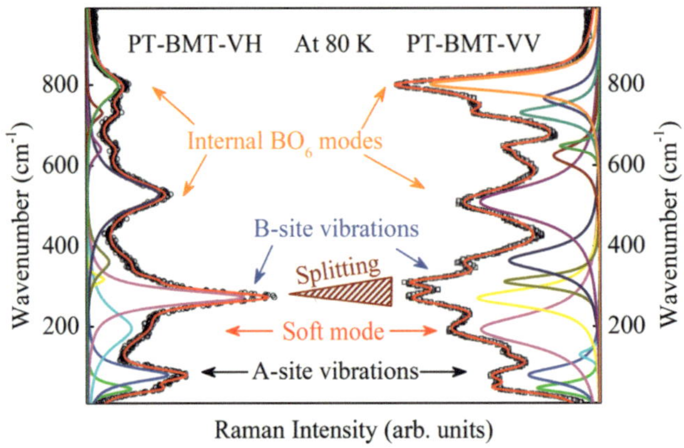

Figure 1. The polarized VH and VV Raman spectra of 0.62PT-0.38BMT single crystal at 80 K. The peak assignment is based on the individual vibration mode, which is fitted by Lorentz function (Figure reproduced with permission from [14]. Copyright 2016, Royal Society of Chemistry).

crystal at representative temperature of 80 K. It can be found that the Raman spectra is consistent with previous reports of the tetragonal crystal structure. Comparing to the theoretical calculations and experimental results, three main regions with different kinds of vibration in the lattice can be distinguished: (1) Low frequency below 150 cm^{-1}, which can be associated with vibrations of the perovskite A-site, involving Pb and Bi ions. (2) Frequency range 150–450 cm^{-1}, which is related to the Ti-O/Mg-O stretching vibrations. (3) High frequency bands above 450 cm^{-1} have all been associated with BO_6 vibrations, namely the breathing and stretching modes of the oxygen octahedra. The Raman spectra suggest a certain degree of local B-site chemical order for the 0.62PT-0.38BMT single crystal with A-site and B-site simultaneously doping.

In order to understand the relationship between negative thermal expansion (NTE) and lattice dynamic behavior for the 0.62PT-0.38BMT crystal, **Figure 2** presents the polarized Raman spectra in different temperature point from 80 to 850 K. It can be found that the peak position and intensity of the polarized Raman phonon modes have a decreasing trend with increasing temperature, while the broadening of the Raman modes is observed, which is consist with the other ferroelectrics [17]. It should be pointed that around the temperature range of 600–700 K, the lowest phonon modes occurs (marked by the dashed rectangles). In XRD results, we can also found the similar anomalies range (different coefficient of thermal expansion values in **Figure 3**). Thus, there are some minor structural changes in the tetragonal PT-BMT crystal, which is corresponding to the change of the Raman phonon modes above 700 K. This phenomenon is the result of interaction by spontaneous volume ferroelectrostriction mechanism and the dynamics of the polar nanoregions (PNRs).

Figure 3 shows the different temperature Raman spectra with different changing trend. As we know, an abnormal phonon softening upon heating can be detected by high resolution Raman spectra if there is any phase transition change. From the 0.62PT-0.38BMT Raman spectra, one clear anomaly is observed at about 670 K, which is quite different from the phase transition temperature. We found the Curie temperature of this single crystal is located at about 800 K through the XRD pattern and dielectric permittivity experiment. The PNRs can play an important role in

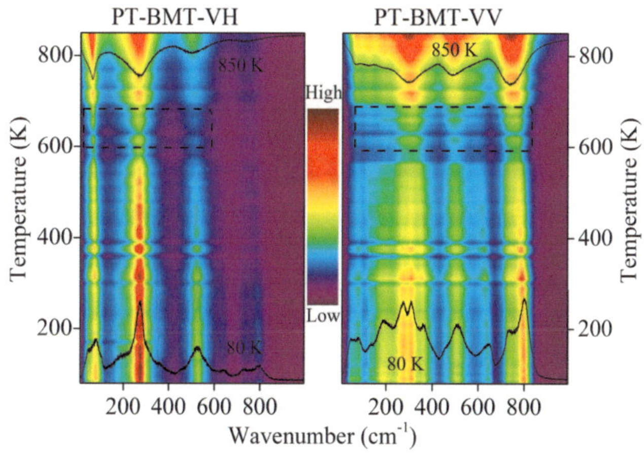

Figure 2. Temperature evolution of Raman intensities for 0.62PT-0.38BMT single crystal. The dashed rectangles mark the temperature ranges of the most pronounced changes (Figure reproduced with permission from [14]. Copyright 2016, Royal Society of Chemistry).

lead-based ferroelectric materials. Note that the physical properties of the single crystal have some linear changes near phase transition, for example the refractive index n, lattice parameters, and unit cell volume V. Thus, the abnormal increasing for the soft mode indicates some structure changes from the crystal Raman spectra. In the whole temperature range, two different slopes of lines are observed (-0.011 and -0.030 cm^{-1} K^{-1}), which can be attributed to the variation of the soft mode frequency induced by the local spontaneous polarization (P_S) inside the PNRs [18–20]. Therefore, the various tendency of the Raman soft mode frequency with temperature agrees well with the PNRs in Pb-based ferroelectrics, in which unit cell volume deviates from linear thermal expansion at Burns temperature according to the XRD results in **Figure 3**. The current Raman study presents a novel method to characterize the relationship between the NTE and P_S of perovskite ferroelectric materials.

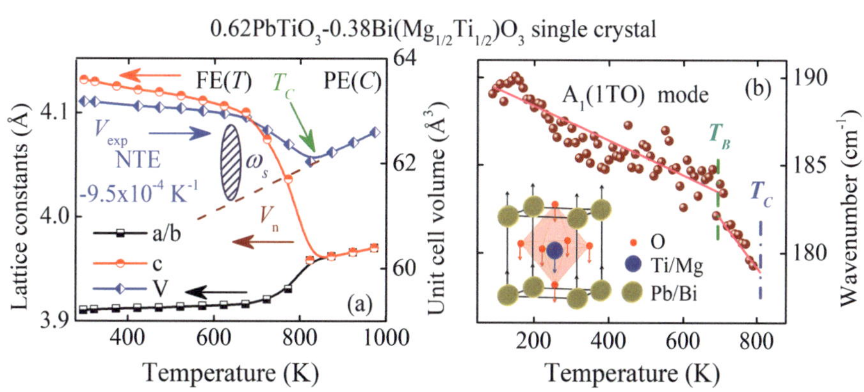

Figure 3. Temperature dependence of the lattice parameters and unit cell volume of the 0.62PT-0.38BMT crystal. FE(T) and PE(C) mean ferroelectric tetragonal phase and paraelectric cubic phase, respectively. The phonon frequency of soft mode as a function of temperature for 0.62PT-0.38BMT crystal. The solid lines indicate the different decrement of the soft mode with temperature (Figure reproduced with permission from [14]. Copyright 2016, Royal Society of Chemistry).

3.2. Tungsten bronze niobate ferroelectrics

Disordered ferroelectric systems are of great interest from the fundamental as well as applications points of view. The uniaxial relaxor ferroelectric SBN, which is of particular interest due to its well-known pyroelectric, electro-optic, and non-linear optical properties, has some potential applications such as holographic data storage, and generation of photorefractive solitons [21]. The aim of this part is to investigate the lattice vibration of the $Sr_xBa_{1-x}Nb_2O_6$ ceramics with different Sr/Ba ratios studied by Raman spectroscopy. The physical mechanism of the ferroelectric to paraelectric phase transition is found from the soft mode variation with the temperature. **Figure 4** shows Raman spectra of the SBN ceramics recorded at 150 K with different Sr compositions. The A_{1g} mode corresponding to a stretch type vibration of the NbO_6 octahedron decrease with increasing the Sr/Ba ratio. The Raman shift of the modes decreases from 643 and 612 cm^{-1} to 636 and 605 cm^{-1}, respectively. Another important feature observed in **Figure 4** is that the lowest modes (labeled with "*LM*") at about 42 cm^{-1} can be detected, which is inferred to the soft mode of the SBN ceramics. The soft mode is eigenvector approximated the ionic displacements occurring at a crystallographic phase transition in the ferroelectric material.

From the temperature dependence of the Raman spectra for all SBN ceramics in the temperature range from 150 to 750 K, the temperature evolution of the soft mode peak position is presented in **Figure 5**. The frequency of the soft modes decreases from 42 to 38 cm^{-1} with the Sr composition at 150 K. The phenomenon can be attributed to the A site substitution originated in the smaller ionic radius of Sr, as compared to Ba element. All frequency of the soft mode first decreases and then disappears in the SBN paraelectric phase. The T_C of the SBN ceramics shifts to lower temperature with increasing Sr composition. It can be well expressed by $T_C(x) = 566 - 371x$, where x is the Sr molar fraction. According to changes of the soft modes, it is clear that increasing Sr composition leads to the shrinking of T_C.

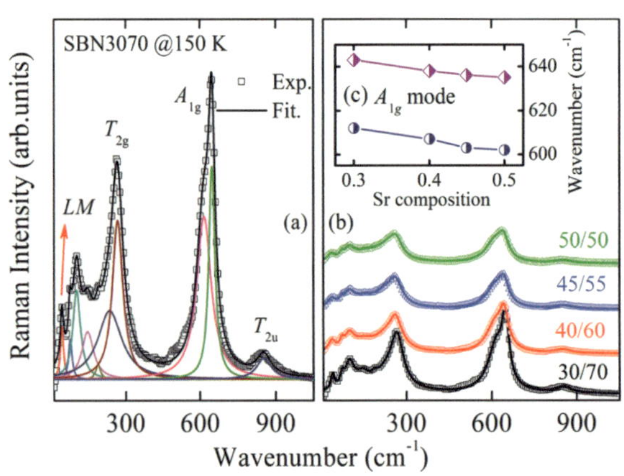

Figure 4. (a) Experimental Raman spectrum (dotted lines) and Lorentzian fitting results (solid lines) of the $Sr_{0.3}Ba_{0.7}Nb_2O_6$ ceramic at 150 K. The arrows indicate that the lowest phonon mode is located at about 40 cm^{-1} (labeled with "LM"). (b) Raman spectra of the SBN ceramics with different Sr composition recorded at 150 K. (c) The A_{1g} Raman-active phonon mode at about 636 and 605 cm^{-1} as a function of Sr composition.

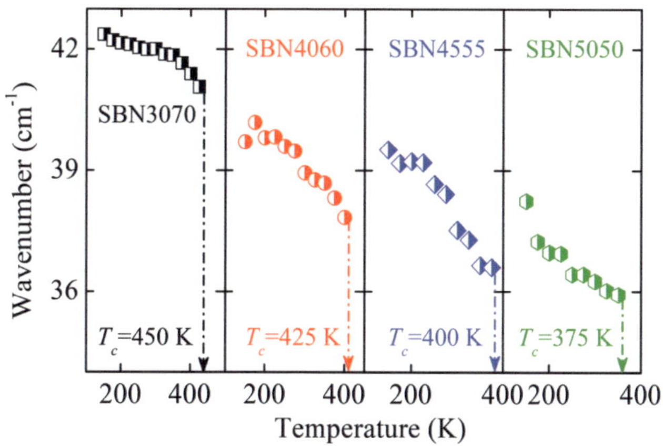

Figure 5. The soft mode frequency as a function of the temperature for the SBN ceramics. The dashed arrows show the temperature points where the soft mode vanishes.

3.3. Aurivillius-type bismuth layer-structured ferroelectrics

Recently, bismuth layer-structured ferroelectrics (BLSFs) have gained increasing attention for the promising applications of non-volatile random access memories (NvRAMs), optical switches and high-temperature piezoelectric devices, because of their relatively high T_c, low dielectric dissipation, and excellent hysteresis behavior [22–25]. Calcium bismuth niobate ($CaBi_2NbO_9$, CBN) is an Aurivillius layered material with the ultra-high T_c and relative higher thermal depoling temperature. The different properties induced by doping in BLSFs are always related to the structure distortions. The spectra of vibration modes in different frequencies could be sensitive to describe the structural distortion. As an example, **Figure 6** shows temperature-dependent Raman spectra of $CaBi_2Nb_{1.97}W_{0.03}O_9$ (W3) ceramic and the CBNW ceramics with different W compositions at 100 K. The results suggest that the peak center of ν5 deviated from the vertical dash line, and the color (in the web version) which represented the intensity of phonon mode faded with increasing temperature. This phenomenon means that the ν5 phonon mode was soften. However, there is no obvious change in the ν6 phonon mode upon heating, as compared to the behaviors of ν5. The similar Raman spectra with different W compositions also confirm that CBN doped with tungsten can keep single phase.

To further investigate the evolution of the ν5 and ν6 phonon modes, Raman spectra are well-fitted with multi-Lorentz oscillators. **Figure 7** shows Raman scattering results and well-fitted peaks with multi-Lorentz oscillators for all samples at 100, 300 and 800 K. The frequency, intensity and full width at half maximum of each phonon mode at different temperature can be derived from the fitting. The frequencies of the ν5 and ν6 phonon modes for W3 are presented in **Figure 8**. The shift of phonon modes can be explained by a simplified Klemens model [26].

From the fitting result according to the model, we can conclude that the change of the Raman phonon mode is dominated by the lattice expansion and thermal evolution. However, the intrinsic anharmonic coupling of phonons is quite weak for ν5. We can observe a slightly blue shift for ν6 phonon mode, while an opposite trend for ν5 in the whole temperature range. The

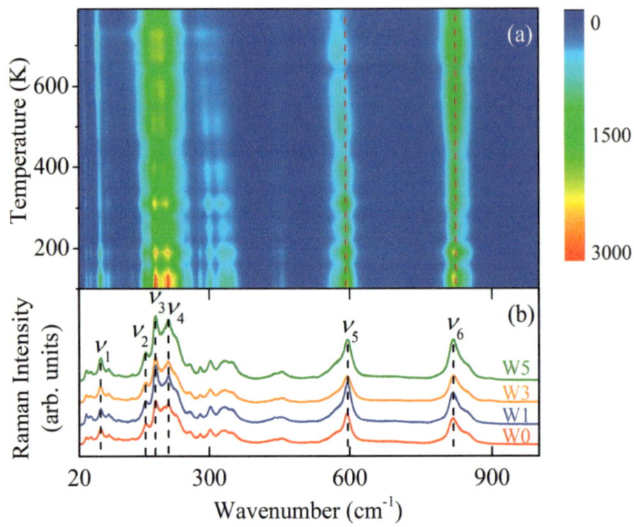

Figure 6. (a) Temperature-dependence of Raman spectra of $CaBi_2Nb_{1.97}W_{0.03}O_9$ (W3) ceramics from 100 to 800 K. (b) Raman scattering of CBNW ceramics with different W compositions at 100 K (Figure reproduced with permission from [22]. Copyright 2015, Elsevier).

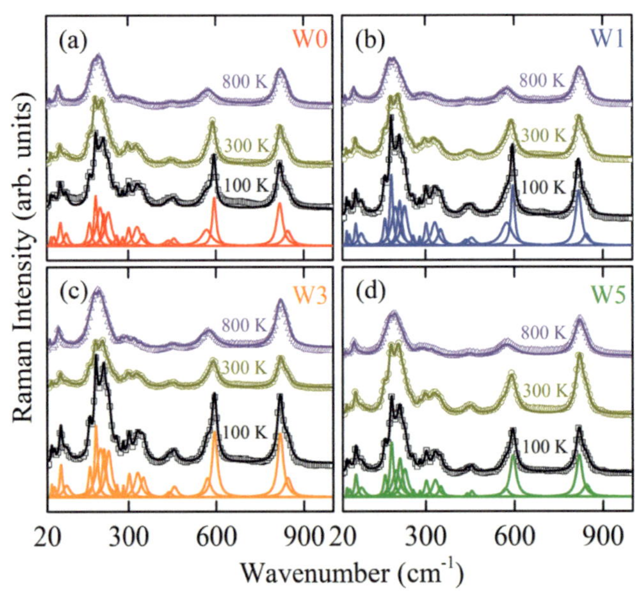

Figure 7. Experimental (dots) and the best fitting (solid lines) Raman spectra of $CaBi_2Nb_{2-x}W_xO_9$ ceramics with (a) 0, (b) 0.01, (c) 0.03, and (d) 0.05 at 100, 300 and 800 K, respectively (Figure reproduced with permission from [22]. Copyright 2015, Elsevier).

different changes between the phonon modes may be ascribed to the unusually strong and positive intrinsic anharmonicity instead of thermal evolution with increasing temperature. On the other hand, for all of the ceramics with increasing W composition, we found that the I(ν6)/I

http://dx.doi.org/10.5772/intechopen.72770

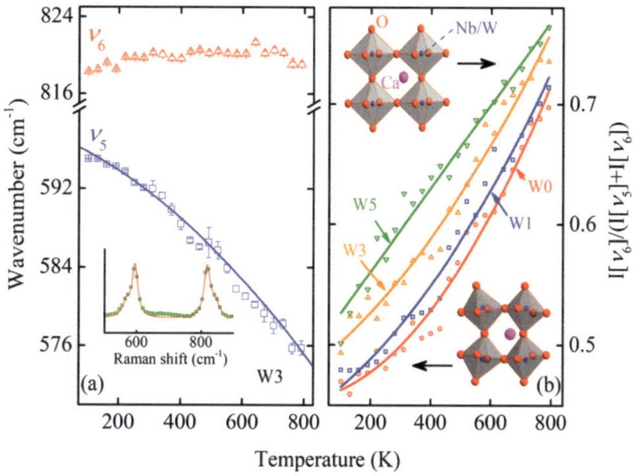

Figure 8. (a) Temperature-dependence of Raman phonon mode of ν5 (squares) and ν6 (trilateral). The inset shows the W3 Raman spectra at 100 K using Lorentz fitting. (b) the ratio of I(ν6)/I[(ν5) + I(ν6)] intensity with different W doping compositions (Figure reproduced with permission from [22]. Copyright 2015, Elsevier).

[(ν5) + I(ν6)] ratio increase with increasing temperature. The result is that the distortion degrees of perovskite (Nb, W)O_6 octahedra decrease with increasing temperature [27], and the BO_6 octahedra distortion predominates the structure distortion with Aurivillius niobate phase [28]. Thus, it can be concluded that the structure distortion in CBNW ceramics decreases with increasing W composition. The variation tendency is consistent with the XRD result in previous study. From the change of the Raman phonon modes with different temperature, we can see the parameter T_C generally decreases with the decrease of structure distortion degree [29, 30]. It was found that the relative peak intensity of I(ν6)/I[(ν5) + I(ν6)] ratio is sensitive for the distortion degree in ferroelectric materials. The Curie temperature of $CaBi_2Nb_{2-x}W_xO_9$ decreases with increasing composition from the Raman results, which is consistent with the results of dielectric permittivity experiments. The results indicate that the Raman spectrum is an effective tool for detecting structure distortion and phase transition of ferroelectric materials.

3.4. ABO_3 perovskite ceramics

In the past few decades, the complex Pb-based ABO_3 perovskite materials have attracted much attentions due to the excellent properties obtained in the compositions close to morphotropic phase boundary (MPB) [31]. The A-site substitution plays an important role in phase transition and more studies are requisite. In this part, the A-site substitution effect on the phase transition near MPB is investigated for PLZST ceramics. Transition temperature region and lattice dynamics are systematically discussed according to the temperature dependent Raman scattering spectroscopy. In addition, a new transient phase called the intermediate phase was found to exist between AFE and PE phase, which could be induced by defects through increasing temperature and doping of foreign ions.

Figure 9 depicts room-temperature Raman scattering results and well-fitted deconvolution peaks for all samples. A sharp increase of relative strength of E(TO2) mode could be observed

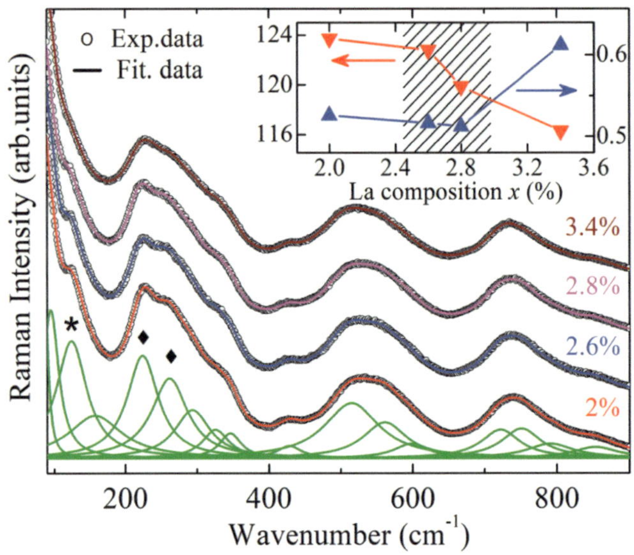

Figure 9. Raman scattering spectra of $Pb_{1-1.5x}La_xZr_{0.42}Sn_{0.4}Ti_{0.18}O_3$ ceramics with different La composition recorded at room temperature. The inset shows the frequency variation of A1(TO1) mode and relative strength of E(TO2) mode as a function of La composition (Figure reproduced with permission from [16]. Copyright 2013, American Institute of Physics).

at La composition of 2.8%. This indicates that PLZST3.4 ceramic is more likely to be tetragonal phase rather than the rhombohedral phase in the PLZST2 ceramic. The softing peak upon La composition is assigned as A1(TO1) symmetry, which stems from splitting of T_{1u} in cubic phase. The mode softs from 123.4 (PLZST2) to 116.3 cm^{-1} (PLZST3.4) with an incommensurate drop at La composition of 2.6%. The fact that both of two dramatic changes occur indicates the phase transition from rhombohedral to tetragonal structure for La composition between 2.6 and 2.8%.

To elucidate the thermal evolution of PLZST ceramics, temperature dependence of Raman spectra from 100 to 650 K are shown in **Figure 10**. We can see all of the Raman modes for the four PLZST ceramics have a blue shift with increasing temperature. Note that some of them disappear above 400 K during the cubic phase appearing. The results indicate that there is a structure transformation at about 400 K. Similar abnormal decrease in Raman intensity of PLZST ceramics has been detected by the spectra with increasing temperature.

To further investigate the phase transition mechanism, we plot temperature dependence of the wavenumber from the Raman modes in **Figure 11**. It can be found that some abrupt variations in the whole temperature range, which is easily divided into two phase transition regions. Note that there is a new intermediate phase at about 300 K for all of samples.

Figure 12 shows the phase diagram of PLZST ceramics according to the temperature dependence of Raman spectra. Three different structure phases can be found: AFE_O phase, intermediate phase, and PE_C phase. We can found that the PLZST ceramics have a transformation from AFEO to the intermediate phase with the composition of Ti exceeds 5.0% at room temperature. However, all of the ceramics remain cubic phase when the temperature upon 450 K. Note that the PLZST ceramics undergo successive phase transitions with increasing

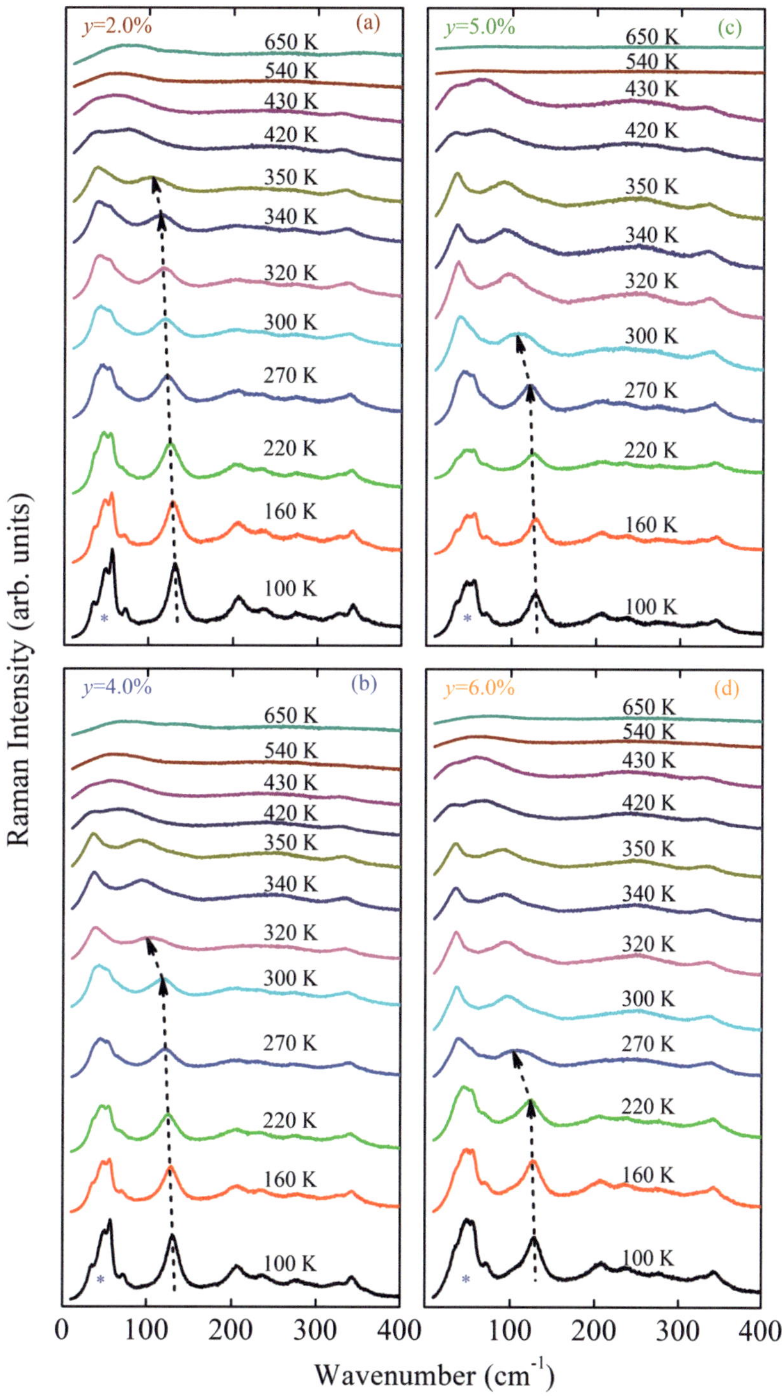

Figure 10. Temperature dependent Raman spectra for PLZST ceramics from 100 to 650 K. The symbol (*) represents the low wavenumber phonon modes (Figure reproduced with permission from [31]. Copyright 2016, Elsevier).

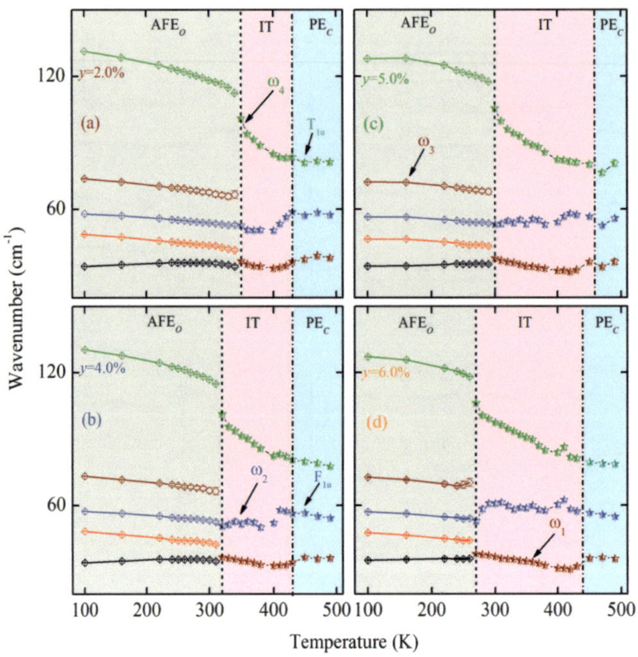

Figure 11. The variations of A1(TO1), AFD, E(TO1), and AHC modes as a function of temperature for PLZST ceramics with different Ti composition. Note that the dashed lines show the boundaries of two adjacent phase structures (Figure reproduced with permission from [31]. Copyright 2016, Elsevier).

temperature: from antiferroelectric phase at low temperature to intermediate phase and paraelectric cubic phase at high temperature. The Curie temperature for PLZST ceramics is lower than that of pure PZT materials because of some ions doping [32, 33]. Note that the intermediate boundaries in the phase diagram have a downward trend with Ti composition. However, the cubic phase boundaries almost remain the same temperature of about 400 K. The intermediate phase temperature ranges gradually become larger with increasing Ti compositions. The Curie temperature detected by Raman spectra are consist with the XRD and dielectric results, which suggests that the Raman spectrum is an effective tool to distinguish different phase structure in PZT ceramics with different temperature and doping level.

3.5. Multiferroics materials

$BiFeO_3$ (BFO) has been hailed as an important material for magnetoelectric devices due to its room-temperature multiferroic properties, in which the electric polarization is coupled to antiferromagnetic (AFM) order, allowing for manipulation of magnetism by applied electric fields and vice versa [34]. During the last several years, intriguing behavior was found in doped BFO compounds. In this part, we present Raman results on La and Ti codoped BFO ceramics to systematically study phase transitions induced by the chemical substitution and temperature.

Figure 13 demonstrates the spectra recorded at several characteristic temperatures and well-fitted deconvolution peaks at 80 K. The frequencies of the phonon modes shift to low energies and the intensity of all major peaks reduces as the temperature increases from 80 to 680 K. The modes in higher-frequency range become severely widening and merge into a broadening

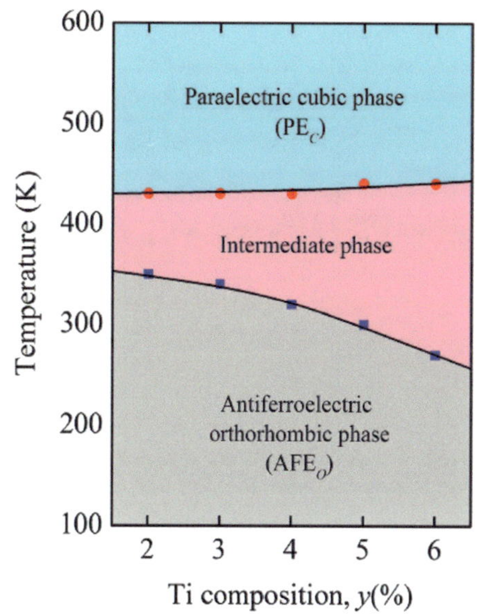

Figure 12. The phase diagram of PLZST ceramics based on the low wavenumber phonon mode variations with the temperature and Ti composition. The phase transformation regions of AFE_O, intermediate phase and PE_C characteristics can be clearly distinguished by the solid dots (Figure reproduced with permission from [31]. Copyright 2016, Elsevier).

peak. From the analysis of the mode position with the temperature, the anomalies in phonon modes suggest a strong magnon-phonon coupling of BLFTO in the present work.

Figure 14 shows composition dependence of the local structural transition temperature (T*) and Neel temperature (T_N). With the increase in the doping compositions of La and Ti, T* and T_N decrease from 580 to 540 K and from 645 to 630 K, respectively. According to La and Ti-codoped in BFO, an external pressure is induced due to size mismatch of host (Bi, Fe) and substitution (La, Ti) cations, which leads to the variation of tolerance factor and structural distortions. On the other hand, the strength of the antiferromagnetic superexchange interaction relies on the Fe-O-Fe angle. These structural effects weaken the magnetic exchange and decrease T_N. Thus, a complicated mechanism induced by the variation of bond length, bond angle, and the exchange interaction between adjacent magnetic moments could substantially contribute to the shrinking of Neel temperature. In summary, the Raman phonon modes have abnormal change around 140 and 205 K, which can be ascribed to the strong magnon-phonon coupling. The structural transition occurred at about 570 K can be detected by the Raman spectra, indicating that the Raman modes are sensitive to the structure changes and spin reorientation.

3.6. Lead-free ferroelectrics

Recently, KNN based lead-free materials were reported to offer comparable piezoelectric properties to that of PZT [35]. The $(K_{0.5}Na_{0.5})NbO_3$-$0.05LiNbO_3$-$yMnO_2$ (y = 0 and 1.0%) (KNN-LN-M) single crystals have been studied by the Raman spectra from the temperature 300–800 K. Moreover, the thermotropic phase boundaries are observed, indicating the existence of the mixed-phase region (i.e. PPT) between orthorhombic (O) and tetragonal (T) phases.

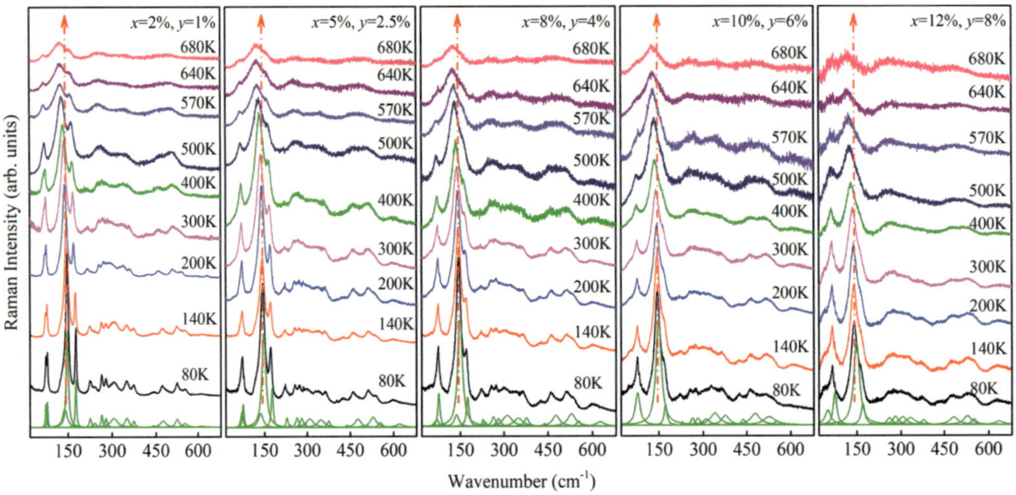

Figure 13. Temperature dependence of Raman scattering for the BLFTO ceramics collected in the temperature range from 80 to 680 K. As an example, Raman spectra recorded at 80 K, which were fitted with independent damped harmonic oscillators, have been indicated on the bottom (Figure reproduced with permission from [34]. Copyright 2014, American Institute of Physics).

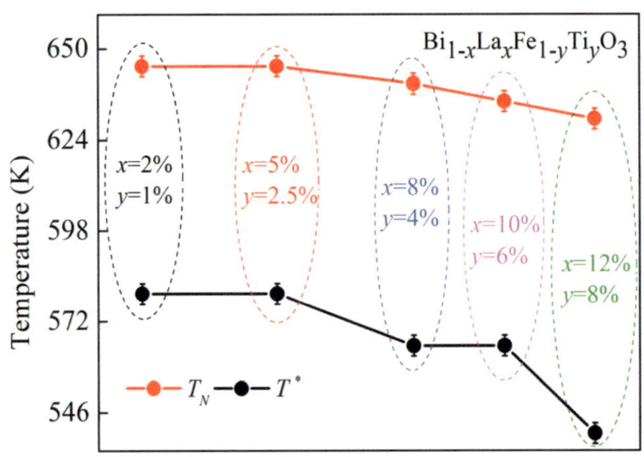

Figure 14. Composition dependence of phase transition temperature T* and Neel temperature TN for the BLFTO ceramics derived from temperature dependent Raman scattering (Figure reproduced with permission from [34]. Copyright 2014, American Institute of Physics).

Figure 15 depicts temperature dependence of Raman scattering for the KNN-LN and KNN-LN-1%MnO_2 crystals, respectively. The phase structure of KNN-LN based crystals transforms from O phase to T phase, then to cubic (C) phase with increasing temperature. The disorder in the O, T, and C phases can be characterized mainly by the Nb central ion allowed positions, and then the framework of the established eight-site model will be adopted [36]. In order to obtain a complete description of the dynamical properties through the successive phase transitions, it is useful to know the classification of the optical modes and their correlations between the various phases [37].

For a closer inspection, the peak positions of all modes in each phase are plotted in **Figure 16** against the temperature between 300 and 800 K. Frequency shifts and transformations of the modes are clear at the O-T and T-C boundaries. The trends within each phase consist primarily of mode softening with increasing temperature, exactly as expected to follow from the lattice thermal expansion. In the light of all the results, displacive and order–disorder mechanisms have to be associated in the description of the phase transitions. Thus, the successive phase transitions in KNN-LN based single crystals derive from competition between a soft phonon mode and a relaxation mode.

In order to obtain the phase fraction, we assume that one phase at thermal phase boundaries is a linear superposition of the spectra below and above phase boundaries. As shown in **Figure 17**, the fitting Raman spectra at 410–440 K in KNN-LN and KNN-LN-1%MnO_2, respectively, are displayed in comparison with the observed spectra at the same temperature. The fitting spectra are obtained from the linear superposition of the Raman spectra below and above phase boundaries. The coincidence of the experimental and the fitted Raman spectra are well for both samples, which strongly confirms the coexistence of the O and T phases at the thermal phase boundary from O to T phase. The mixed structure state at the thermal phase boundaries is due to sufficient competing mechanical and dipolar interactions between domains in multi-domain configurations [38]. In single domain case, the system undergoes a series of first-order ferroelectric transitions upon heating, sequentially adopting the O and T ferroelectric phases before reverting to the C parent phase.

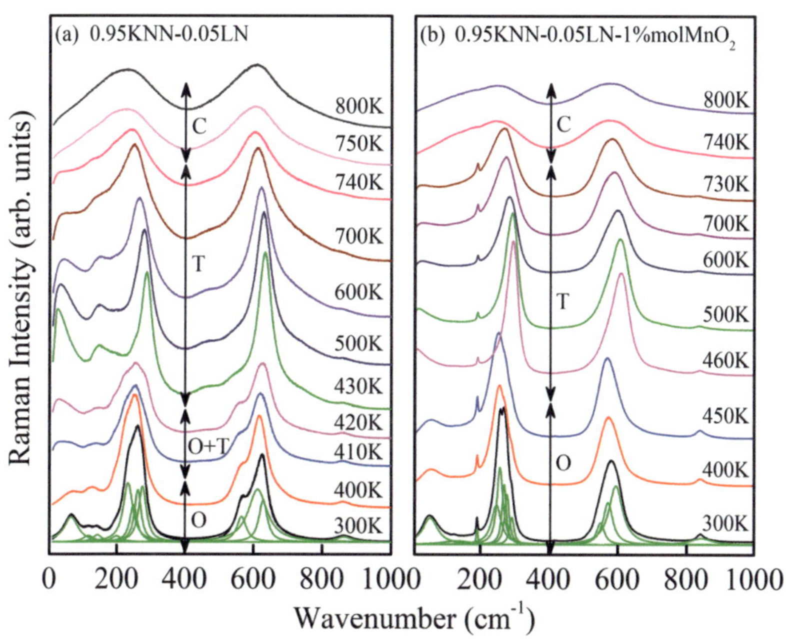

Figure 15. Temperature dependence of Raman scattering for the KNN-LN based single crystals collected in the temperature range from 300 to 800 K and Lorentzian-shaped deconvolution at the temperature of 300 K. The arrows are applied to separate different phase transition (Figure reproduced with permission from [35]. Copyright 2015, American Institute of Physics).

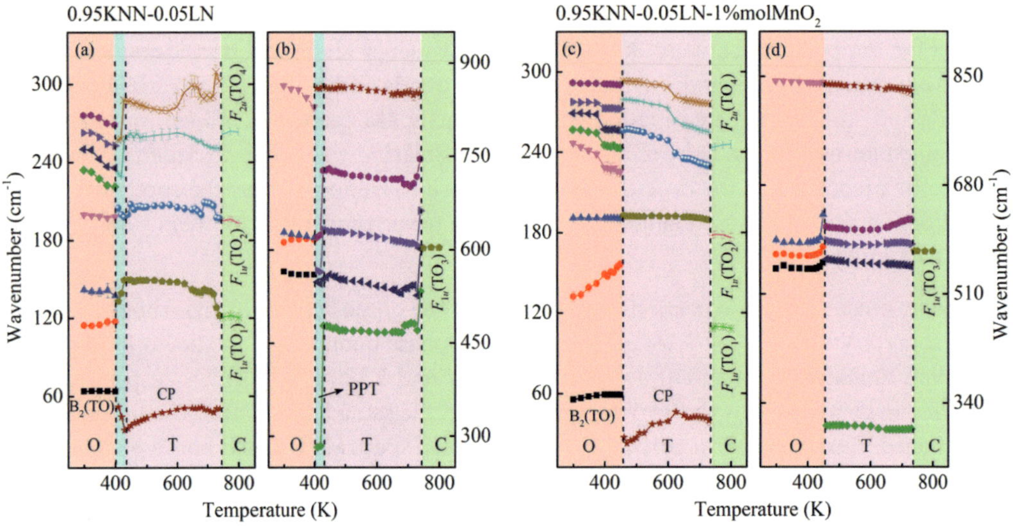

Figure 16. Temperature dependence of the phonon frequency from KNN-LN based crystals. (a) and (c): The low frequency range of 10–300 cm^{-1}, (b) and (d): The high frequency range of 400–900 cm^{-1}. Note that different shade regions indicate that the crystals are located in diverse phase (Figure reproduced with permission from [35]. Copyright 2015, American Institute of Physics).

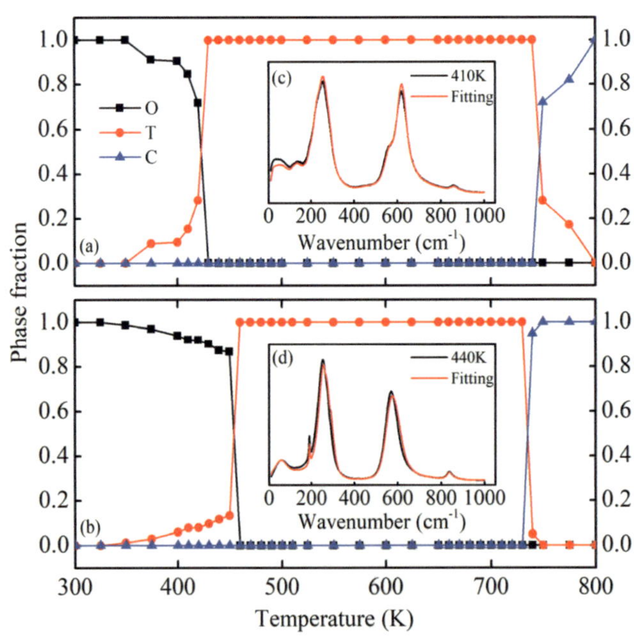

Figure 17. Phase fraction of (a) KNN-LN, and (b) KNN-LN-1%MnO_2 single crystals as a function of temperature. Note that (c) and (d) is the fitting Raman spectra at 410–440 K for KNN-LN and KNN-LN-1%MnO_2, respectively (Figure reproduced with permission from [35]. Copyright 2015, American Institute of Physics).

4. Conclusions and outlook

This chapter reports a systematic micro-Raman scattering study on A- and B-site doped lead-based, tungsten bronze niobate, Bi-layered, ABO_3 perovskites, multiferroics and lead-free ferroelectrics. The processing conditions, substitution, temperature-dependent Raman spectra and the structure–property correlations are discussed in the ceramics and single crystal forms of these materials. It can be concluded that the Raman spectra can provide us with a lot of important physical parameters for application in the future, such as spontaneous polarization, Curie temperature, structure distortion degree, mix-phase region and phase diagram and so on. The excellent agreement between the Raman, XRD or dielectric observations of merging of phase transition temperatures in all of the ferroelectric materials with different doping compositions suggested the powerful tool for detecting phase transition with solid state spectroscopy.

It should be emphasized that the domains in these ferroelectric materials and their related phenomena with different conditions, such as electric field, magnetic field and pressure have not been well investigated. The Raman spectra can be used to characterize the species of domain (including a/c domains, 90°/180° domains), critical sizes and domain wall in ferroelectrics. Thus, we can check the domain status by recording the high-resolution Raman response. Our next goal is to characterize the domain information in these ferroelectric ceramics and films and identify the factors which can affect the domain structure. We believe that our research results will be of great significance to the development of microelectronic physics in the future.

Acknowledgements

The authors thank Dr. X. Chen, K. Shi and X. J. Ding for helpful discussions. This work was financially supported by Major State Basic Research Development Program of China (Grant Nos. 2013CB922300), Natural Science Foundation of China (Grant Nos. 61674057, 11374097, 61376129, 61504156, and 61227902), Projects of Science and Technology Commission of Shanghai Municipality (Grant Nos. 15JC1401600 and 14XD1401500), and the Program for Professor of Special Appointment (Eastern Scholar) at Shanghai Institutions of Higher Learning.

Author details

Kai Jiang, Liping Xu, Jinzhong Zhang, Zhigao Hu* and Junhao Chu

*Address all correspondence to: zghu@ee.ecnu.edu.cn

Technical Center for Multifunctional Magneto-Optical Spectroscopy (ECNU), Shanghai and Department of Electronic Engineering, East China Normal University, Shanghai, China

References

[1] Haertling GH. Ferroelectric ceramics: History and technology. Journal of the American Ceramic Society. 1999;**82**:797-818. DOI: 10.1111/j.1151-2916.1999.tb01840.x

[2] Dobal PS, Katiyar RS. Studies on ferroelectric perovskites and bi-layered compounds using micro-Raman spectroscopy. Journal of Raman Specroscopy. 2002;**33**:405-423. DOI: 10.1002/jrs.876

[3] Yang Y, Wang X, Sun C, Li L. Phase transition in $BaTiO_3$ nanotube arrays. Journal of Applied Physics. 2011;**109**:014109. DOI: 10.1063/1.3530614

[4] Pandey CS, Schreuer J, Burianek M, Muhlberg M. Relaxor behavior of ferroelectric $Ca_{0.22}Sr_{0.12}Ba_{0.66}Nb_2O_6$. Applied Physics Letters. 2013;**102**:022903. DOI: 10.1063/1.4775686

[5] Fan L, Chen J, Ren Y, Pan Z, Zhang L, Xing X. Unique piezoelectric properties of the monoclinic phase in $Pb(Zr,Ti)O_3$ ceramics: Large lattice strain and negligible domain switching. Physical Review Letters. 2016;**116**:027601. DOI: 10.1103/PhysRevLett.116.027601

[6] Schütz D, Deluca M, Krauss W, Feteira A, Jackson T, Reichmann K. Lone-pair-induced covalency as the cause of temperature- and field-induced instabilities in bismuth sodium Titanate. Advanced Functional Materials. 2012;**22**:2285-2294. DOI: 10.1002/adfm.201102758

[7] Grinberg I, Cooper VR, Rappe AM. Relationship between local structure and phase transitions of a disordered solid solution. Nature. 2002;**419**:909-911. DOI: 10.1038/nature01115

[8] Sun Y, Welch GC, Leong WL, Takacs CJ, Bazan GC, Heeger AJ. Solution-processed small-molecule solar cells with 6.7% efficiency. Nature Materials. 2012;**11**:44-48. DOI: 10.1038/nmat3160

[9] Dupe B, Prosandeev S, Geneste G, Dkhil B, Bellaiche L. $BiFeO_3$ films under tensile epitaxial strain from first principles. Physical Review Letters. 2011;**106**:237601. DOI: 10.1103/PhysRevLett.106.237601

[10] Khalyavin DD, Salak AN, Vyshatko NP, Lopes AB, Olekhnovich NM, Pushkarev AV, Maroz II, Radyush YV. Crystal structure of metastable perovskite $bi(Mg_{1/2}Ti_{1/2})O_3$: Bi-based structural analogue of antiferroelectric $PbZrO_3$. Chemistry of Materials. 2006;**18**:5104-5110. DOI: 10.1021/cm061129w

[11] Rusakov DA, Abakumov AM, Yamaura K, Belik AA, Van Tendeloo G, Takayama-Muromachi E. Structural evolution of the $BiFeO_3-LaFeO_3$ system. Chemistry of Materials. 2010;**23**:285-292. DOI: 10.1021/cm1030975

[12] Damodaran AR, Liang C-W, He Q, Peng C-Y, Chang L, Chu Y-H, Martin LW. Nanoscale structure and mechanism for enhanced electromechanical response of highly strained $BiFeO_3$ thin films. Advanced Materials. 2011;**23**:3170-3175. DOI: 10.1002/adma.201101164

[13] Liu J, Chen X, Xu G, Yang D, Tian Y, Zhu X. Novel high-temperature ferroelectric single crystals $0.38Bi(Mg_{1/2}Ti_{1/2})O_3$-$0.62PbTiO_3$ with good and temperature-stable piezoelectric properties. CrystEngComm. 2015;**17**:5605-5608. DOI: 10.1039/C5CE01114K

[14] Jiang K, Zhang P, Zhang J, Xu G, Li W, Hu Z, Chu J. Relationship between negative thermal expansion and lattice dynamics in a tetragonal $PbTiO_3$-bi($Mg_{1/2}Ti_{1/2}$)O_3 perovskite single crystal. RSC Advances. 2016;**6**:3159-3164. DOI: 10.1039/C5RA24408K

[15] Liang P, Kai J, Jinzhong Z, Zhigao H, Genshui W, Xianlin D, Junhao C. Temperature-dependent phonon Raman scattering and spectroscopic ellipsometry of pure and Ca-doped $Sr_xBa_{1-x}Nb_2O_6$ ferroelectric ceramics across the phase transition region. Journal of Physics D: Applied Physics. 2016;**49**:035307. DOI: 10.1088/0022-3727/49/3/035307

[16] Chen X, Hu ZG, Duan ZH, Chen XF, Wang GS, Dong XL, Chu JH. Effects from A-site substitution on morphotropic phase boundary and phonon modes of ($Pb_{1-1.5x}La_x$) ($Zr_{0.42}Sn_{0.40}Ti_{0.18}$)$O_3$ ceramics by temperature dependent Raman spectroscopy. Journal of Applied Physics. 2013;**114**:043507. DOI: 10.1063/1.4816093

[17] Zhang J, Tong W-Y, Zhu J, Xu J, Duan Z, Xu L, Hu Z, Duan C-G, Meng X, Zhu Z, Chu J. Temperature-dependent lattice dynamics and electronic transitions in 0.93Pb($Zn_{1/3}Nb_{2/3}$)O_3-0.07$PbTiO_3$ single crystals: Experiment and theory. Physical Review B. 2015;**91**:085201. DOI: 10.1103/PhysRevB.91.085201

[18] Chen J, Hu L, Deng J, Xing X. Negative thermal expansion in functional materials: Controllable thermal expansion by chemical modifications. Chemical Society Reviews. 2015;**44**:3522-3567. DOI: 10.1039/C4CS00461B

[19] Zhang Q, Li Z, Xu Z. Phase transition in (1-x)Bi($Mg_{1/2}Ti_{1/2}$)O_3-x$PbTiO_3$ ceramics. Materials Letters. 2011;**65**:3143-3145. DOI: 10.1016/j.matlet.2011.06.078

[20] Bokov AA, Ye ZG. Recent progress in relaxor ferroelectrics with perovskite structure. Journal of Materials Science. 2006;**41**:31-52. DOI: 10.1007/s10853-005-5915-7

[21] Dec J, Kleemann W, Miga S, Shvartsman VV, Łukasiewicz T, ŚWirkowicz M. Aging, rejuvenation, and memory effects in the domain state of $Sr_{0.75}Ba_{0.25}Nb_2O_6$. Phase Transitions. 2007;**80**:131-140. DOI: 10.1080/01411590701315492

[22] Shi K, Peng L, Li M, Zhou Z, Jiang K, Zhang J, Hu Z, Dong X, Chu J. Structural distortion, phonon behavior and electronic transition of Aurivillius layered ferroelectric $CaBi_2Nb_{2-x}W_xO_9$ ceramics. Journal of Alloys and Compounds. 2015;**653**:168-174. DOI: 10.1016/j.jallcom.2015.09.037

[23] Kleemann W, Dec J, Shvartsman VV, Kutnjak Z, Braun T. Two-dimensional ising model criticality in a three-dimensional uniaxial relaxor ferroelectric with frozen polar nanoregions. Physical Review Letters. 2006;**97**:065702. DOI: 10.1103/PhysRevLett.97.065702

[24] Zhang J, Wang G, Gao F, Mao C, Cao F, Dong X. Influence of Sr/Ba ratio on the dielectric, ferroelectric and pyroelectric properties of strontium barium niobate ceramics. Ceramics International. 2013;**39**:1971-1976. DOI: 10.1016/j.ceramint.2012.08.048

[25] Kitaev YE, Aroyo MI, Perez-Mato JM. Site symmetry approach to phase transitions in perovskite-related ferroelectric compounds. Physical Review B. 2007;**75**:064110. DOI: 10.1103/PhysRevB.75.064110

[26] Balkanski M, Wallis RF, Haro E. Anharmonic effects in light scattering due to optical phonons in silicon. Physical Review B. 1983;**28**:1928-1934. DOI: 10.1103/PhysRevB.28.1928

[27] Shimakawa Y, Imai H, Kimura H, Kimura S, Kubo Y, Nishibori E, Takata M, Sakata M, Kato K, Hiroi Z. Orbital hybridization and covalency in paraelectric and ferroelectric $SrBi_2Nb_2O_9$. Physical Review B. 2002;**66**:144110. DOI: 10.1103/PhysRevB.66.144110

[28] Shimakawa Y, Kubo Y, Tauchi Y, Kamiyama T, Asano H, Izumi F. Structural distortion and ferroelectric properties of $SrBi_2(Ta_{1-x}Nb_x)_2O_9$. Applied Physics Letters. 2000;**77**:2749-2751. DOI: 10.1063/1.1319509

[29] Sun L, Chu J, Feng C, Chen L. Analysis of relaxor mechanism and structural distortion for $SrBi_{1.6}Nd_{0.4}Nb_2O_9$ bismuth-layer-structured ceramics. Applied Physics Letters. 2007;**91**: 242902. DOI: 10.1063/1.2824383

[30] Blake SM, Falconer MJ, McCreedy M, Lightfoot P. Cation disorder in ferroelectric Aurivillius phases of the type $Bi_2ANb_2O_9$(A=Ba, Sr, Ca). Journal of Materials Chemistry. 1997;**7**:1609-1613. DOI: 10.1039/A608059F

[31] Ding X, Guo S, Hu Z, Chen X, Wang G, Dong X, Chu J. The intermediate phase and low wavenumber phonon modes in antiferroelectric $(Pb_{0.97}La_{0.02})(Zr_{0.60}Sn_{0.40-y}Ti_y)O_3$ ceramics discovered from temperature dependent Raman spectra. Journal of Alloys and Compounds. 2016;**667**:310-316. DOI: 10.1016/j.jallcom.2016.01.188

[32] Buixaderas E, Gregora I, Kamba S, Petzelt J, Kosec M. Raman spectroscopy and effective dielectric function in PLZTx/40/60. Journal of Physics: Condensed Matter. 2008;**20**:345229. DOI: 10.1088/0953-8984/20/34/345229

[33] Jankowska-Sumara I. Antiferroelectric phase transitions in single crystals $PbZrO_3$:Sn revisited. Phase Transitions. 2014;**87**:685-728. DOI: 10.1080/01411594.2014.900554

[34] Xu LP, Zhang LL, Zhang XL, Zhang JZ, Hu ZG, Yu J, Chu JH. Phase transformations in multiferroic $Bi_{1-x}La_xFe_{1-y}Ti_yO_3$ ceramics probed by temperature dependent Raman scattering. Journal of Applied Physics. 2014;**116**:164103. DOI: 10.1063/1.4900439

[35] Xu LP, Jiang K, Zhang JZ, Xu GS, Hu ZG, Chu JH. Phase transitions and thermotropic phase boundaries in MnO_2-doped $(K_{0.5}Na_{0.5})NbO_3$-$0.05LiNbO_3$ single crystals: Raman scattering evidence at elevated temperatures. Applied Physics Letters. 2015;**106**:122901. DOI: 10.1063/1.4916226

[36] Pruzan P, Gourdain D, Chervin JC. Vibrational dynamics and phase diagram of $KNbO_3$ up to 30 GPa and from 20 to ~500 K. Phase Transitions. 2007;**80**:1103-1130. DOI: 10.1080/01411590701473176

[37] Fontana MD, Metrat G, Servoin JL, Gervais F. Infrared spectroscopy in $KNbO_3$ through the successive ferroelectric phase transitions. Journal of Physics C: Solid State Physics. 1984;**17**:483. DOI: 10.1088/0022-3719/17/3/020

[38] Lummen TTA, Gu Y, Wang J, Lei S, Xue F, Kumar A, Barnes AT, Barnes E, Denev S, Belianinov A, Holt M, Morozovska AN, Kalinin SV, Chen L-Q, Gopalan V. Thermotropic phase boundaries in classic ferroelectrics. Nature Communications. 2014;**5**:3172. DOI: 10.1038/ncomms4172

Section 5

Raman Spectroscopy Applied to Biomedical Sciences

Chapter 13

Raman Spectroscopy Applied to Health Sciences

Alexandra Nunes and Sandra Magalhães

Additional information is available at the end of the chapter

http://dx.doi.org/10.5772/intechopen.73087

Abstract

Raman spectroscopy has remarkable analytical abilities to scientists who want to study biological samples. The use of Raman spectroscopy within biologic samples has been increasing in the last years because it can provide biochemical information, allows discrimination between two or more sample groups, and, contrary to what happens with other spectroscopic techniques, water has no interference in the spectra. Biological samples typically do not require extensive preparation, and biochemical and structural information extracted from spectroscopic data can be used to characterize different groups. This chapter presents the general features of Raman spectroscopy and Raman spectroscopic tools relevant to the application in health sciences. In order to emphasize the potential of Raman in this research field, examples of its application in oncology, in bacterial identification and in dementia diagnosis are given.

Keywords: Raman spectroscopy, metabolomics, diagnosis, health and disease

1. Introduction

Spectroscopy is a relevant tool for biomedical analysis. Significant progresses in the application of spectroscopy in clinical field were done in the last years. Spectroscopic techniques provide information at the molecular level, and it is possible to evaluate functional groups, bond types and molecular conformations of the biological components of a sample, once spectral signals in vibrational spectra are specific to each molecule and act like a fingerprint.

A spectroscopic-based metabolomic study includes sample collection, sample analysis, statistical analysis and identification of altered metabolites. The resulting data can be translated into defining disease biomarkers/pathways, with the generation of a disease metabolic fingerprint.

Vibrational spectroscopy has been proposed as an approach to diagnosis. Raman has a past research regarding its potential as a diagnostic tool of a wide range of pathologies using a wide range of sample types. Although being recognized as not as specific and sensitive as other metabolomics techniques, several works already demonstrated the potential of Raman applied to health sciences for metabolic fingerprinting because it is possible with only one spectra to simultaneously analyze carbohydrates, amino acids, fatty acids, lipids, proteins, nucleic acids and polysaccharides with a minimum sample preparation. As Raman is a scattering technique and it is not perturbed by aqueous media, it is suitable to analyze biological samples.

For diagnostic purposes, it is expected that Raman spectra of biological samples result in quantitative data, so it is essential to define some categorical differentiable classes for data by dividing samples in healthy or disease sample classes. For these purposes, chemometric data processing is a valuable tool.

Due to Raman spectroscopy features, it is currently widely used in health sciences for spectral imaging of cells and tissues, for the *in-vivo* and *ex-vivo* diagnosis of tissues, where fiber probes can be used, and for biofluid analysis, contributing to a better knowledge of the disease and disease diagnosis at the molecular level. This chapter describes the most relevant application of Raman in biomedical field.

2. Raman features and sampling

Raman spectroscopy is an optical technique based on inelastic scattering of light due to the vibration of the molecules that can provide chemical fingerprints of several samples. In health sciences, Raman can be used in DNA analysis, lipids, proteins and amino acids identification, bacteria classification and recognition, cell responses, cancer diagnosis and prognosis, and dental prosthesis, among others. This wide range of application is due to Raman's capabilities that go far beyond of being a noninvasive and nondestructive method that does not require samples preparation. Raman is compatible with aqueous solutions and is also a technique that produces results in a short time, requiring less than a minute to obtain a good quality spectrum, and no sample preparation is needed. This spectroscopic technique is sensitive to identify many different functional groups that produce weak signals in infrared spectroscopy (C═C; S─S; C─S). Besides that, it has a highly selective fingerprint that allows the discrimination of similar molecules and has high spatial resolution that allows single cell analysis and intracellular imaging.

The ability to use advanced optical technologies in the visible or near-infrared spectral range (lasers, microscopes, and optic fibers) is responsible for the growth of Raman spectroscopy in medical diagnostic. As molecular changes in cells, tissues or biofluids can be detected and quantified by Raman spectroscopy, it is possible to use this technique in the diagnosis of diseases and to study the effects of drugs (treatments) in biosystems.

In fact, Raman spectroscopy can offer chemical fingerprints of cells, tissues (*in vivo* or *in-vitro*), or biofluids. A large range of samples can be used for Raman analysis: formalin-fixed and fresh frozen mammalian tissue, fixed cells and biofluids.

Whether a sample is a solid, liquid, powder, slurry, or gas, no sample preparation is normally required, and there is no need to dissolution, grinding, glass formation, or pressing in order to obtain good spectroscopic outcome. Raman spectroscopy is also a nondestructive method, and it is possible to evaluate the chemical composition of a sample, and the same aliquot can be used further to extract additional biological information by using other methodologies [1].

3. Raman spectroscopic variations

In recent past, Raman spectroscopy was known as a technique with low signals requiring longer acquisition times. However, recent developments were made to overcome this limitation in the last years. Nonlinear optical effects and metallic nanoparticles are currently used to improve Raman signals, fiber-optic Raman probes were introduced and are used for real-time *in-vivo* experiments, and multimodal integration with other optical techniques increased the acquisition speed and spatial accuracy. These advances in the accuracy allow the application of Raman spectroscopy into clinical diagnosis, and time of analysis allows its clinical use.

Surface enhanced Raman spectroscopy (SERS), Tip Enhanced Raman Scattering (TERS) or nano-Raman and resonance effects increase significantly Raman sensitivity to study biological samples. SERS increases Raman intensity compared to the usual and weak Raman scattering. These improvement features are sufficient to allow even single molecule detection using Raman. SERS is useful in trace material analysis, flow cytometry and other applications where the traditional sensitivity/speed of a Raman measurement is insufficient [2]. Resonance Raman spectroscopy is a variant of Raman spectroscopy that instead of using laser excitation at any wavelength to measure Raman scattering of the laser light, the excitation wavelength is used to overlap with an electronic transition. The overlap results in an extraordinary increase in scattering intensities, thus detection limits and measurement times can be significantly decreased. It is also possible to couple Raman to an optical microscope. Raman microspectroscopy uses visible and near-infrared excitation lasers and allows to extract molecular properties of the samples with diffraction-limited spatial resolution. The typical method to obtain Raman spectral images is by scanning the sample with the laser spot and then applying a uni- or multivariate spectral model to each Raman spectrum [3]. In order to decrease the time of analysis, Raman spectral imaging can be based on line-mapping (laser beam is expanded to form a line spot on the sample surface) [4].

TERS is a chemical imaging technique that is label-free and have enhanced-resolution. TERS imaging is performed with a Raman spectrometer, a scanning probe microscope (SPM) integrated with an optical microspectrometer. The scanning probe microscope provides the means for nanoscale imaging and the optical microscope provides the resources to bring the light to a functionalized probe, and the spectrometer is the sensor analyzing the light output providing chemical specificity. It is possible to increase the signal to obtain high-spatial resolution spectral images for large samples using selective-sampling Raman microspectroscopy. In this approach, it is possible to (1) obtain information about sample spatial features by other optical technique [5] or (2) estimate information in real-time from the Raman spectra [6]. When traditional variations of Raman spectroscopy are used to study tissues, the results are not good due to insufficient penetration depth. The advance of spatially offset Raman spectroscopy (SORS) overcame this limitation enabling spectral measurements until 10–20 mm of the sample and, with this modification, the application of Raman in clinical fields increases by collecting the scattered light away from the point of laser illumination [7]. Transmission Raman spectroscopy (TRS) is the term used when the collection and illumination points are on opposite sides of the sample, and it is quite useful to analyze opaque materials. Besides, it was proved to be very useful to read many millimeters of tissues [8]. To increase the resolution of Raman microspectroscopy, it is possible to use Coherent anti-Stokes Raman spectroscopy (CARS) or stimulated Raman spectroscopy (SRS), both based on nonlinear optical effects that upgrade the spatial resolution. Coherent Raman spectroscopy techniques are based on non-linear effects to increase the speed and spatial resolution of Raman spectroscopy by inducing coherent molecular vibrations in the sample increasing resolution. CARS is used to obtain image from cells and tissues by exciting the CH stretching vibrations of lipids and proteins and SRS allows to obtain high-speed images [9, 10]. Raman spectroscopy is also adaptable for fiber-optic probes, making it valuable for medical diagnosis *in vivo*, for instance in hollow organs. The probe should have very reduced dimensions to permit the access to body cavities, and the spectra acquisition time should be short to allow accurate measurement [11].

4. Multivariate analysis applied to spectroscopic data analysis

Multivariate analysis tools are used to extract information from spectral data set and to assist in biomedical interpretation. Chemometric methods are diverse and offer different approaches to extract specific information from the data. These methods used for data interpretation and to extract information from complex datasets were usually univariate and therefore they were not suitable to describe the sample variation or composition. A range of processing methods can be applied to a Raman data set; usually baseline corrections should be performed as a first approach, prior to normalization methods and then a wide range of multivariate analysis tolls can be applied [12].

Multivariate methods include techniques of multivariate classification (or pattern recognition techniques) and multivariate regression. Techniques of multivariate classification can be divided into unsupervised and supervised learning procedures. In unsupervised pattern recognition techniques such as principal-component analysis (PCA) and cluster analysis (CA), there is no need for *a priori* knowledge about the training set samples (spectra). These methods are used to consider differences and similarities between spectra. By contrast, supervised pattern recognition techniques such as linear discriminant analysis (LDA) and artificial neural networks (ANNS) require some *a priori* knowledge, for example, undoubtedly identifying samples from disease and samples from healthy cases. In this way, supervised procedures allow a more precise classification, while unsupervised methods are useful for an exploratory analysis of data.

On the other hand, multivariate regression techniques (or multivariate calibration methods) are usually applied to analyze one or multiple molecules of a complex sample that possess overlapping spectroscopic signals. These techniques include principal components regression (PCR) and partial least squares regression (PLS) (**Figure 1**).

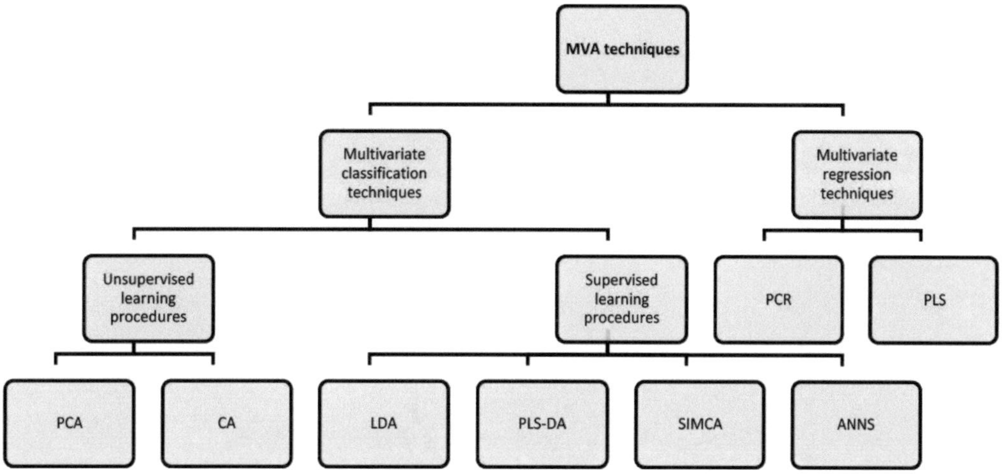

Figure 1. Most frequently applied multivariate data analysis techniques in combination with spectroscopic methods. MVA: Multivariate data analysis; PCR: Principal component regression; PLS: Partial least-squares regression; PCA: Principal-component analysis; CA: Cluster analysis; LDA: Linear discriminant analysis; PLS-DA: Partial least-squares discriminant analysis; SIMCA: Soft independent modeling of class analogy; ANNS: Artificial neural networks. Adapted from [13, 14].

5. Raman spectroscopy in cancer diagnosis

A preeminent application of Raman spectroscopy in health sciences is its use in cancer diagnosis. According to Cancer Research UK, in 2012, an estimated 14.1 million new cancer cases occurred worldwide, resulting in 8.2 million people died [15]. Despite the majority of new cancer cases was registered in less-developed countries, a significative number (more than 6 million cases) occurs in developed regions, with access to advanced medical care and treatments [16]. These data reveal the urgent need of reliable diagnostic tools to reduce cancer cases and cancer mortality.

The common methodologies for cancer diagnosis are based on invasive histological analysis and biomedical imaging, which are expensive, time-consuming and can give rise to subjective diagnosis [17]. Raman spectroscopy has been extensively studied in an attempt to replace or complement current methods and increase sensitivity and specificity of the diagnosis [18]. In this section, we discuss the advances in Raman spectroscopy applied to biofluids and tissue for diagnosis of different cancer types (**Figure 2**).

5.1. Skin cancers

Raman has been widely investigated to discriminate benign skin lesions from malignant lesions, mainly basal cell carcinoma and squamous cell carcinoma (**Table 1**). Despite the high complexity of skin tissue, studies *in vivo* show huge potential as a routine procedure in hospitals for cancer screening [19]. In 2008, a preliminary report used 289 patients with different

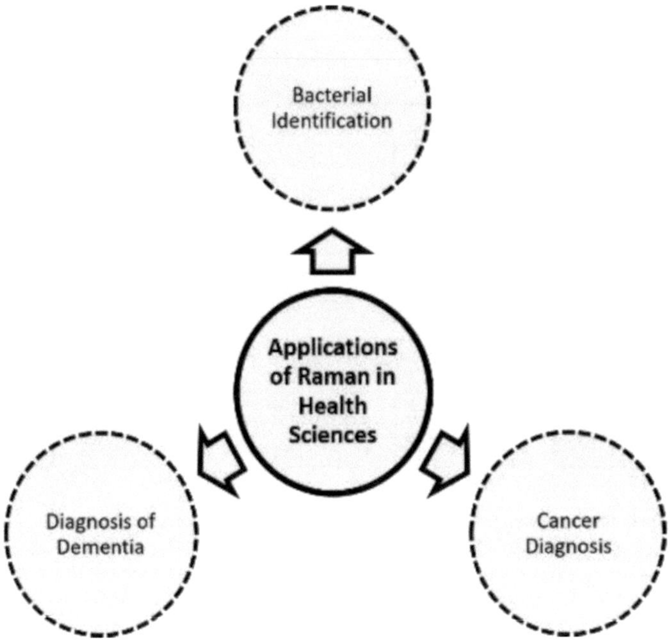

Figure 2. Main application areas of Raman spectroscopy in health science.

types of malignant and benign skin lesions [20]. The authors applied real-time *in vivo* Raman spectroscopy to acquire spectra of all samples and analyzed data using PLS and LDA analysis. Direct analysis of spectra showed distinct biomolecular signatures and ROC curves allowed to discriminate skin cancers from benign lesions with an area under curve greater than 0.9 [20]. Later, the same authors published another study *in vivo* using a higher cohort of patients and 518 skin lesions (both malignant and benign conditions that are visually similar to skin cancer) [21]. Using the same instrumentation as before [20], the authors analyzed data using PC-GDA and PLS analysis, once visual inspection of spectra did not show distinctive Raman peaks assigned to skin cancer [21]. Although Raman peaks are almost the same between normal and cancer samples, the intensity of the signals allows the discrimination between different types of lesions and are compatible with those obtained using histopathological methods. To complement and validate these results, in 2015, Zhao et al. [22] added 127 samples to the previous cohort, creating a consolidated group. Using real-time Raman spectroscopy and PC-GDA, an PLS analysis obtained the same discrimination as before, proving the ability of Raman spectroscopy for *in vivo* skin cancer detection. In 2014, other *in vivo* study successfully used Raman microspectroscopy to study 20 skin samples and detect basal cell carcinoma in tissue removed during surgery [21].

Besides *in vivo* studies, there are also some studies *in vitro* to assess the use of Raman spectroscopy for skin cancer detection. In 2010, Bodanese et al. used dispersive Raman spectroscopy and successfully distinguished normal samples from basocellular cell carcinoma [23]. They found spectral differences in the region between 800 and 1000 cm^{-1} and between 1200 and

Cancer type	Study type	Statistical analysis	Sensitivity and specificity (%)	Spectral differences (cancer *vs.* normal)	Refs.
Skin	*In vivo*	PC-GDA, PLS	95, 54	NA	[22]
Skin	*Ex vivo*	LDA, MLR	100, 92	NA	[44]
Skin	*Ex vivo*	PCA	99.1, 93.3	Proteins, lipids and melanin	[24]
Skin	*In vitro*	GDA	~90, ~90	Phenylalanine, tryptophan and DNA bases	[25]
Skin	*In vivo*	PC-GDA, PLS	90, 64	NA	[21]
Skin	*In vitro*	PCA	89, 93	Lipids and proteins	[23]
Skin	*In vivo*	PLS, LDA	91, 75	NA	[20]
Nasopharyngeal	*In vivo*	PLS	91, 95	NA	[33]
Oral	*Ex vivo*	PCA, LDA	80.7, 84.1	Nucleic acids, proteins, lipids	[28]
Laryngeal	*In vivo*	PLS-DA, LOPCV	93.3, 90.1	Nucleic acids, proteins, lipids	[32]
Esophageal	*In vivo*	PLS-DA, LOPCV	97, 97.4	Lipids, proteins	[31]
Esophageal	*Ex vivo*	PCA	90.5, 95	NA	[45]
Esophageal	*Ex vivo*	PCA, LDA	93, 95	NA	[30]
Esophageal	*Ex vivo*	PCA, LDA	71–81, 81–98	NA	[29]
Nasopharyngeal	*Ex vivo*	PCA, LDA	90.7, 100	Nucleic acids, collagen, phospholipids, phenylalanine	[27]
Gastric intestinal pre-cancer	*In vivo*	PCA, LDA, LOPCV	89.3, 92.2	Collagen, lipids, phenylalanine, proteins	[36]
Gastric	*Ex vivo*	PCA-DA	90, 90.9	NA	[46]
Gastric	*Ex vivo*	PCA, LDA	100, 94.1	RNA bases, ribose	[35]
Gastric	*Ex vivo*	PCA, LDA	100, 97	Tyrosine, adenine, coenzyme A	[34]
Breast	*Ex vivo*	Direct analysis	NA	Carotenoids, lipids, carbohydrates, proteins	[40]
Breast	*Ex vivo*	PLS-DA, LOPCV	74.2, 86.4	Collagen, amino acids	[39]
Breast	*Ex vivo*	LDA, MNLR	95.6, 96.2	Nucleic acids, collagen, lipids	[38]
Leukemia	*Ex vivo*	PCA, LDA	100, 100	Lipids, phospholipids, amino acids, carotenes	[43]
Colorectal	*Ex vivo*	PCA, LDA	97.4, 100	Nucleic acids, saccharides, proteins	[42]
Bladder	*In vitro*	PCA	98, 95	NA	[41]

GDA: General discriminant analysis; LDA: Linear discriminant analysis; LOPCV: Leave-one patient-out cross validation; MLR: Multinomial logistic regression; MNLR: Multinomial logistic regression; NA: Not available; PCA: Principal component analysis; PC-GDA Principal component with generalized discriminant analysis; PLS: Partial least square; PLS-DA: Partial least square discriminant analysis.

Table 1. Raman spectroscopy in cancer diagnosis.

1300 cm^{-1}, assigned to proteins and lipids, using PCA multivariate analysis, with high sensitivity and specificity [23]. Later, in 2012, the same author reported new results corroborating the previous study [24]. Using a Raman spectrometer attached to a fiber optic and PCA analysis, they analyzed 145 different samples of basocellular cell carcinoma, melanoma and without malignant lesions and were able to discriminate cancer and normal samples with sensitivity and specificity values over 90% [24]. In the same year, a different study was performed by Wang H et al. using HaCaT cells, melanocytes and their malignant derivatives [25]. They tested the ability of micro-Raman spectroscopy to separate different cell lines and found significant spectral differences between HaCaT cells and squamous cell carcinoma, melanocytes and melanoma cells as well as between all normal cells *versus* all tumor cells [25] (**Table 1**). The results of these *in vitro* studies are of extreme importance and can help the interpretation spectra of *in vivo* samples for cancer skin diagnosis.

5.2. Oral cavity, nasopharyngeal and laryngeal cancers

In what concerns diagnostic of oral cancers, Raman spectroscopy has been widely used in biofluids like blood, urine and saliva and using fiber optic probes for *in vivo* diagnosis (**Table 1**). The advantage of using biofluids instead of tissue is that they can be collected using noninvasive and painless methods [26].

In 2010, Feng et al. described for the first time the use of a surface-enhanced Raman spectroscopy (SERS) method for plasma analysis for nasopharyngeal cancer detection using silver nanoparticles [27]. This type of spectroscopy produces strong signals and has a low associated cost. Besides, it is suitable for use with intact tissue, biofluids and during endoscopies [26]. A total of 76 samples were used in that study and using PCA and LDA multivariate analysis, the authors found a distinct biological signature between control and cancer samples mainly due to an increase in nucleic acids, collagen, phospholipids and phenylalanine and also a decrease in amino acids and saccharide in cancer samples compared to control samples [27]. These results gave new insights about the use of surface-enhanced Raman spectroscopy for noninvasive diagnostic methods. SERS using gold nanoparticles was also applied to blood serum to diagnose oral squamous cell carcinoma [28]. Analysis of 370 Raman spectra using PCA and LDA multivariate analysis allowed to discriminate serum samples of patients with and without neoplasia with high sensitivity and specificity [28].

The use of Raman spectroscopy in oral cancer diagnosis relies mainly on fiber optic probes, and data published so far present promising results. In 2010, a study showed the applicability of fiber optic Raman probed spectroscopy to target biopsies at endoscopy [29]. This type of approaches can be useful to avoid excision of normal tissue for biopsy. In this way, there is a reduction of the burden in histopathology departments and in the number of invasive procedures for the patient. Later, in 2012, Almond et al. also tested a fiber optic Raman probe to discriminate between benign, metaplastic and neoplastic esophageal *ex vivo* tissue [30]. Results showed high sensitivity and specificity, so it is suitable to think in this approach as a new technique for clinical diagnosis although *in vivo* clinical trials are needed to confirm the accuracy of this probe. In 2015, fiber optic Raman spectroscopy was successfully applied in 48 patients during endoscopy [31]. The authors were able to simultaneously acquire both fingerprint and high wavenumber Raman spectra to extract the maximum biological information and obtained sensitivity and specificity values about 97% for the diagnosis of esophageal squamous cell carcinoma [31]. Similar results were obtained with a probe designed to diagnose laryngeal cancer [32]. Analysis of 2124 Raman spectra of 60 patients during endoscopy showed sensitivity and specificity above 90% for the identification of laryngeal cancer when combined fingerprint and high-wavenumber spectra [32]. Recently, Ming et al. [33] performed a pilot study in 79 patients with and without nasopharyngeal cancer and in post-irradiated patients. They detect a specific signature for each one of the three cohorts, which may indicate that Raman could not only be used for diagnostic purposes but also for surveillance in post-treated patients. Furthermore, the authors used a probe with only 1.8 mm, which is the smallest probe used in Raman diagnostics and is more suitable to be used in clinical endoscopies [33].

5.3. Gastric cancers

Diagnosis of gastric cancer using Raman technologies relies either on the use of fiber optic probes or on SERS (**Table 1**). In fact, SERS was applied to plasma samples to detect gastric cancer in a noninvasive way [34], similar to what was done to diagnose nasopharyngeal cancer [27]. The authors use two cohorts, with a total of 65 samples (32 patients with confirmed gastric cancer and 33 control patients). Using PCA and LDA multivariate analysis, it was observed discrimination between cancer and normal samples with sensitivity and specificity of 100 and 97%, respectively. In 2012, the same methodology was applied to discriminate gastric cancer from normal controls based on serum RNAs, also achieving high sensitivity and specificity (100 and 94.1%, respectively) [35]. SERS seems to be a useful technique to apply in routine clinical diagnosis coupled, for instance, with endoscopy. Fiber optic probes can be used for *in vivo* identification of gastric metaplasia. Lin et al. [36] coupled fingerprinting and high-wavenumber Raman spectroscopy with a fiber optic Raman probe and were able to detect, in real-time, pre-cancerous gastric lesions. They acquired 4520 spectra in real time, during gastroscopy, and by using PCA and LDA analysis, they were able to identify precancerous lesions with high sensitivity and specificity [36]. This can improve early diagnosis of neoplasia and significantly improve the efficacy of treatments.

5.4. Breast cancers

Breast cancer is the second most prevalent cancer in the world and, it is the most common cancer in women, causing more than 500,000 deaths every year [37]. According to these statistics, it is not surprising that Raman spectroscopy has been used as a diagnostic tool for this disease. Kong et al. used Raman microspectrometry to detect ductal carcinoma in tissue excised during breast-conserving surgery [38]. They developed a model that allowed to discriminate normal and cancerous tissue in approximately 17 min, with sensitivity and specificity above 95% [38]. A different approach was used by Feng et al. in 2015 [39]. Similar to what this group did for other types of cancer (see **Table 1** for detailed information), they applied SERS to saliva proteins of 97 patients and were able to discriminate between control, benign tumors and malignant tumors with sensitivities and specificities between 72.7–75.8% and 81.2–93.4%, respectively, using PLS-DA analysis [39]. These results give good perspectives for new noninvasive diagnostic tools. Depciuch et al. acquired Raman spectra of breast biopsies and did direct spectral analysis without performing multivariate analysis [40]. They reported differences in spectral regions assigned to the principal biomolecules: lipids, sugars and proteins between normal and cancerous samples [40]. All these reports give new light in understanding the molecular mechanisms involved in breast cancer.

5.5. Other cancers

Besides skin, oral, gastric and breast cancers, Raman spectroscopy techniques are also applied to diagnose other types of neoplasia (**Table 1**), mainly *in vitro*. For instance, in 2011, there was a study that used modulated Raman spectroscopy to detect the presence of human urothelial cells and bladder cancer cells after cell lines were incubated with urine [41]. The results achieved high sensitivity and specificity and, in the future, this approach may be applied in routine urine exams to detect bladder cancer or to monitor patients under treatment for bladder cancer.

Blood samples are widely used in the context of diagnosis. In the case of colorectal cancer, Lin et al. used gold nanoparticle-based SERS in blood serum samples of 83 patients to differentiate Raman spectra of healthy and disease samples [42]. This approach allowed to achieve sensitivity and specificity of 97.4 and 100%, respectively. Besides, the authors detected spectral differences between normal and cancer samples, mainly an increase in the relative amount of nucleic acids and a decrease in the amount of proteins in colorectal cancer patients, compared to control healthy subjects [42]. Serum blood samples can also be used to monitor the efficacy of treatments. In the specific case of leukemia, Gonzalez et al. used standard Raman spectroscopy and multivariate analysis to distinguish normal samples from leukemia samples with 100% of both sensitivity and specificity [43]. The authors detect some molecular changes between both groups of samples, mainly in the regions of lipids, phospholipids and β-carotene. Therefore, this Raman-PCA technique can be easily applied as a noninvasive tool to diagnosis and progression evaluation of leukemia.

In the last decade, there was an increase in the use of Raman spectroscopy in the field of cancer diagnostic and monitoring. As it was possible to see in this section, the improvement in the algorithms to process Raman signals as well as the development of new SERS techniques and fiber optic probes allowed to produce results with high sensitivity and specificity and to apply Raman-based approaches in *in vivo*, *ex vivo* and *in vitro* clinical diagnosis of several cancer types in different biological samples.

6. Raman spectroscopy for bacterial identification

Correct and in-time identification of microorganisms is crucial in clinical diagnosis. Nowadays, despite the advances of technology and methodologies of bacterial identification, most of the hospitals use bacterial culture as a standard method [47, 48]. However, these approaches are time-consuming, and sometimes it requires more than a day until the results are available. This can have serious implications to patients, mainly to those with severe infections. In this way, there are several investigations that evaluate the potential of Raman-based approaches to bacterial identification and typing and producing fast and accurate results. Raman spectra of bacteria are like a fingerprint, since it represents the molecular composition and it is specific for each sample. The following studies are examples of the use of Raman in the field of clinical microbiology.

In 2009, Willemse et al. used Raman spectroscopy to type methicillin-resistant and methicillin-sensible *Staphylococcus aureus* and compare it with traditional DNA typing methods, which are time- and labor-consuming [49]. After spectra collection and cluster analysis, results showed that Raman spectroscopy has reproducibility and discrimination ability, and all Raman clusters were in accordance with epidemiologic data of the isolates [49]. Similar results were obtained by the same group using *Escherichia coli* and *Klebsiella pneumoniae* isolates [50]. Using SpectraCell analyzer (River Diagnostics), they obtained high reproducible spectra and a discriminatory power similar to traditional DNA typing methods [50]. However, Raman spectroscopy was not able to detect ESBL-producing *E. coli* transmission events even when coupled to High-throughput MultiLocus Sequence Typing [51]. SERS is also used in the field of microbiology to identify pathogens. For instance, malaria parasite can be detected using a SERS nanoplatform [52]. Besides, it is also possible to discriminate wild-type malaria DNA from mutant malaria DNA using this technique [52]. SERS was also used in milk samples to identify *Salmonella enterica* serotype Enteritidis [53]. In this study, SERS was used with Au nanoprobes, and the results were almost 100x more sensitive than those obtained by PCR [53].

In clinical microbiology, it would be of particular interest to apply bacterial identification approaches directly to biological samples. One of the possible drawbacks of using spectroscopy in this field is that biological fluids can have complex matrixes that may mask the specific spectral signature of a given pathogen. To verify the applicability of Raman to bacterial identification in biofluids, Harz et al. used micro-Raman to directly analyze cerebrospinal fluid (CFS) of patients with bacterial meningitis [54]. Since lethality of the disease depends on the pathogen involved, time is crucial and it is necessary to properly identify the bacteria and initiate the adequate antibiotic therapy in a short period of time. In this study, the authors showed that CFS did not affect Raman spectra of bacteria, and it was possible to identify it with accuracy [54]. This corroborates the idea that Raman can be used as a diagnostic assay. In hospitals, it is important not only to identify the pathogen but also to understand the antimicrobial susceptibility profile of the microorganism in order to choose the right antibiotic to mitigate and treat the infection. Raman spectroscopy has been recently successfully used with this purpose [55]. The authors were able to discriminate Raman spectra of 67 antibiotic-susceptible strains isolated from positive blood cultures in the presence of different concentrations of antibiotic in only 5 h [55]. Further development of this technology could produce results with robustness similar to current methods used in hospitals, and therefore in the future, it can be applied to clinical diagnosis.

Raman-based approaches can also be used in the field of virology. It is possible to identify rotavirus with an accuracy above 96% using SERS fingerprinting, and the detection of the virus was possible even using a complex cellular matrix, although the results were not as sensitive as those obtained with purified samples [56]. A similar procedure was used to detect respiratory syncytial virus [57]. In this study, the authors applied SERS enzyme-catalyzed immunoassay of respiratory syncytial virus in cell lysates, and the results showed a linear correlation between the intensity of spectra and the amount of virus with a detection limit lower than traditional methods [57].

The studies discussed in this section suggest that Raman-based approaches are a good alternative for real-time clinical usage, since they are easy to use, fast (it is possible to have results within 45 min after positive culture), reliable and can be easily applied to a wide variety of microorganisms, since specific dyes or labels are no need.

7. Raman spectroscopy for the diagnosis of dementia

Alzheimer's disease is the most prevalent type of neurodegenerative dementia in the elderly population worldwide. The key for efficiently and accurately diagnosing dementia occurs during the early stages. Scientists struggle to find a blood-based method to perform an accurate diagnosis of this type of dementia, which affects 35.6 million people worldwide [58].

In this context, Raman microspectroscopy was already used to analyze serum of Alzheimer's disease patients, patients with other dementias and healthy controls. Data were analyzed with multivariate statistics for the differential identification of Alzheimer's disease patients. The study was a proof of concept; however, it proves that Raman microspectroscopy and artificial neural network classification were able to differentiate patients with more than 95% sensitivity and specificity, demonstrating that Raman has potential to become a blood-based tests that may help clinical assessments for effective and accurate differential diagnosis of Alzheimer's disease [59].

8. Conclusions and outlook

In the last years, there was a significant improvement in instrumentation of Raman spectroscopy, leading to an enhancement of sensitivity and accuracy of this approach. In this way, Raman has been broadly used in the context of medical investigation, mainly in the oncologic field. Coupled to adequate chemometric analysis tools, Raman spectroscopy, mainly SERS, was already applied to successfully diagnose several types of cancer, infections and preliminary results are also promising in the context of Alzheimer's disease. It is expected that in the next years, Raman spectroscopy can be routinely used in hospitals to *in vivo* diagnose and surveillance of several diseases, producing fast and accurate results, decreasing waiting times and, in this way, improving efficacy of treatments.

Author details

Alexandra Nunes* and Sandra Magalhães

*Address all correspondence to: alexandranunes@ua.pt

Department of Medical Sciences, iBiMED – Institute of Biomedicine, University of Aveiro, Aveiro, Portugal

References

[1] Kong K, Kendall C, Stone N, Notingher I. Raman spectroscopy for medical diagnostics — From in-vitro biofluid assays to in-vivo cancer detection. Advanced Drug Delivery Reviews. 2015;**89**:121-134. DOI: 10.1016/J.ADDR.2015.03.009

[2] Schütz M, Steinigeweg D, Salehi M, Kömpe K, Schlücker S. Hydrophilically stabilized gold nanostars as SERS labels for tissue imaging of the tumor suppressor p63 by immuno-SERS microscopy. Chemical Communications. 2011;**47**:4216-4218

[3] Delhaye M, Dhamelincourt P. Raman microprobe and microscope with laser excitation. Journal of Raman Specroscopy. 1975;**3**:33-43

[4] De Grauw CJ, Otto C, Greve J. Line-scan Raman microspectrometry for biological applications. Applied Spectroscopy. 1997;**51**:1607-1612

[5] Rowlands CJ, Varma S, Perkins W, Leach I, Williams H, Notingher I. Rapid acquisition of Raman spectral maps through minimal sampling: Applications in tissue imaging. Journal of Biophotonics. 2012;**5**:220-229. DOI: 10.1002/jbio.201100098

[6] Kong K, Rowlands CJ, Varma S, Perkins W, Leach IH, Koloydenko AA, et al. Increasing the speed of tumour diagnosis during surgery with selective scanning Raman microscopy. Journal of Molecular Structure. 2014;**1073**:58-65

[7] Matousek P, Clark IP, Draper ERC, Morris MD, Goodship AE, Everall N, et al. Subsurface probing in diffusely scattering media using spatially offset Raman spectroscopy. Applied Spectroscopy. 2005;**59**:393-400

[8] Matousek P, Parker AW. Bulk Raman analysis of pharmaceutical tablets. Applied Spectroscopy. 2006;**60**:1353-1357

[9] Saar BG, Freudiger CW, Reichman J, Stanley CM, Holtom GR, Xie XS. Video-rate molecular imaging in vivo with stimulated Raman scattering. Science (80-). 2010;**330**:1368-1370

[10] Evans CL, Potma EO, Puoris' haag M, Côté D, Lin CP, Xie XS. Chemical imaging of tissue in vivo with video-rate coherent anti-stokes Raman scattering microscopy. Proceedings of the National Academy of Sciences of the United States of America. 2005; **102**:16807-16812

[11] Matousek P, Stone N. Recent advances in the development of Raman spectroscopy for deep non-invasive medical diagnosis. Journal of Biophotonics. 2013;**6**:7-19. DOI: 10.1002/jbio.201200141

[12] Afseth NK, Segtnan VH, Wold JP. Raman spectra of biological samples: A study of preprocessing methods. Applied Spectroscopy. 2006;**60**:1358-1367

[13] Wentzell PD, Montoto LV. Comparison of principal components regression and partial least squares regression through generic simulations of complex mixtures. Chemometrics and Intelligent Laboratory Systems. 2003;**65**:257-279

[14] Ellis DI, Goodacre R. Metabolic fingerprinting in disease diagnosis: Biomedical applications of infrared and Raman spectroscopy. Analyst. 2006;**131**:875-885. DOI: 10.1039/b602376m

[15] Cancer Research UK. Worldwide Cancer Statistics n.d. Available from: http://www.cancerresearchuk.org/health-professional/cancer-statistics/worldwide-cancer [Accessed: Aug 21, 2017]

[16] Ferlay J, Soerjomataram I, Ervik M, Dikshit R, Eser S, Mathers C, et al. GLOBOCAN 2012: Estimated cancer incidence, mortality and prevalence worldwide in 2012: IARC CancerBase No. 11 International Agency for Research on Cancer 2014. Available from: http://globocan.iarc.fr [Accessed: Mar 16, 2016]

[17] Almond LM, Hutchings J, Shepherd N, Barr H, Stone N, Kendall C. Raman spectroscopy: A potential tool for early objective diagnosis of neoplasia in the oesophagus. Journal of Biophotonics. 2011;**4**:685-695. DOI: 10.1002/jbio.201100041

[18] Pence I, Mahadevan-Jansen A. Clinical instrumentation and applications of Raman spectroscopy. Chemical Society Reviews. 2016;**45**:1958-1979. DOI: 10.1039/C5CS00581G

[19] Kourkoumelis N, Balatsoukas I, Moulia V, Elka A, Gaitanis G, Bassukas ID. Advances in the in vivo Raman spectroscopy of malignant skin tumors using portable instrumentation. International Journal of Molecular Sciences. 2015;**16**:14554-14570. DOI: 10.3390/ijms160714554

[20] Zhao J, Lui H, McLean DI, Zeng H. Real-time raman spectroscopy for non-invasive skin cancer detection - preliminary results. Vol. 2008, 30th Annual International Conference of the IEEE Engineering in Medicine and Biology Society, Vancouver, BC. 2008; pp. 3107-3109. DOI: 10.1109/IEMBS.2008.4649861

[21] Lui H, Zhao J, McLean D, Zeng H. Real-time Raman spectroscopy for in vivo skin cancer diagnosis. Cancer Research. 2012;**72**:2491-2500. DOI: 10.1158/0008-5472.CAN-11-4061

[22] Zhao J, Lui H, Kalia S, Zeng H. Real-time Raman spectroscopy for automatic in vivo skin cancer detection: An independent validation. Analytical and Bioanalytical Chemistry. 2015;**407**:8373-8379. DOI: 10.1007/s00216-015-8914-9

[23] Bodanese B, Silveira LJ, Albertini R, Zangaro RA, Pacheco MTT. Differentiating normal and basal cell carcinoma human skin tissues in vitro using dispersive Raman spectroscopy: A comparison between principal components analysis and simplified biochemical models. Photomedicine and Laser Surgery. 2010;**28**(Suppl 1):S119-S127. DOI: 10.1089/pho.2009.2565

[24] Bodanese B, Silveira FL, Zangaro RA, Pacheco MTT, Pasqualucci CA, Silveira LJ. Discrimination of basal cell carcinoma and melanoma from normal skin biopsies in vitro through Raman spectroscopy and principal component analysis. Photomedicine and Laser Surgery. 2012;**30**(7):381. DOI: 10.1089/pho.2011.3191

[25] Wang H, Tsai T-H, Zhao J, Lee AMD, Lo BKK, Yu M, et al. Differentiation of HaCaT cell and melanocyte from their malignant counterparts using micro-Raman spectroscopy

guided by confocal imaging. Photodermatology, Photoimmunology & Photomedicine. 2012;**28**:147-152. DOI: 10.1111/j.1600-0781.2012.00660.x

[26] Austin LA, Osseiran S, Evans CL. Raman technologies in cancer diagnostics. Analyst. 2016;**141**:476-503. DOI: 10.1039/c5an01786f

[27] Feng S, Chen R, Lin J, Pan J, Chen G, Li Y, et al. Nasopharyngeal cancer detection based on blood plasma surface-enhanced Raman spectroscopy and multivariate analysis. Biosensors & Bioelectronics. 2010;**25**:2414-2419. DOI: 10.1016/j.bios.2010.03.033

[28] Tan Y, Yan B, Xue L, Li Y, Luo X, Ji P. Surface-enhanced Raman spectroscopy of blood serum based on gold nanoparticles for the diagnosis of the oral squamous cell carcinoma. Lipids in Health and Disease. 2017;**16**:73. DOI: 10.1186/s12944-017-0465-y

[29] Kendall C, Day J, Hutchings J, Smith B, Shepherd N, Barr H, et al. Evaluation of Raman probe for oesophageal cancer diagnostics. Analyst. 2010;**135**:3038-3041. DOI: 10.1039/c0an00536c

[30] Almond LM, Hutchings J, Kendall C, Day JCC, Stevens OAC, Lloyd GR, et al. Assessment of a custom-built Raman spectroscopic probe for diagnosis of early oesophageal neoplasia. Journal of Biomedical Optics. 2012;**17**:81421. DOI: 10.1117/1.JBO.17.8.081421

[31] Wang J, Lin K, Zheng W, Ho KY, Teh M, Yeoh KG, et al. Simultaneous fingerprint and high-wavenumber fiber-optic Raman spectroscopy improves in vivo diagnosis of esophageal squamous cell carcinoma at endoscopy. Scientific Reports. 2015;**5**:12957. DOI: 10.1038/srep12957

[32] Lin K, Zheng W, Lim CM, Huang Z. Real-time in vivo diagnosis of laryngeal carcinoma with rapid fiber-optic Raman spectroscopy. Biomedical Optics Express. 2016;**7**:3705-3715. DOI: 10.1364/BOE.7.003705

[33] Ming LC, Gangodu NR, Loh T, Zheng W, Wang J, Lin K, et al. Real time near-infrared Raman spectroscopy for the diagnosis of nasopharyngeal cancer. Oncotarget. 2017;**8**:49443-49450. DOI: 10.18632/oncotarget.17703

[34] Feng S, Chen R, Lin J, Pan J, Wu Y, Li Y, et al. Gastric cancer detection based on blood plasma surface-enhanced Raman spectroscopy excited by polarized laser light. Biosensors & Bioelectronics. 2011;**26**:3167-3174. DOI: 10.1016/j.bios.2010.12.020

[35] Chen Y, Chen G, Zheng X, He C, Feng S, Chen Y, et al. Discrimination of gastric cancer from normal by serum RNA based on surface-enhanced Raman spectroscopy (SERS) and multivariate analysis. Medical Physics. 2012;**39**:5664-5668. DOI: 10.1118/1.4747269

[36] Lin K, Wang J, Zheng W, Ho KY, Teh M, Yeoh KG, et al. Rapid fiber-optic Raman spectroscopy for real-time in vivo detection of gastric intestinal metaplasia during clinical gastroscopy. Cancer Prevention Research (Philadelphia, PA). 2016;**9**:476-483. DOI: 10.1158/1940-6207.CAPR-15-0213

[37] International Agency for Research on Cancer. GLOBOCAN 2012: Estimated Cancer Incidence, Mortality and Prevalence Worldwide in 2012. n.d. Available from: http://globocan.iarc.fr/Pages/fact_sheets_cancer.aspx [Accessed: Aug 24, 2017]

[38] Kong K, Zaabar F, Rakha E, Ellis I, Koloydenko A, Notingher I. Towards intra-operative diagnosis of tumours during breast conserving surgery by selective-sampling Raman micro-spectroscopy. Physics in Medicine and Biology. 2014;**59**:6141-6152. DOI: 10.1088/0031-9155/59/20/6141

[39] Feng S, Huang S, Lin D, Chen G, Xu Y, Li Y, et al. Surface-enhanced Raman spectroscopy of saliva proteins for the noninvasive differentiation of benign and malignant breast tumors. International Journal of Nanomedicine. 2015;**10**:537-547. DOI: 10.2147/IJN.S71811

[40] Depciuch J, Kaznowska E, Zawlik I, Wojnarowska R, Cholewa M, Heraud P, et al. Application of Raman spectroscopy and infrared spectroscopy in the identification of breast cancer. Applied Spectroscopy. 2016;**70**:251-263. DOI: 10.1177/0003702815620127

[41] Canetta E, Mazilu M, De Luca AC, Carruthers AE, Dholakia K, Neilson S, et al. Modulated Raman spectroscopy for enhanced identification of bladder tumor cells in urine samples. Journal of Biomedical Optics. 2011;**16**:37002. DOI: 10.1117/1.3556722

[42] Lin D, Feng S, Pan J, Chen Y, Lin J, Chen G, et al. Colorectal cancer detection by gold nanoparticle based surface-enhanced Raman spectroscopy of blood serum and statistical analysis. Optics Express. 2011;**19**:13565-13577

[43] Gonzalez-Solis JL, Martinez-Espinosa JC, Salgado-Roman JM, Palomares-Anda P. Monitoring of chemotherapy leukemia treatment using Raman spectroscopy and principal component analysis. Lasers in Medical Science. 2014;**29**:1241-1249. DOI: 10.1007/s10103-013-1515-y

[44] Takamori S, Kong K, Varma S, Leach I, Williams HC, Notingher I. Optimization of multimodal spectral imaging for assessment of resection margins during Mohs micrographic surgery for basal cell carcinoma. Biomedical Optics Express. 2015;**6**:98-111. DOI: 10.1364/BOE.6.000098

[45] Zhou X, Chen G-Y, Zhang J-M, Fan C-Z, Liang E-J, Guo Y, et al. Research on early diagnosis of esophageal cancer by Raman spectroscopy of human hemoglobin. Guang Pu Xue Yu Guang Pu Fen Xi. 2013;**33**:2989-2992

[46] Wang W, Pan Z, Tang W, Li Y, Fan C. Research on early diagnosis of gastric cancer by the surface enhanced Raman spectroscopy of human hemoglobin. Guang Pu Xue Yu Guang Pu Fen Xi. 2015;**35**:3402-3405

[47] Pahlow S, Meisel S, Cialla-May D, Weber K, Rösch P, Popp J. Isolation and identification of bacteria by means of Raman spectroscopy. Advanced Drug Delivery Reviews. 2015;**89**:105-120. DOI: 10.1016/j.addr.2015.04.006

[48] Krafft C, Popp J. The many facets of Raman spectroscopy for biomedical analysis. Analytical and Bioanalytical Chemistry. 2015;**407**:699-717. DOI: 10.1007/s00216-014-8311-9

[49] Willemse-Erix DFM, Scholtes-Timmerman MJ, Jachtenberg J-W, van Leeuwen WB, Horst-Kreft D, Bakker Schut TC, et al. Optical fingerprinting in bacterial epidemiology: Raman spectroscopy as a real-time typing method. Journal of Clinical Microbiology. 2009;**47**:652-659. DOI: 10.1128/JCM.01900-08

[50] Willemse-Erix D, Bakker-Schut T, Slagboom-Bax F, Jachtenberg J, Lemmens-den Toom N, Papagiannitsis CC, et al. Rapid typing of extended-spectrum beta-lactamase- and carbapenemase-producing *Escherichia coli* and *Klebsiella pneumoniae* isolates by use of SpectraCell RA. Journal of Clinical Microbiology. 2012;**50**:1370-1375. DOI: 10.1128/JCM.05423-11

[51] Voor In't Holt AF, Wattel AA, Boers SA, Jansen R, Hays JP, Goessens WHF, et al. Detection of healthcare-related extended-Spectrum Beta-lactamase-producing Escherichia Coli transmission events using combined genetic and phenotypic epidemiology. PLoS One 2016;**11**:e0160156. DOI: 10.1371/journal.pone.0160156

[52] Ngo HT, Gandra N, Fales AM, Taylor SM, Vo-Dinh T, Sensitive DNA. Detection and SNP discrimination using ultrabright SERS nanorattles and magnetic beads for malaria diagnostics. Biosensors & Bioelectronics. 2016;**81**:8-14. DOI: 10.1016/j.bios.2016.01.073

[53] Draz MS, Lu X. Development of a loop mediated isothermal amplification (LAMP) – surface enhanced Raman spectroscopy (SERS) assay for the detection of *Salmonella enterica* serotype Enteritidis. Theranostics. 2016;**6**:522-532. DOI: 10.7150/thno.14391

[54] Harz M, Kiehntopf M, Stockel S, Rosch P, Straube E, Deufel T, et al. Direct analysis of clinical relevant single bacterial cells from cerebrospinal fluid during bacterial meningitis by means of micro-Raman spectroscopy. Journal of Biophotonics. 2009;**2**:70-80. DOI: 10.1002/jbio.200810068

[55] Dekter HE, Orelio CC, Morsink MC, Tektas S, Vis B, Te Witt R, et al. Antimicrobial susceptibility testing of gram-positive and -negative bacterial isolates directly from spiked blood culture media with Raman spectroscopy. European Journal of Clinical Microbiology & Infectious Diseases. 2017;**36**:81-89. DOI: 10.1007/s10096-016-2773-y

[56] Driskell JD, Zhu Y, Kirkwood CD, Zhao Y, Dluhy RA, Tripp RA. Rapid and sensitive detection of rotavirus molecular signatures using surface enhanced Raman spectroscopy. PLoS One. 2010;**5**:e10222. DOI: 10.1371/journal.pone.0010222

[57] Zhan L, Zhen SJ, Wan XY, Gao PF, Huang CZ. A sensitive surface-enhanced Raman scattering enzyme-catalyzed immunoassay of respiratory syncytial virus. Talanta. 2016;**148**:308-312. DOI: 10.1016/j.talanta.2015.10.081

[58] Brookmeyer R, Evans DA, Hebert L, Langa KM, Heeringa SG, Plassman BL, et al. National estimates of the prevalence of Alzheimer's disease in the United States. Alzheimer's Dement. 2011;**7**:61-73. DOI: 10.1016/J.JALZ.2010.11.007

[59] Ryzhikova E, Kazakov O, Halamkova L, Celmins D, Malone P, Molho E, et al. Raman spectroscopy of blood serum for Alzheimer's disease diagnostics: Specificity relative to other types of dementia. Journal of Biophotonics. 2015;**8**:584-596

Chapter 14

Raman Spectroscopy for In Vivo Medical Diagnosis

Miguel Ghebré Ramírez-Elías and
Francisco Javier González

Additional information is available at the end of the chapter

http://dx.doi.org/10.5772/intechopen.72933

Abstract

Raman spectroscopy is a noninvasive optical technique that can be used as an aid in diagnosing certain diseases and as an alternative to more invasive diagnostic techniques such as the biopsy. Due to these characteristics, Raman spectroscopy is also known as an optical biopsy technique. The success of Raman spectroscopy in biomedical applications is based on the fact that the molecular composition of healthy tissue is different from diseased tissue; also, several disease biomarkers can be identified in Raman spectra, which can be used to diagnose or monitor the progress of certain medical conditions. This chapter outlines an overview of the use of Raman spectroscopy for *in vivo* medical diagnostics and demonstrates the potential of this technique to address biomedical issues related to human health.

Keywords: Raman spectroscopy, biomedical, chemometrics

1. Introduction

Raman spectroscopy is based on the inelastic scattering of photons, also known as Raman effect, discovered by C. V. Raman in 1928 [1]. When a sample is illuminated with a light source, the incoming photons are absorbed or scattered. If absorbed, the photon energy is transferred to the molecules, whereas if a photon is scattered and the energy is conserved, it is called elastic scattering. However, a small portion of scattered photons (1 in every 10 billion photons) can be scattered inelastically, which means a slight change in the photon energy. This small energy difference between the incident and the scattered photon is the Raman effect. Raman spectroscopy has several advantages for biomedical applications, including being nondestructive and relatively fast to acquire, and provides information at the molecular level. Additionally, water produces weak Raman scattering, which means the presence of water in the sample does not interfere with the spectrum that is being analyzed. The main disadvantages of Raman spectroscopy include the extremely weak Raman signal and the presence of undesirable noise sources such as the intense fluorescence background present in biological samples.

2. Instrumentation

A Raman spectrometer useful for *in vivo* measurements should be an integrated system that can provide real-time spectral acquisition and analysis [1]. A Raman system for *in vivo* measurements includes a light source, sample light delivery and collection, spectrograph with detector, and the computer interface. Lasers are the excitation source for Raman spectroscopy due to the fact they can provide sufficient power to the sample in order to detect Raman spectra in a reasonable integration time. However, it is necessary to consider important issues such as power, integration time, and wavelength of the laser to optimize the Raman system for in vivo biomedical applications. For example, to avoid tissue damage, the maximum permissible exposure (defined by ANSI) and temperature increase must be considered. Therefore, a correct laser power selection depends on achieving a good signal to noise and to minimize tissue damage. In biological tissue, the fluorophores can generate signals that mask or overwhelm the weak Raman signal, and to avoid fluorescence background, multiple approaches have been proposed including the excitation in the near infrared (NIR) [2]. It is known that most biological fluorophores have no peak emission in this region of the spectrum, which results in lower fluorescence background compared to visible or UV excitation. Due to these advantages, most of the Raman spectroscopy systems for skin diagnosis use a 785-nm diode laser as the excitation source, since it provides low-cost light source that generates low fluorescence and can penetrate deep into human tissue. In sample light delivery and collection, the most used method for clinical applications is optical fibers. The Raman fiber probe design varies depending on the clinical application. In the case of Raman spectroscopy of the skin, the probe consists of a single central delivery fiber surrounded by several collection fibers. The selection of a suitable detection system is an important issue for Raman spectroscopy. The

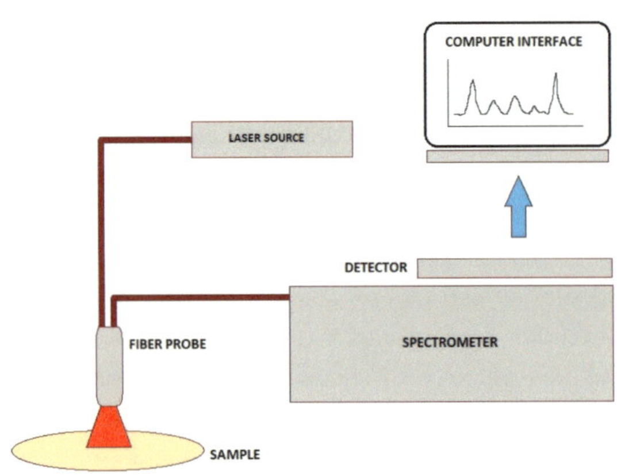

Figure 1. Schematic of a typical Raman system for *in vivo* biomedical applications.

typical Raman detection system used for biomedical applications consists of a spectrograph attached to a cooled charge coupled device (CCD). Most CCDs use a thermoelectric (TE) system to cool the detector down to −70°C in order to reduce thermal noise. The detection system also requires a spectrograph coupled to the Raman probe and to the CCD. It is recommended the spectrograph have a spectral resolution of 8–10 cm^{-1} in order to provide detailed information of biological Raman bands. The spectral resolution depends on spectrograph optical parameters, the diffraction grating, and the CCD pixel size. A schematic of the typical arrangement of these components is shown in **Figure 1**.

3. Data preprocessing

A big issue in biological Raman spectroscopy is the presence of undesirable background elements related to different sources such as intrinsic fluorescence, noise introduced by the equipment used, and the noise generated by external sources.

3.1. Smoothing and denoising

The main sources of noise present in Raman spectra from biological samples are the shot noise, fluorescence background, flicker noise, dark current, and thermal noise. One alternative to reduce the thermal noise and dark signal is the use of a Raman system with high quality, thermoelectric cooled spectrometers. In Raman spectra, most of the time, the shot noise is the predominant noise associated with the particle nature of light. The approximate shot noise associated with measurement of n counts is $n^{1/2}$. Thus the signal to noise ratio (S/N) can be improved incrementing the number of counts n. In other words, S/N can be improved by increasing averaging time due to the fact the signal increases proportionally with time. There are several multitude noise removal techniques that can be applied to Raman spectra. Smoothing is often employed for the removal of high-frequency components from Raman spectra, based on the fact that noise appears as high-frequency fluctuations, whereas signals are assumed to be low frequency. One smoothing technique is Fourier filtering [3]. In this technique, the higher frequency fluctuations, which are considered only noise, can be removed and the lower frequency ones can be used to reconstruct Raman spectra without noise. One drawback of this method is that the removal of the higher frequency noise may often introduce artifacts and distortion in Raman spectra. A commonly used smoothing technique is Savitzky-Golay (SG) filtering. The SG filter is a moving window–based local polynomial fitting procedure [4]. As the moving window size increases, some of the Raman bands may disappear. Therefore, it is very important to choose the appropriate parameters such as the polynomial order and the moving window size to avoid loss of Raman data. Other smoothing methods are locally weighted scatter plot smoothing (LOWESS) [5] and wavelet filtering [6] whereby the spectrum is decomposed using the discrete wavelet transform in order to isolate the noise by localizing it in space and frequency. Once it is isolated, it can be set to zero and the inverse wavelet transform is used to reconstruct the data. In all the mentioned methods, parameters have to be chosen carefully to avoid the important Raman bands being eliminated during smoothing.

3.2. Background removal

As mentioned is the last section, one noise source in biological Raman spectra is the fluorescence background. This intrinsic fluorescence emission is several orders of magnitude greater than the Raman scattering intensity of biological tissues; therefore, fluorescence appears as a strong band that obscures Raman signals and must be removed in order to perform the analysis on the Raman spectra. Background elimination has been performed using two approaches: experimental and computational. The experimental methods are related to changes in the instrumentation and those include shifted excitation [7], photo bleaching [8], and time gating [9]. One drawback of these methods is the relatively complex instrumentation, the long acquisition times, and alterations in the sample that could make the analysis of biological samples difficult. On the other hand, background removing by using computational approaches has the advantages such as easy to implement, inexpensive, and fast. Such methods include polynomial fitting [10–12], Fourier transform [13], wavelet transform [13], first- and second-order differentiation [14], multiplicative signal correction [15], linear programming [16], geometric approach [17], asymmetric least squares [18], methods based on iterative reweighted quantile regression [19], iterative exponential smoothing [20], and morphology operators [21, 22]. However, the most used method is polynomial fitting due to simplicity. In this method, a polynomial is fitted and subsequently subtracted from the Raman spectrum to eliminate background effects. The selection of polynomial order is extremely important, because a higher order polynomial fitting may consider Raman bands as background and may be affected by high frequency noise. To solve this issue, some modified polynomial fitting methods were proposed. **Figure 2** shows the Raman spectra of *in vivo* mouse skin tissue with and without fluorescence removal using the polynomial fitting method.

For example, the algorithm proposed by Zhao et al. [11] also known as the Vancouver Raman algorithm (VRA) is widely used for baseline correction in biomedical applications due to effectiveness and simplicity. The main advantage of this method is that it accounts for noise effects and Raman signal contribution.

3.3. Normalization

Raman spectra from the same sample could have different intensity levels if they were acquired at different times or under different experimental parameters such as changes in laser power levels. Normalization process deals with these differences in intensity levels by making that the intensity of a specific Raman band of the same material is the same or similar possible in all the spectra recorded under the same experimental parameters. One approach is the normalization to area. In this method, the intensity at each frequency in the spectrum is divided by the square root of the sum of the squares of all intensities. This normalization is useful when the spectra do not share a common band and it is better to normalize the spectra so that the total area under the spectrum is 1.0. This method has the advantage that is not dependent on any single band but one disadvantage is that the background can contribute to the normalization [1]. Another approach is the peak normalization, which uses intensity corresponding to the central frequency of a particular Raman band as reference (internal or external). The 1660 cm^{-1} (amide I) and the 1450 cm^{-1} band (C—H vibrations) are commonly

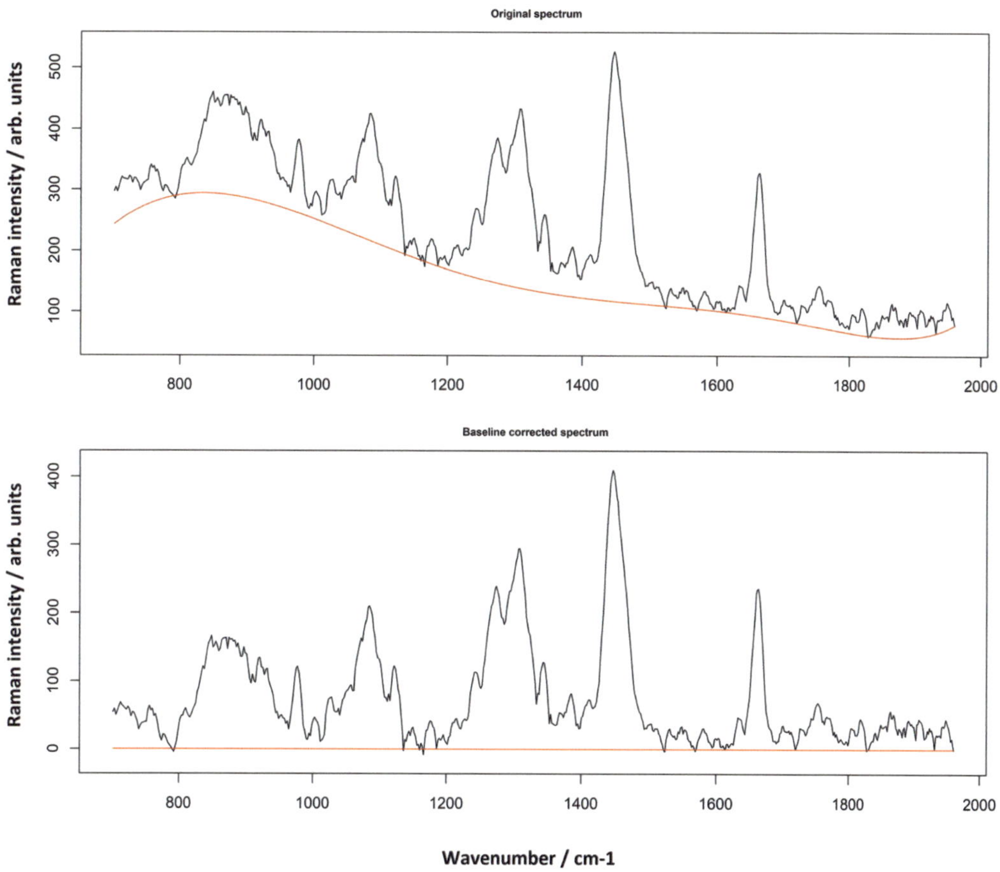

Figure 2. In vivo Raman spectra of skin with fluorescence (top) and without fluorescence (bottom) using the polynomial fitting method.

used as reference due to their intensities that are not significantly affected by other changes in the sample [23]. This method assumes the reference does not change from one spectrum to other and therefore is not suitable when the nature of the samples could lead to a shift in the band position.

4. Chemometrics

Chemometrics uses mathematical and statistical methods to provide chemical/physical information from chemical data or for the subject under consideration, spectroscopic data. In order to identify components in a sample, one possibility is to use individual bands, but this approach is not the best option because one band is not specific for a molecule, as many molecules have the a band in the same localization. A more precise identification is to use multiple bands or the complete spectrum. Such approach considers each point in a spectrum as a variable and spectroscopic data can be displayed as a matrix where columns represent the variables (Raman shift or wavenumber) and the rows represent observations (Raman spectra). To analyze data with more than one variable, multivariate data analysis is used. There are many multivariate data analysis techniques available and their correct use depends on the objective of the analysis. The objective can be data description or exploratory analysis, discrimination, classification, clustering, regression, and prediction. Also, the data analysis methods can be divided into unsupervised and supervised methods. The supervised methods are used when there is no a priori knowledge available and are very useful to find hidden structures in the unlabeled data and sometimes are used as a first step to supervised methods. Hierarchical cluster analysis (HCA) and principal component analysis (PCA) are examples of unsupervised methods. On the other hand, supervised methods need a priori information such as class labels and the analysis involves the use of a training data set to find the patterns in the data and later validate the model using a test set. One example of the supervised method is partial least squares (PLS).

4.1. Principal component analysis (PCA)

Principal component analysis (PCA) is an unsupervised method often used to reduce the number of variables [24] and exploratory analysis of data. PCA is based on the eigenvector decomposition of the covariance matrix of the spectra matrix into eigenvectors and eigenvalues. The eigenvectors (or principal components) are orthogonal along n-dimensional axes and are ordered by decreasing value of each associated eigenvalue. This means the principal components are independent of each other and uncorrelated, as opposed to the original ones, which may be correlated. Also, their decreasing order means that the first principal component explains the maximum amount of variance of the original data, and the second one explains more variance than the third, and so on. The original data can be considered as an $M \times N$ matrix of M spectra sampled at N wavenumbers. Applying the PCA to this matrix, PCA yields three results: N principal components, an $N \times N$ matrix containing the coefficients for the transformation between the original data and the principal components, and N eigenvalues describing the importance of the corresponding principal components. The original N experimental spectra are transformed into a new set of N 'synthetic' spectra called principal components. In summary, one advantage of PCA is that by evaluating the relative importance of the consecutive principal components, it is possible to reduce the dimension of the original dataset by finding a smaller collection of variables that explain the highest amount of variance. Additionally, because changes in Raman signal are uncorrelated with the noise in the spectra, the random noise and the significant spectral changes will be separated into different principal components. Therefore, many principal components can be discarded, removing noise without losing useful information from Raman signal.

4.2. Partial least squares (PLS)

PLS is one of the most widely used multivariate data analysis techniques along with vibrational spectroscopy to estimate and quantify components in a sample [25]. As a supervised method, the concentrations of all constituents in the calibration samples are known. As with PCA, the noise observed in the spectra is isolated into separate latent variables (LVs), which are left out of the calibration, improving prediction precision, and nonlinear relationships between the properties of interest and intensity can be accommodated in a PLS model by including multiple LVs.

4.3. Classification and clustering models

Several data analysis methods are focused on looking for differences between the spectra so that groups of spectra can be identified and classified. The most common methods used in biomedical Raman spectroscopy are k-nearest neighbors (KNN), hierarchical cluster analysis (HCA), artificial neural networks (ANN), discriminant analysis (DA), and support vector machines (SVM). The KNN method compares all spectra in the dataset through the use of the metrics of similarity between spectra like the Euclidean distance. This method has been used in combination with PCA and Raman spectroscopy for the diagnosis of colon cancer [26]. HCA uses a variety of multivariate distance calculations such as Euclidean and Mahalanobis metrics to identify similar spectra and is one of the used methods in Raman and IR imaging [27]. Similarly, artificial neural networks can be used to identify clusters or to find patterns in complex data. ANNs are computational models inspired by the functionality and structure of the central nervous system and the networks consist of interconnected group of nodes or neurons, which have different functions such data input, output, storage, or forwarding. The layout of ANN is composed of a number of layers and a number of neurons per layer. The use of ANN in the data analysis of blood serum Raman spectra allows for the differentiation between patients with Alzheimer's disease, other types of dementia, and healthy individuals [28]. DA is a supervised data analysis technique, which requires a priori knowledge of each sample group membership. DA computes a set of discriminant functions based on linear combinations of variables that maximize the variance between groups and minimize the variance within groups according to Fisher's criterion. Sometimes it is very useful to combine both PCA and LDA approaches (called PC-LDA model), which improves the efficiency of classification as it automatically finds the most diagnostically significant features [29–31]. SVMs are kernel-based algorithms that transform data into a high-dimensional space and construct a hyperplane that maximizes the distance to the nearest data point of any of the input classes. Raman spectroscopy and SVM have been used as methods for cancer screening [32].

5. Applications

The importance of the *in vivo* Raman spectroscopy is the number of potential biomedical applications. One application is the *in vivo* noninvasive diagnosis, and most research papers focus on cancer and skin diagnosis. In this section, a wide overview over applications in cancer and skin diagnosis is given, with a focus on developments over the past 5 years.

5.1. Cancer diagnosis

One of the most common clinical targets under investigation with Raman spectroscopy is cancer due to the possibility to measure biological samples minimally invasive, *in vivo,* and without labeling. One important step that enables the introduction of *in vivo* measurements of cancer in hollow organs is the development of fiber-optic Raman probes that can be implemented during endoscopy [33].

5.1.1. Lung cancer

Short et al. designed a Raman probe for *in vivo* detection of lung cancer during autofluorescence bronchoscopy [34], and they demonstrated the potential of Raman for *in vivo* diagnosis of lung cancer by reducing the false positives of autofluorescence bronchoscopy [35].

5.1.2. Gastrointestinal cancer

In 2014, Bergholt et al. [36] performed an *in vivo* diagnostic trial to classify dysplasia in Barrett's esophagus (BE). They reported a diagnostic sensitivity of 87.0% and a specificity of 84.7%, which demonstrate that real-time Raman spectroscopy can be performed prospectively in screening of the patients with suspicious BE *in vivo*. In a study conducted on mice with colon cancer, Taketani et al. [37] identified alterations in its molecular composition of lipids and collagen type I, along with its advancement. The tumor lesion was discriminated from normal tissues of the control mouse with an accuracy of 86.8%. Stomach cancer diagnosis has been another application in biomedical Raman spectroscopy [38]. Bergholt et al. have also reported a statistically robust study where 450 patients underwent Raman endoscopy for identifying gastric precancer based on PLS-DA [39]. The same group used *in vivo* Raman spectroscopy to characterize the properties of normal colorectal tissues and to assess distinctive biomolecular variations of different anatomical locations in the colorectum for cancer diagnosis. They conclude that interanatomical Raman spectral variability of normal colorectal tissue is subtle compared to cancer tissue. Their PLS-DA model provided a diagnostic accuracy of 88.8%, a sensitivity of 93.9% and a specificity of 88.3% for colorectal cancer detection [40].

5.1.3. Oral cancer

In a study conducted by Guze et al. [41], Raman spectra of oral diseases from 18 patients were classified into a benign or malignant category using PCA-LDA, and the method provided 100% specificity with 77% sensitivity. Murali Krishna et al. reported the potential for Raman spectroscopy to identify early changes in oral mucosa and the efficacy of this approach in oral cancer applications [42]. Comparing noncancer locations in a smoking and nonsmoking population demonstrated prediction accuracies from 75 to 98%. Another group reported the discrimination of normal oral tissue from different lesion categories with accuracies ranging from 82 to 89% [43]. Recently, Lin et al. [44] reported the utility of fiber-optic–based Raman spectroscopy for real-time *in vivo* diagnosis of nasopharyngeal carcinoma (NPC) at endoscopy. A total of 3731 *in vivo* Raman spectra were acquired in real time from 95 subjects. Raman spectra differ significantly between normal and cancerous nasopharyngeal tissues. Using PCA-LDA, their method provided a diagnostic accuracy of 93.1% (sensitivity of 93.6%; specificity of 92.6%) for nasopharyngeal cancer identification. The Raman spectra of the diseased tissue include oral squamous cell carcinoma (OSCC), oral submucosa fibrosis (OSMF), and oral leukoplakia (OLK). The study achieved good diagnostic accuracy for the three diseased groups and the normal group, which were 89, 85, 82, and 85%, respectively.

5.1.4. Skin cancer

A clinical study of 453 patients to investigate different types of skin cancer was published in 2012 by Lui et al. [45]. The instrument used by the authors allowed an acquisition time of approximately 1s and the software preprocessed the spectra immediately, which allowed to investigate skin lesions in real time. Benign and malignant skin lesions including melanomas, basal cell carcinomas, squamous cell carcinomas, actinic keratoses, atypical nevi, melanocytic nevi, blue nevi, and seborrheic keratosis were investigated and discriminated by multivariate analysis tools with sensitivities between 95 and 99%. Lim et al. determined the diagnostic capability of a multimodal spectral diagnosis for *in vivo* noninvasive disease diagnosis of melanoma and nonmelanoma skin cancers [46]. They acquired reflectance, fluorescence, and Raman spectra from 137 lesions in 76 patients using optical fiber–based systems. They obtained the best classification for nonmelanoma skin cancers when using multimodal approach. On the other hand, the best melanoma classification occurred when using Raman spectroscopy alone. A Raman probe to detect invasive brain cancer *in situ* in real time in patients was developed by Jermyn et al. [47]. They demonstrated that Raman spectroscopy can accurately detect grade 2–4 gliomas *in vivo* during human brain cancer surgery and it was possible to differentiate between cancer cell–invaded brain and normal brain, with sensitivity and specificity greater than 90%. Additionally, this approach can classify in real time, making it an invaluable tool for surgical procedure and decision making.

5.2. Skin diseases

5.2.1. Atopic dermatitis

Several published works have used Raman spectroscopy to analyze the molecular composition of skin and correlate it with history of atopic dermatitis (AD) and filaggrin gene (FLG) mutations; Kezic et al. measured NMFs noninvasively on the skin of 137 Irish children with a history of moderate to severe AD [48]. González et al. detected the presence of the protein filaggrin in the skin of newborns using Raman spectroscopy and PCA as an early detection procedure for filaggrin-related AD [49]. In order to detect the presence of filaggrin in the Raman spectra, the coefficients of the principal components for each of the skin spectra from newborns were calculated. The first and second principal components accounted for 93.86% of all the explained variance of the original data. **Figure 3** shows a graph of these two principal components, also known as scores plot. In the figure, the gray solid circles correspond to those infants who developed AD; the rest of the subjects are grouped together around the location of the filaggrin spectrum, represented as a black solid circle. The geometrical distance of each Raman spectra to the spectrum of filaggrin in the principal component plane indicates the amount of filaggrin in the subjects. Lower distances indicate higher amount of filaggrin and higher distances indicate lees amount of filaggrin or a filaggrin with a different molecular structure than the molecule that was taken as a reference spectrum.

This result indicates that this approach can be used to identify the persons who are more susceptible to develop AD, making it possible to use this technique as a method for early detection of AD. González et al. validated the use of Raman spectroscopy as a noninvasive tool to detect filaggrin gene mutations [50]. In this study, the amount of filaggrin was estimated by performing the correlation between the pure filaggrin Raman spectrum and the skin spectra obtained from Mexican patients with AD; the genetic analysis showed that 8 out of the 19 patients (42%) presented an FLG mutation. These 8 patients presented the 2282del4 FLG mutation, 2 of which (10.5%) were homozygous and 6 (31.5%) heterozygous, whereas 1 (5.2%) resulted in a compound heterozygote for the 2282del4 and the R501X mutations. These genetic results were compared to the filaggrin amount estimated; a lower correlation value of the spectra with the filaggrin spectrum indicates a lower filaggrin concentration. **Figure 4** shows the results of the correlation for the patients with an FLG mutation (FLG –) and without an FLG mutation (FLG +). The patients with an FLG mutation presented an average correlation of 0.286, while the patients without an FLG mutation showed an average correlation of 0.4. Their results show that the correlation of the filaggrin Raman spectrum with the Raman

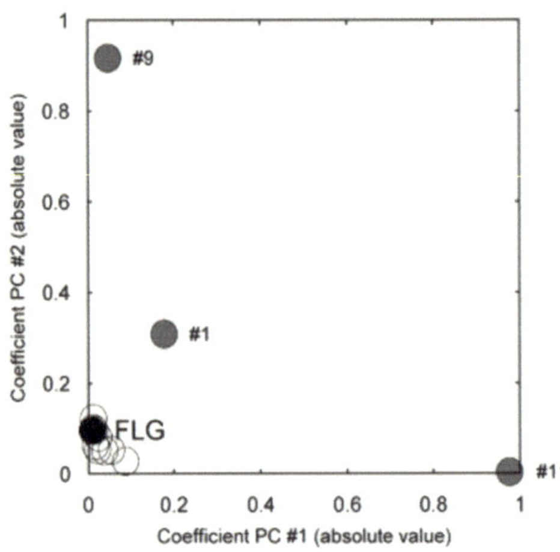

Figure 3. Plot of the two first principal components of the Raman spectra for each newborn (white circles) and the Raman spectrum of filaggrin (black circle). The infants identified with the numbers 1, 9, and 11 developed AD (gray circles) [49].

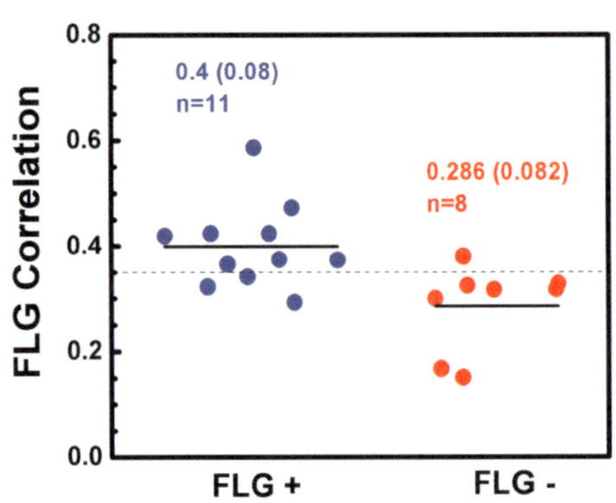

Figure 4. Correlation between the filaggrin Raman spectrum and the skin spectrum of subjects with (FLG –) and without (FLG +) filaggrin gene mutations [50].

spectra of skin can be an indicator of filaggrin gene mutations. In another work, Baclig et al. used a genetic algorithm to demonstrate that strongly reduced Raman spectral information is sufficient for clinical diagnosis of atopic dermatitis [51].

5.2.2. *Skin aging*

Tfayli et al. reported slight variability in skin lipids upon aging [52]. The Raman spectral features of the skin lipids shifted in lateral packing with increasing age of the volunteers. González et al. differentiated between chronological aging and photoinduced skin damage by PCA of *in vivo* Raman spectra from sun-protected and sun-exposed skin [53].

5.2.3. *Nickel allergy*

Alda et al. [54] detected biochemical differences in the structure of the skin of subjects with nickel allergy when comparing with healthy subjects. The Raman spectral differences between groups were classified using PCA.

5.2.4. *Melasma*

Moncada et al. [55] used Raman spectroscopy in melasma patients treated with a triple combination cream (Tretinoin, Fluocinolona, and Hydroquinone) and found that the Raman skin spectra of the melasma patients showed differences in the peaks associated to melanin at 1352 and 1580 cm^{-1} (**Figure 5**). The Raman skin spectrum of patients who did not respond to treatment (**Figure 1B**) showed peaks that are not well defined, which are consistent with molecule degradation and protein breakdown. These results are consistent with the results reported previously by González et al. [56].

5.2.5. *Other in vivo applications: UV/Vis Raman, Raman imaging, and SERS*

In most of the *in vivo* Raman applications, near infrared (NIR) excitation sources are preferred. NIR wavelengths in the range of 780–1100 nm result in lower fluorescence background in the tissue and simplify the analysis of the Raman bands in comparison to visible or UV excitation. The visible excitation sources have been used in various biomedical Raman applications [57]. However, the use of visible wavelengths has several disadvantages for *in vivo* biomedical Raman applications such as the decrease of penetration depth, autofluorescence, and heat generation. The UV radiation is not used for *in vivo* measurements due to the mutagenicity. In Raman imaging [58], a laser spot scans the sample area and acquires Raman spectra at every set point. The intensity of a specific Raman band or bands is used to build an image from cells and tissues. Also the Raman spectra can be discriminated by chemometric analysis and the result is an image of the sample that contains chemical information, also known as Raman chemical image. Other methods of Raman imaging include coherent anti-Stokes Raman spectroscopy (CARS) and stimulated Raman scattering. These methods have been applied to study biochemical interactions in cells and tissues. However, the *in vivo* applications have been limited to animal models. [59–61]. The Raman imaging has the disadvantage that long integration times are needed, which limit its use for *in vivo* measurement in humans. Surface-enhanced Raman spectroscopy (SERS)

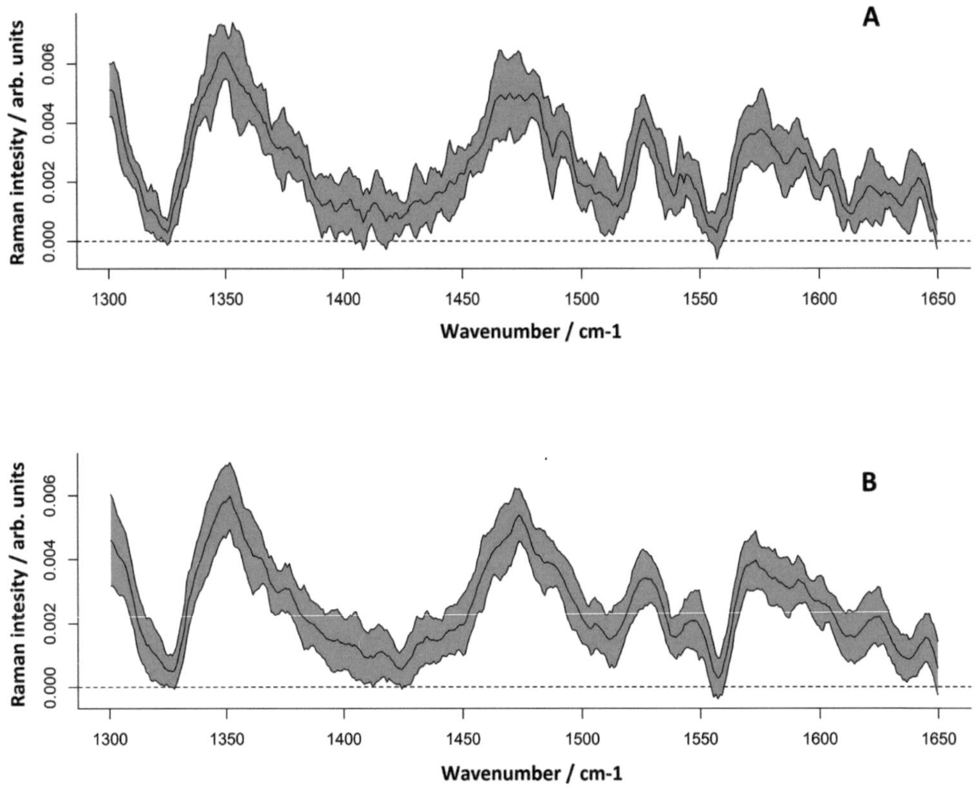

Figure 5. Raman skin spectra of all melasma patients, grouped by patients who responded to the treatment (A) and patients who did not respond to treatment (B). For each group, the central solid line corresponds to the mean of the spectra of each group, and the gray shadow around this line represents the standard deviation [55].

was used for *in vivo* biomedical applications to detect biomarkers in animal models [62–64]. However, the toxicity issues related to nanoparticles used for SERS make this method infeasible for *in vivo* Raman measurements of human tissue.

Other alternatives for *in vivo* biomedical applications are to combine Raman spectroscopy with other optical methods. For example, Raman spectroscopy has been combined with optical coherence tomography [65, 66], confocal reflectance microscopy [67, 68], diffuse reflectance, and fluorescence spectroscopy [46, 69]. The disadvantages of the multimodal approach are the higher cost and complexity of the system needed to perform the measurements. However, the multimodal approach has the advantage, when comparing with Raman spectroscopy alone that provides complementary and more detailed information about the disease and more accurate diagnosis in terms of both sensitivity and specificity.

6. Limitations

Among the disadvantages of Raman spectroscopy for biomedical applications is the weakness of the Raman effect, which most of the time is often accompanied by a stronger background signal particularly in biological samples. The background removing includes changes in instrumentation, which means high-complexity and high-cost systems. One alternative is the algorithm-based methods for fluorescence background removing. However, these methods cannot deal with all types of fluorescence without user intervention to adjust algorithm parameters. Additionally, the complexity of the fitting algorithms makes it difficult to use by nonexperts. Other limitation is that not all the molecules are Raman active, which means that some molecules do not give Raman signal. The potential of damaging the sample due to the laser exposure, which depends on the excitation wavelength, has to be taken for *in vivo* measurements. To solve this problem, lower energy excitation sources in the NIR range are preferred. Demonstrating the safety of these devices to regulatory agencies is a very important step for clinical implementation. For the *in vivo* diagnosis applications, larger studies are needed in order to test the reliability of the results. To date, a short number of studies involving a sufficient number of patients are reported. The lack of standardized and reliable methods for data analysis is an important limitation. Thus, standardization of measurement procedures, instrument calibration, processing, and evaluation of data is needed. Also the information provided by Raman spectra must be displayed in user-friendly, simple format, including clinically relevant information for diagnosis.

7. Conclusions and outlook

From the applications described in this chapter, it is clear that Raman spectroscopy has a great potential for *in vivo* measurements and identification of disease markers, which would make this technique a viable option for noninvasive medical diagnosis. Among the advantages of using Raman spectroscopy as a noninvasive tool for medical diagnosis is the fact that water is not Raman-active; therefore, it does not interfere with measurements. Also, the technique is noninvasive and fast and gives specific information about the structure and biochemical composition of samples, making it a viable option to identify molecules that are associated with disease.

Raman spectroscopy is likely to become a key player for *in vivo* and noninvasive medical diagnosis; however, in order to become a useful and reliable technique, it is important to use it along with signal processing methods and chemometrics in order to automatize and increase the reliability of the measurements and the identification of the molecules of interest.

An area of development that would accelerate the use of Raman spectroscopy in a clinical environment is the design of low-cost and portable Raman spectrometers, which would make their use more appealing for the medical community. Research in this area could also lead to an integrated optics Raman spectrometer, which would make the use of this technique useful in wearable health devices and monitoring of health parameters in a clinical environment.

It is the authors' belief that the combination of optimized instrumentation, standardized measurement procedures, preprocessing, and data analysis will allow Raman spectroscopy to become a powerful tool for disease diagnostics and a common clinical tool in a hospital environment.

Acknowledgements

This work was supported by the Consejo Nacional de Ciencia y Tecnología (CONACyT), Mexico, through Grants CEMIE-Sol 32, and by the National Laboratory program from CONACYT, through the Terahertz Science and Technology National Lab (LANCYTT).

Author details

Miguel Ghebré Ramírez-Elías[1]* and Francisco Javier González[2]

*Address all correspondence to: miguel.ghebre@uaslp.mx

1 School of Sciences, Autonomous University of San Luis Potosi, San Luis Potosi, Mexico

2 Terahertz Science and Technology National Lab (LANCYTT), Autonomous University of San Luis Potosi, San Luis Potosi, Mexico

References

[1] Ferraro JR, Nakamoto K, Brown CW. Introductory Raman Spectroscopy. 2nd ed. San Diego, California: Academic Press; 2003

[2] Zhao J, Lui H, McLean DI, Zeng H. Real-time raman spectroscopy for noninvasive in vivo skin analysis and diagnosis. In: New Developments in Biomedical Engineering. InTech; 2010

[3] Mosier-Boss PA, Lieberman SH, Newbery R. Fluorescence rejection in Raman spectroscopy by shifted-spectra, edge detection, and FFT filtering techniques. Applied Spectroscopy. 1995;**49**(5):630-638

[4] Gautam R, Deobagkar-Lele M, Majumdar S, Chandrasekar B, Victor E, Ahmed SM, Wadhwa N, Verma T, Kumar S, Sundaresan NR, Umapathy S. Molecular profiling of sepsis in mice using Fourier transform infrared microspectroscopy. Journal of Biophotonics. 2016;**9**(1–2):67-82

[5] Cleveland WS, Grosse E. Computational methods for local regression. Statistics and Computing. 1991;**1**(1):47-62

[6] Barclay VJ, Bonner RF, Hamilton IP. Application of wavelet transforms to experimental spectra: Smoothing, denoising, and data set compression. Analytical Chemistry. 1997 Jan 1;**69**(1):78-90

[7] Zhao J, Carrabba MM, Allen FS. Automated fluorescence rejection using shifted excitation Raman difference spectroscopy. Applied Spectroscopy. 2002 Jul 1;**56**(7):834-845

[8] Wang H, Zhao J, Lee AM, Lui H, Zeng H. Improving skin Raman spectral quality by fluorescence photobleaching. Photodiagnosis and Photodynamic Therapy. 2012 Dec 31;**9**(4): 299-302

[9] Knorr F, Smith ZJ, Wachsmann-Hogiu S. Development of a time-gated system for Raman spectroscopy of biological samples. Optics Express. 2010 Sep 13;**18**(19):20049-20058

[10] Lieber CA, Mahadevan-Jansen A. Automated method for subtraction of fluorescence from biological Raman spectra. Applied Spectroscopy. 2003 Nov 1;**57**(11):1363-1367

[11] Zhao J, Lui H, McLean DI, Zeng H. Automated autofluorescence background subtraction algorithm for biomedical Raman spectroscopy. Applied Spectroscopy. 2007 Nov;**61**(11): 1225-1232

[12] Gallo C, Capozzi V, Lasalvia M, Perna G. An algorithm for estimation of background signal of Raman spectra from biological cell samples using polynomial functions of different degrees. Vibrational Spectroscopy. 2016 Mar 31;**83**:132-137

[13] Zhang ZM, Chen S, Liang YZ, Liu ZX, Zhang QM, Ding LX, Ye F, Zhou H. An intelligent background-correction algorithm for highly fluorescent samples in Raman spectroscopy. Journal of Raman Spectroscopy. 2010 Jun 1;**41**(6):659-669

[14] O'Grady A, Dennis AC, Denvir D, McGarvey JJ, Bell SE. Quantitative Raman spectroscopy of highly fluorescent samples using pseudosecond derivatives and multivariate analysis. Analytical Chemistry. 2001 May 1;**73**(9):2058-2065

[15] Candeloro P, Grande E, Raimondo R, Di Mascolo D, Gentile F, Coluccio ML, Perozziello G, Malara N, Francardi M, Di Fabrizio E. Raman database of amino acids solutions: A critical study of extended multiplicative signal correction. The Analyst. 2013;**138**(24):7331-7340

[16] Baek SJ, Park A, Shen A, Hu J. A background elimination method based on linear programming for Raman spectra. Journal of Raman Spectroscopy. 2011 Nov 15;**42**(11):1987-1993

[17] Kourkoumelis N, Polymeros A, Tzaphlidou M. Background estimation of biomedical Raman spectra using a geometric approach. Journal of Spectroscopy. 2012 Jul 11;**27**(5–6):441-447

[18] He S, Zhang W, Liu L, Huang Y, He J, Xie W, Wu P, Baseline DC. Correction for Raman spectra using an improved asymmetric least squares method. Analytical Methods. 2014;**6**(12): 4402-4407

[19] Liu X, Zhang Z, Sousa PF, Chen C, Ouyang M, Wei Y, Liang Y, Chen Y, Zhang C. Selective iteratively reweighted quantile regression for baseline correction. Analytical and Bioanalytical Chemistry. 2014 Mar 1;**406**(7):1985-1998

[20] Liu X, Zhang Z, Liang Y, Sousa PF, Yun Y, Yu L. Baseline correction of high resolution spectral profile data based on exponential smoothing. Chemometrics and Intelligent Laboratory Systems. 2014 Dec 15;**139**:97-108

[21] Koch M, Suhr C, Roth B, Meinhardt-Wollweber M. Iterative morphological and mollifier-based baseline correction for Raman spectra. Journal of Raman Spectroscopy. 2017 Feb 1;**48**(2): 336-342

[22] González-Vidal JJ, Pérez-Pueyo R, Soneira MJ. Automatic morphology-based cubic p-spline fitting methodology for smoothing and baseline-removal of Raman spectra. Journal of Raman Spectroscopy. 2017 Jun 1;**48**(6):878-883

[23] Butler HJ, Ashton L, Bird B, Cinque G, Curtis K, Dorney J, Esmonde-White K, Fullwood NJ, Gardner B, Martin-Hirsch PL, Walsh MJ. Using Raman spectroscopy to characterize biological materials. Nature Protocols. 2016 Apr 1;**11**(4):664-687

[24] Jolliffe IT. Principal component analysis and factor analysis. In: Principal Component Analysis. New York: Springer; 1986. pp. 115-128

[25] Luinge HJ, Van der Maas JH, Visser T. Partial least squares regression as a multivariate tool for the interpretation of infrared spectra. Chemometrics and Intelligent Laboratory Systems. 1995 Apr 1;**28**(1):129-138

[26] Li X, Yang T, Li S, Wang D, Song Y, Zhang S. Raman spectroscopy combined with principal component analysis and k nearest neighbour analysis for non-invasive detection of colon cancer. Laser Physics. 2016;**26**(3):035702

[27] Pacia MZ, Mateuszuk L, Buczek E, Chlopicki S, Blazejczyk A, Wietrzyk J, Baranska M, Kaczor A. Rapid biochemical profiling of endothelial dysfunction in diabetes, hypertension and cancer metastasis by hierarchical cluster analysis of Raman spectra. Journal of Raman Spectroscopy. 2016 Nov 1;**47**(11):1310-1317

[28] Elena R, Kazakov O, Halamkova L, Celmins D, Malone P, Molho E, Zimmerman EA, Lednev IK. Raman spectroscopy of blood serum for Alzheimer's disease diagnostics: specificity relative to other types of dementia. Journal of Biophotonics. 2015;**8**(7):584-596

[29] Chen L, Wang Y, Liu N, Lin D, Weng C, Zhang J, Zhu L, Chen W, Chen R, Feng S. Near-infrared confocal micro-Raman spectroscopy combined with PCA–LDA multivariate analysis for detection of esophageal cancer. Laser Physics. 2013 Apr 30;**23**(6):065601

[30] Ramírez-Elías MG, Guevara E, Zamora-Pedraza C, Rogelio Aguirre RJ, Juárez F, Bertha I, Bárcenas P, Guadalupe M, Ruiz F, González FJ. Assessment of mezcal aging combining Raman spectroscopy and multivariate analysis techniques. Biomedical Spectroscopy and Imaging. 2017 Jan 1;**6**(1–2):75-81

[31] Pérez MR, Mendoza MG, Elías MG, González FJ, Contreras HR, Servín CC. Raman spectroscopy an option for the early detection of citrus Huanglongbing. Applied Spectroscopy. 2016 May 1;**70**(5):829-839

[32] Li S, Zhang Y, Xu J, Li L, Zeng Q, Lin L, Guo Z, Liu Z, Xiong H, Liu S. Noninvasive prostate cancer screening based on serum surface-enhanced Raman spectroscopy and support vector machine. Applied Physics Letters. 2014 Sep 1;**105**(9):091104

[33] Duraipandian S, Zheng W, Ng J, Low JJ, Ilancheran A, Huang Z. Simultaneous fingerprint and high-wavenumber confocal Raman spectroscopy enhances early detection of cervical precancer in vivo. Analytical Chemistry. 2012 Jun 20;**84**(14):5913-5919

[34] Eberhardt K, Stiebing C, Matthäus C, Schmitt M, Popp J. Advantages and limitations of Raman spectroscopy for molecular diagnostics: An update. Expert Review of Molecular Diagnostics. 2015 Jun 3;**15**(6):773-787

[35] Short MA, Lam S, McWilliams AM, Ionescu DN, Zeng H. Using laser Raman spectroscopy to reduce false positives of autofluorescence bronchoscopies: A pilot study. Journal of Thoracic Oncology. 2011 Jul 31;**6**(7):1206-1214

[36] Bergholt MS, Zheng W, Ho KY, Teh M, Yeoh KG, So JB, Shabbir A, Huang Z. Fiberoptic confocal raman spectroscopy for real-time in vivo diagnosis of dysplasia in Barrett's esophagus. Gastroenterology. 2014 Jan 1;**146**(1):27-32

[37] Taketani A, Hariyani R, Ishigaki M, Andriana BB, Sato H. Raman endoscopy for the in situ investigation of advancing colorectal tumors in live model mice. The Analyst. 2013;**138**(14):4183-4190

[38] Wang J, Lin K, Zheng W, Ho KY, Teh M, Yeoh KG, Huang Z. Comparative study of the endoscope-based bevelled and volume fiber-optic Raman probes for optical diagnosis of gastric dysplasia in vivo at endoscopy. Analytical and Bioanalytical Chemistry. 2015 Nov 1;**407**(27):8303-8310

[39] Bergholt MS, Zheng W, Ho KY, Yeoh KG, Teh M, So JB, Huang Z. Real-time depth-resolved fiber optic Raman endoscopy for in vivo diagnosis of gastric precancer. Proceedings of SPIE. 2014 Mar 4;**8939**:89390M-1

[40] Bergholt MS, Zheng W, Lin K, Wang J, Xu H, Ren JL, Ho KY, Teh M, Yeoh KG, Huang Z. Characterizing variability of in vivo Raman spectroscopic properties of different anatomical sites of normal colorectal tissue towards cancer diagnosis at colonoscopy. Analytical Chemistry. 2014 Dec 26;**87**(2):960-966

[41] Guze K, Pawluk HC, Short M, Zeng H, Lorch J, Norris C, Sonis S. Pilot study: Raman spectroscopy in differentiating premalignant and malignant oral lesions from normal mucosa and benign lesions in humans. Head & Neck. 2015 Apr 1;**37**(4):511-517

[42] Singh SP, Sahu A, Deshmukh A, Chaturvedi P, Krishna CM. In vivo Raman spectroscopy of oral buccal mucosa: A study on malignancy associated changes (MAC)/cancer field effects (CFE). The Analyst. 2013;**138**(14):4175-4182

[43] Krishna H, Majumder SK, Chaturvedi P, Sidramesh M, Gupta PK. In vivo Raman spectroscopy for detection of oral neoplasia: A pilot clinical study. Journal of Biophotonics. 2014 Sep 1;**7**(9):690-702

[44] Lin K, Zheng W, Lim CM, Huang Z. Real-time in vivo diagnosis of nasopharyngeal carcinoma using rapid fiber-optic Raman spectroscopy. Theranostics. 2017 Jan 1;7(14): 3517-3526

[45] Lui H, Zhao J, McLean D, Zeng H. Real-time Raman spectroscopy for in vivo skin cancer diagnosis. Cancer Research. 2012 May 15;**72**(10):2491-2500

[46] Lim L, Nichols B, Migden MR, Rajaram N, Reichenberg JS, Markey MK, Ross MI, Tunnell JW. Clinical study of noninvasive in vivo melanoma and nonmelanoma skin cancers using multimodal spectral diagnosis. Journal of Biomedical Optics. 2014 Nov 1;**19**(11):117003

[47] Jermyn M, Mok K, Mercier J, Desroches J, Pichette J, Saint-Arnaud K, Bernstein L, Guiot MC, Petrecca K, Leblond F. Intraoperative brain cancer detection with Raman spectroscopy in humans. Science Translational Medicine. 2015 Feb 11;**7**(274):274ra19

[48] Kezic S, O'Regan GM, Lutter R, Jakasa I, Koster ES, Saunders S, Caspers P, Kemperman PM, Puppels GJ, Sandilands A, Chen H. Filaggrin loss-of-function mutations are associated

with enhanced expression of IL-1 cytokines in the stratum corneum of patients with atopic dermatitis and in a murine model of filaggrin deficiency. Journal of Allergy and Clinical Immunology. 2012 Apr 30;**129**(4):1031-1039

[49] González FJ, Alda J, Moreno-Cruz B, Martínez-Escanamé M, Ramírez-Elías MG, Torres-Álvarez B, Moncada B. Use of Raman spectroscopy for the early detection of filaggrin-related atopic dermatitis. Skin Research and Technology. 2011 Feb 1;**17**(1):45-50

[50] González FJ, Valdes-Rodríguez R, Ramírez-Elías MG, Castillo-Martínez C, Saavedra-Alanis VM, Moncada B. Noninvasive detection of filaggrin gene mutations using Raman spectroscopy. Biomedical Optics Express. 2011 Dec 1;**2**(12):3363-3366

[51] Baclig AC, Bakker Schut TC, O'Regan GM, Irvine AD, McLean WH, Puppels GJ, Caspers PJ. Possibilities for human skin characterization based on strongly reduced Raman spectroscopic information. Journal of Raman Spectroscopy. 2013 Mar 1;**44**(3):340-345

[52] Tfayli A, Guillard E, Manfait M, Baillet-Guffroy A. Raman spectroscopy: Feasibility of in vivo survey of stratum corneum lipids, effect of natural aging. European Journal of Dermatology. 2011 Dec;**22**(1):36-41

[53] González FJ, Castillo-Martínez C, Martínez-Escanamé M, Ramírez-Elías MG, Gaitan-Gaona FI, Oros-Ovalle C, Moncada B. Noninvasive estimation of chronological and photoinduced skin damage using Raman spectroscopy and principal component analysis. Skin Research and Technology. 2012 Nov 1;**18**(4):442-446

[54] Alda J, Castillo-Martinez C, Valdes-Rodriguez R, Hernández-Blanco D, Moncada B, González FJ. Use of Raman spectroscopy in the analysis of nickel allergy. Journal of Biomedical Optics. 2013 Jun 1;**18**(6):061206

[55] Moncada B, Castillo-Martínez C, Arenas E, León-Bejarano F, Ramírez-Elías MG, González FJ. Raman spectroscopy analysis of the skin of patients with melasma before standard treatment with topical corticosteroids, retinoic acid, and hydroquinone mixture. Skin Research and Technology. 2016 May 1;**22**(2):170-173

[56] Moncada B, Sahagún-Sánchez LK, Torres-Álvarez B, Castanedo-Cázares JP, Martínez-Ramírez JD, González FJ. Molecular structure and concentration of melanin in the stratum corneum of patients with melasma. Photodermatology, Photoimmunology & Photomedicine. 2009 Jun 1;**25**(3):159-160

[57] Liu CH, Zhou Y, Sun Y, Li JY, Zhou LX, Boydston-White S, Masilamani V, Zhu K, Pu Y, Alfano RR. Resonance Raman and Raman spectroscopy for breast cancer detection. Technology in Cancer Research & Treatment. 2013 Aug;**12**(4):371-382

[58] Klein K, Gigler AM, Aschenbrenner T, Monetti R, Bunk W, Jamitzky F, Morfill G, Stark RW, Schlegel J. Label-free live-cell imaging with confocal Raman microscopy. Biophysical Journal. 2012 Jan 18;**102**(2):360-368

[59] Saar BG, Freudiger CW, Reichman J, Stanley CM, Holtom GR, Xie XS. Video-rate molecular imaging in vivo with stimulated Raman scattering. Science. 2010 Dec 3;**330**(6009): 1368-1370

[60] Patel II, Steuwe C, Reichelt S, Mahajan S. Coherent anti-stokes Raman scattering for label-free biomedical imaging. Journal of Optics. 2013 Sep 10;**15**(9):094006

[61] Imitola J, Côté D, Rasmussen S, Xie XS, Liu Y, Chitnis T, Sidman RL, Lin CP, Khoury SJ. Multimodal coherent anti-stokes Raman scattering microscopy reveals microglia-associated myelin and axonal dysfunction in multiple sclerosis-like lesions in mice. Journal of Biomedical Optics. 2011 Feb 1;**16**(2):021109

[62] Maiti KK, Dinish US, Samanta A, Vendrell M, Soh KS, Park SJ, Olivo M, Chang YT. Multiplex targeted in vivo cancer detection using sensitive near-infrared SERS nanotags. Nano Today. 2012 Apr 30;**7**(2):85-93

[63] Dinish US, Balasundaram G, Chang YT, Olivo M. Sensitive multiplex detection of serological liver cancer biomarkers using SERS-active photonic crystal fiber probe. Journal of Biophotonics. 2014 Nov 1;**7**(11–12):956-965

[64] Dinish US, Balasundaram G, Chang YT, Olivo M. Actively targeted in vivo multiplex detection of intrinsic cancer biomarkers using biocompatible SERS nanotags. Scientific Reports. 2014;**4**:4075

[65] Khan KM, Krishna H, Majumder SK, Rao KD, Gupta PK. Depth-sensitive Raman spectroscopy combined with optical coherence tomography for layered tissue analysis. Journal of Biophotonics. 2014 Jan 1;**7**(1–2):77-85

[66] Wang J, Zheng W, Lin K, Huang Z. Development of a hybrid Raman spectroscopy and optical coherence tomography technique for real-time in vivo tissue measurements. Optics Letters. 2016 Jul 1;**41**(13):3045-3048

[67] Patil CA, Arrasmith CL, Mackanos MA, Dickensheets DL, Mahadevan-Jansen AA. Handheld laser scanning confocal reflectance imaging–confocal Raman microspectroscopy system. Biomedical Optics Express. 2012 Mar 1;**3**(3):488-502

[68] Wang H, Lee AM, Lui H, McLean DI, Zeng H. A method for accurate in vivo micro-Raman spectroscopic measurements under guidance of advanced microscopy imaging. Scientific Reports. 2013 May 28;**3**:1890

[69] Shaikh R, Prabitha VG, Dora TK, Chopra S, Maheshwari A, Deodhar K, Rekhi B, Sukumar N, Krishna CM, Subhash NA. Comparative evaluation of diffuse reflectance and Raman spectroscopy in the detection of cervical cancer. Journal of Biophotonics. 2017 Feb 1;**10**(2):242-252